THE WABASH RIVER ECOSYSTEM

by

James R. Gammon

Published by

Cinergy Corporation, Plainfield, Indiana
Eli Lilly and Company, Indianapolis, Indiana
and
DePauw University, Greencastle, Indiana

Distributed by

Indiana University Press, Bloomington, Indiana

This book is a publication of

Indiana University Press
601 North Morton Street
Bloomington, IN   47404-3797   USA

http://www.indiana.edu/~iupress

Telephone orders   800-842-6796
Fax orders   812-855-7931
Orders by e-mail   iuporder@indiana.edu

The paper used in this publication meets the minimum
requirements of American National Standard for Information
Sciences--Permanence of Paper for Printed Library
Materials, ANSI Z39.48-1984.

Manufactured in the United States of America

Library of Congress Cataloging-in-Publication Data

Gammon, James R. (James Robert), date.
The Wabash River ecosystem / James R. Gammon.
    p.   cm.
Includes bibliographical references (p.   ).
ISBN 0-253-21272-3 (pbk. : alk. paper)
1. Water--Pollution--Environmental aspects--Wabash River.
2. Freshwater fishes--Effect of water quality on--Wabash River.
3. Stream ecology--Wabash River. I. Title.
QH545.W3G36   1999
577.6'4'097724--dc21   98-7311

1 2 3 4 5   04 03 02 01 00 99

ACKNOWLEDGEMENTS

Beginning in 1967 with an evaluation of thermal effects at the Wabash Electrical Generating Station near Terre Haute, Indiana, the initial thrust of the research was to evaluate the effects of heated effluents from electric generating stations. The emphasis has changed over time and many other different kinds of questions have been posed, some of which have been answered, at least in part. The information gleaned has offered glimpses into the structure of this river's ecosystem. The emphasis has always been to examine the biotic components of the aquatic community, especially fish populations, and how they respond to natural and man-produced constraints and perturbations.

The data was collected under conditions which were sometimes difficult by an experienced and knowledgeable field crew under the direction of John Riggs, who relinquished writing about Garth Ryland and working at the DePauw University Archives to become a "river rat". A long list of students assisted in a variety of ways in the river research program during the past generation: Bob Poppe, Russ Stullken, David and Sue Allard, David H. Veach, David S. White, Candyce Moring, Mark McKee, Ronald Burk, Susan Bell, Stephen Bowen, Michael Harves, Eugene Mancini, Jay Hatch, Ron VanSeventer, Steven Pierce, David McCammack, Terry Teppen, Jerry Rud, Douglas Meikle, David Dee, David Gammon, Jim Thayer, Mike Stroup, Joe Reidy, Louisa Witten, Sue Gilbertson, Randy Jones, David Petree, Robert Gammon, Douglas Bauer, Chris Yoder, Brandon Kulik, Ann Kohlstaedt, Andy Hickman, Clifford Gammon, Ed Snizek, Greg Seketa, James Cooper, Brad Pearman, Kathy Mohar, Neil Parke, Greg Willhite, Cliff Jones, Steve Dawson, Julie Heyward, Todd Sellers, Cole Remsburg, Brian Pickens, Julie Ankenbrook, Mike Myers, Chris Hansen, Neil Masten, John Hecko, Bradley Garner, Dean Wallace, Mike Giesecke, Mark Davis, Shawn Riggs, Ben Anderson, Simon Krughoff, and others.

The research directly and indirectly benefitted over the years from the concerns, suggestions, criticisms, and active involvement of many interested colleagues. John Bell, Peter Howe, Pete Redmond, Gary Milburne, Anne Spacie, Jerry Hamelink, John Winters, Lee Bridges, Steve Boswell, Jack Gakstatter, Bob Hughes, Ed Herricks, Harold McReynolds, John Whitaker, Charlie Crawford, and Mike Lydy come immediately to mind, but many others in IDEM, IDNR, USEPA, and USGS have also had a hand.

Colleagues from various Departments at DePauw University have also provided assistance and I am especially appreciative for the support of my colleagues in the Department of Biological Sciences. Betty McKee provided outstanding clerical assistance over the decades. More recently, Lisa Portune, Debora Steele, David R. Richards, and Randy Lewis have labored to produce a much better manuscript. Most of the original line drawings of fish were produced by David M. Gammon over the past several years, but John Hecko also contributed.

i

I am most appreciative of the inputs of David Hoffman, Randy Lewis, Vince Griffin, Wayne Swallow, Bob Christian, and Bill Nelson of PSI Energy and also of Stanley Parka, Neil Parke, Roger Meyerhoff, John Federmann, George Herr, Jerry Hamelink, Don Brannon and others of Eli Lilly and Company.

Support of this research since its inception has been underwritten largely by Cinergy Corporation (formerly Public Service Indiana) and, for more than a decade and a half, by Eli Lilly and Company. At times support was also secured through the Office of Water Resources Research. The initial verbal agreement with the corporate sponsors was that the results of the research, whatever they happened to be, would be freely available to any and all interested agencies, organizations, and professionals. That agreement remains intact today.

The catch data on which much of the report is based is too extensive to include in Appendix Tables. Those interested in securing summaries of the catch data are urged to request it from the author. Several forms of output are capable of being generated and placed as ASCI files on computer diskette.

TABLE OF CONTENTS

LIST OF TABLES

LIST OF FIGURES

vii

LIST OF FIGURES
(Continued)

LIST OF FIGURES
(Continued)

LIST OF FIGURES
(Continued)

LIST OF FIGURES
(Continued)

LIST OF FIGURES
(Continued)

EXECUTIVE SUMMARY

This book summarizes three decades of ecological study of the middle Wabash River, its post-glacial history, physical and chemical environment, and its rich and abundant plant and animal life - fish, mussels, macroinvertebrates, and phytoplankton.

Some rivers, the Wabash included, inspire poets and musical composers to levels of creative eloquence. Rivers, together with mountains, lakes, and oceans, are among the few natural features which bestow a permanence to an otherwise transitory world. There is something satisfying in the thought that a familiar nearby river once floated a birchbark canoe and will continue to be a source of enjoyment to our children and grandchildren. The large challenge here is to find ways to moderate and resolve multiple demands and uses.

As with our children, pets, and automobiles, all large rivers have certain needs which must be understood and satisfied if they are to be maintained in a healthy and worthwhile condition. One essential set of requirements is to keep the ecological components in good working condition. Describing those components is the primary focus of this volume.

Prior to 1965 or so our aspirations for usefulness was satisfied if a river had sufficient water to float a barge, dilute waste products or at least carry them away from us, and provide some recreational fishing. There was no pressing need to understand the subsurface performers or their ecological

roles as long as diseases and bad odors weren't produced in the immediate vicinity.

A large river also serves a variety of public needs. Their use as corridors for travel was discovered long ago by native Americans, itchy-footed European explorers, and the settlers who followed. Large rivers are still an economical means for cheaply transporting materials and goods as long as commercial users can convince a sympathetic government to fund and maintain locks and dams. Rivers also provide food and recreation, roles which are sure to grow in importance in the future.

A river also has institutional requirements which are vague, but increasingly important. The Clean Water Acts of 1967 and subsequent years set into motion an enormous investigative effort which continues to yield positive fruits. Multistate organizations such as ORSANCO have evolved into complex mega-agencies with nightmarish coordinating challenges. State water laws created during a simpler era have become jurisdictional quagmires which ensure an unacceptable status quo. On the positive side is the growing public clamor for cleaner, more usable running waters with environmental organizations watch-dogging both the river and the legislature.

The studies summarized here were initiated in 1967 and continue to the present time. The fish community has always been the primary focus although other biotic groups have been examined sporadically. Collection methods have included hoop nets and seines, but D.C. electrofishing has been the principle method. The cumulative length of electrofishing over the years is about 1875 miles (3002 km), a distance equivalent to boating from the beginning of the Wabash River in Ohio, down the lower Ohio River into the Mississippi River, then continuing down the entire lower Mississippi River and out into the Gulf of Mexico some 200-300 miles.

The long-term studies have permitted us to:

(a) relate changes in the fish communities to natural events and man-induced factors,

(b) identify problem sections of the river,

(c) evaluate ecological changes associated with operational modifications by industry, and

(d) clarify ecological interactions among the major biotic components of the ecosystem.

The overall fish community was mediocre until about 1983 after which it improved markedly. Using an Index of Well-being as an indication of community quality, the upper five reaches improved from poor-fair to fair and good while the lower three reaches improved from Poor to Fair.

Most species populations improved noticeably. Some populations (eg. blue sucker, mooneye, and spotted bass) expanded into previously unoccupied areas of the river. Many other species populations also increased, especially in the upper portion of the study segment, those which reproduced

and lived in the mainstem as well as species which entered the mainstem from tributaries and offstream reservoirs. At the same time, the carp population remained stable and gizzard shad declined because of a quadrupling of the piscivore population. Despite greater numbers of young recruits for many species populations, there was also an increase in the average size.

This remarkable recovery probably resulted from a combination of long-term, 50% reduction in BOD loading through improved point-source waste treatment and low-flow summers in 1983, 1988, and 1991 which facilitated good reproduction and survival through the first year. Point source pollution from the Lafayette\West Lafayette and Terre Haute areas has been reduced in recent years with demonstrable improvement in local fish communities.

A 25% reduction in agricultural loadings to the river during the 1983 Payment-in-Kind (PIK) program also may have contributed. No-till agriculture is rapidly being adopted throughout the basin and may become an important positive force in the future.

Droughts in 1988, 1991, and 1994 probably reduced nonpoint-source pollution from strip-mining areas as well as agricultural fields and promoted good reproduction and survivorship. Prolonged midsummer periods of high water not only destroyed reproductive production in 1992 and 1993, but also devastated older year classes and, therefore, reduced populations of most larger species to lower levels than ever before observed. The fish populations are once again poised for another positive surge because of the benevolent flows of the mid-1990s.

Clams (Mussels) are the most commercially important biotic component of the Wabash River ecosystem. Depletion of the clam population because of over-harvesting during drought years forced the state to close the harvesting of clams indefinitely. The impending arrival of zebra mussels (*Dreissena polymorpha*) to the Wabash River ecosystem could lead to further declines in native mussels, although it remains to be seen if this species can achieve the high population densities in turbid rivers such as the Wabash as they have in the Great Lakes.

The riparian corridor of the Wabash River mainstem and its tributaries has become critically narrow, and even nonexistent, in far too many places to buffer the river from the flanking agricultural fields. In 1994 more than 11% of the river corridor consisted of bare banks or banks supporting only 1-2 trees between water's edge and the flanking corn and soybean fields. Significant changes of the riparian corridor were noted during the 1970s when agriculture expanded. Willow groves have invaded many formerly denuded areas during the last decade, but accelerated lateral erosion has accompanied higher rates of discharge with resulting depressive effects on the fish community. Efforts should be made to restore green belts along the river corridors of the Wabash River and its tributaries. Many of its former backwaters and oxbos no longer connect with the main Wabash River channel, thus reducing habitat diversity and eliminating valuable spawning habitat for some desirable species of fishes.

The positive changes in water quality and fish communities demonstrate that the Wabash River is amenable to pollution abatement efforts and could become a highly valuable recreational asset. Continued improvement depends partly upon reducing both nutrient delivery and resulting algal densities.

Phytoplankton are undoubtedly the most important member of the biotic community, yet they have been all but neglected by both science and government. It is essential that they be examined much more closely since they are the most important single biotic determinant of water quality in the river because of its effect on turbidity and dissolved oxygen during the summer.

Nonpoint-source pollution from active and derelict strip-mines is slowly being addressed and may ultimately benefit the Wabash River south of the Big Vermilion River.

There has been improvement in rivers throughout the United States during the past three decades. However, the recovery has required an extraordinary effort from industry, government, and environmental organizations. There is, however, little recognition from governmental units that clean rivers are just as important as sound bridges and safe highways. It is likely that our elected officials will join Indiana's citizens in seeking better environmental conditions only through the application of sustained pressure.

THE WABASH RIVER ECOSYSTEM

HISTORIC INFLUENCES
THE INDIAN ERA (8,000BC - 1780AD)

The pre-glacial Wabash River ran through a large valley extending from about Lafayette to its mouth and dates back to the Devonian Period more than 350 million years ago. As part of the ancient Mississippi River system, its aquatic fauna has evolved over an immense period of time. This ancient valley, however, is buried in sand and gravel to a depth of 60 or 70 feet, a legacy of the Illinoian and Wisconsin glaciers of 18,000 to 20,000 years ago. During this period, the Wabash River valley was covered by ice nearly to its junction with the Ohio River.

Geological and natural biological processes influenced the Wabash River and its valley for millions of years before the advent of glaciation, but after the glacial retreat a new ecological element - Man - entered the valley.

During the period of glaciation, perhaps 18,000 to 28,000 years ago, the first humans are believed to have entered into the North American continent from Asia, probably during an interglacial period (Bray, Swanson, and Farrington 1973). From what is now western Alaska, they may have proceeded eastward and southward, perhaps between the Coast Range and the Rocky Mountains and almost certainly east of the Rockies into the upper Saskatchewan River valley. They left behind evidence of their presence in the form of bone tools and the bones of elephants and camels which they hunted and ate, but evidence is scattered and meager.

The glaciers reached their most southerly extent by 20,000 to 18,000 years ago and pushed these early "colonists" southward into what is now western U.S., an area which was unglaciated except at higher elevations. From 18,000 to 9,000 years ago, as the glaciers retreated to the north, these **Clovis** people spread not only further southward, but also eastward into present midwestern U.S. In present southern Indiana, they may have killed deer and remnants of caribou herds and expanded in numbers and diversity.

From 10,000 to 8,000 years ago there was a continent-wide change in prehistoric cultures which coincided with a rise in temperature. Perhaps this was necessitated by the spread of prairies eastward and by oak forests northward. During this period native camps and villages shifted from upland locations to river valleys where freshwater shellfish were gathered, whitetail deer hunted, and berries, nuts, and roots gathered. With this increased skill in the utilization of natural resources, specialized tools and distinctive regional cultures developed.

Southern Indiana lies in the northern part of the **Indian Knoll** culture, part of the **Archaic** culture which prevailed from 10,000 to 3,000 years ago and extended throughout southeast North America from Florida to West Virginia west to Louisiana and Arkansas. The northern part of the **Archaic** culture, the **Laurentian** culture, extended from Wisconsin to Laborador and may have included extreme northern Indiana. Extensive trading was carried on, as evidenced by the appearance of artifacts such as copper tools from the upper Great Lakes in burial sites in the south, as well as milling stones, atlatls, awls, and baskets.

The river floodplain provided soil enriched each year by floods. Agriculture began to be practiced about 3,000 years ago, and by 300 A.D., most river valleys throughout North America, including the lower Wabash River valley, had farming settlements. This included the **Adena-Hopewell** culture of the Ohio and Illinois River valleys, a culture which constructed many large and small rounded burial mounds, used tobacco and copper ornaments, and built wood and wattle houses. The **Adena** people reached a rather high population density and formed large settled areas consisting of scattered small villages of two to five huts. Intertribal trading was extensive. Two main features of the **Adena** culture were cremation of the dead and the entombment of their remains in log or clay-lined chambers covered by rounded mounds (Hyde, 1962).

From 200 B.C. to 500 A.D. the **Hopewell** culture appeared in the Ohio Valley borrowing from the **Adena** and elaborating mound-building practices to their cultural zenith and introducing pottery. By 400 to 500 A.D. the **Hopewell** arts and crafts declined for unknown reasons. Large burial chambers were no longer built and lowland mounds were abandoned for fortified hilltops.

From 700 A.D. to 1700 A.D. the **Mississippi** culture expanded northward from its southern nucleus, bringing with it flat-topped mounds bearing temples, a new strain of corn, flint hoes, and elaborate fortifications which protected fields and homes from warring invaders. This culture probably reached the lower Wabash River valley near the end of its development around 1400 A.D. to 1500 A.D.. A great fortress was constructed east of the Wabash River and north of the Ohio River near the present Evansville (Hyde 1962, pg. 149). This new culture was transitory, however, and declined shortly thereafter. Early Spanish explorers arrived in the southeast just in time to briefly describe the last of the temples.

From the foregoing summary it is evident that the Wabash River valley has been occupied by humans for thousands of years. For at least 10,000 years the rich land and water resources first sustained hunting and gathering **Archaic** cultures and then a sequence of agricultural cultures dating from about 3,000 years ago to the time Europeans first entered the area.

Events during the 1500s and 1600s are confused and poorly understood. In 1608 Samuel de Champlain, following the route traveled in 1535-1543 by Jacque Cartier up the St. Lawrence River, built a fort at Quebec and ushered in the era of trading, which included guns as well as fur and other items. The **Iroquois** had been raiding into the Ohio Valley before the year 1600, but after obtaining European weapons, particularly firearms, they eliminated many neighboring tribes and drove survivors westward and southward. Hyde (1962, pg. 166) suggests that the **Iroquois** had driven most other Indian tribes out of the Ohio River valley by the time Marquette and Joliet reached the Mississippi River in 1673 and "from 1670 on their war parties continued to rove in the Ohio Valley, through lands desolate and silent, stripped of all inhabitants by Iroquois attacks." During the next decade or so the **Iroquois** continued to drive other tribes westward even beyond the Mississippi River, but never themselves settled in the Ohio River valley.

At the close of the 17th century tribes from the north began to recolonize the Ohio

3

River valley. **Delawares** and **Shawnees** returned to the upper Ohio River Valley and the Algonquian tribes of Wisconsin, the **Miamis, Mascoutens, Weas,** and **Piankashaws**, moved into the Wabash River Valley (Figure 1). The **Miamis** established themselves in the upper Wabash River valley where their main village was Kekionga, the present Fort Wayne (Figure 2). The **Wea**

moved to the Wabash River near present West Lafayette and established Ouiatenon. In 1750, a branch of the Kickapoos joined a **Piankashaw** village at the mouth of the Big Vermillion River and another village called Chippekawkay near present Vincennes. Other tribes occupied many other tributaries of the Wabash River.

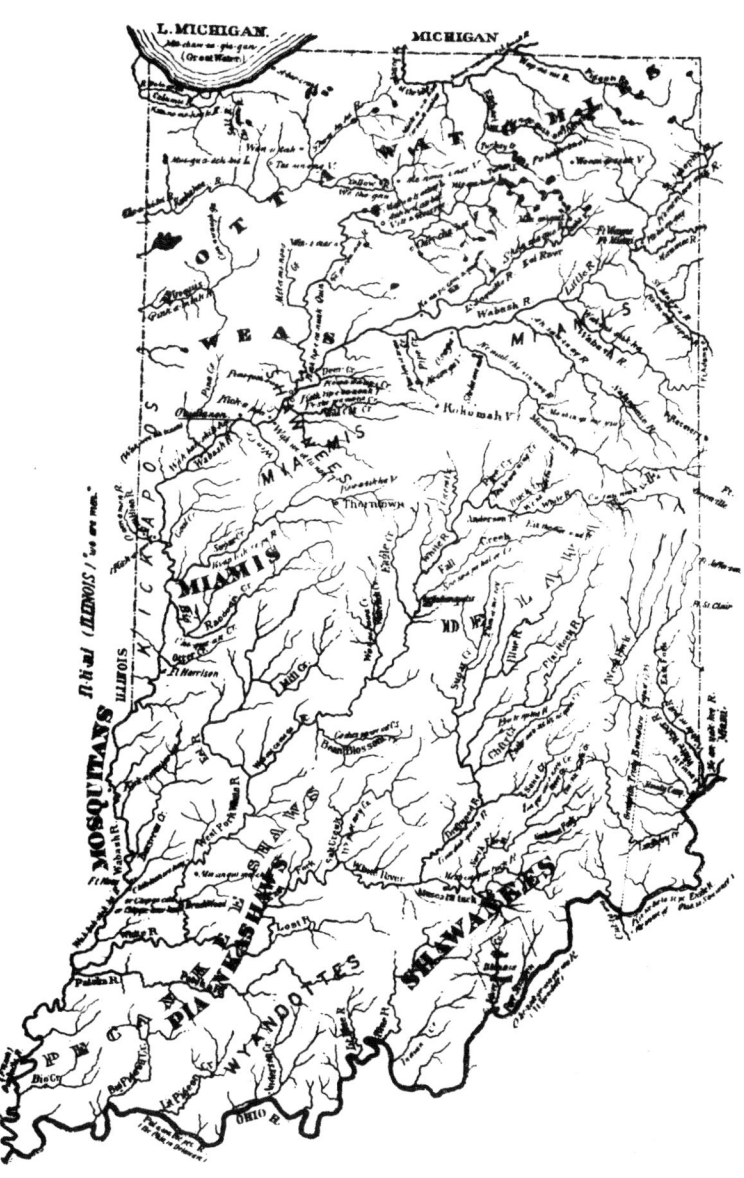

Figure 1: Locations of Indian tribes and villages 1750 - 1850. (After State Geological Report, 1882)

4

Throughout this period and into the 1700's the French and English expanded their trading activities. The French built forts and a chain of trading posts to connect French settlements and military posts in Louisiana with fur trading centers in Chicago, Detroit, and Canada. They also established Fort de Chartres near Kaskaskia in Illinois and sent Frenchmen, the *coureurs de bois*, to establish themselves among the Indians on the Wabash River. When Francois-Marie Bissot founded a military and trading post at Vincennes in 1732-33 (McCord 1970) "the Wabash (was) composed of five nations who compose four villages of which the least has sixty men carrying arms, and all of them could furnish from six to seven hundred men if it were necessary to assemble them."

During the middle 1700's the French and English struggled for control over native tribes, but not until 1765 was the issue finally settled in favor of the British. George Rogers Clark defeated the British at Vincennes in 1779 and created the first American seat of government located northwest of the Ohio River. Originally called the County of Illinois under the government of Virginia, it was known later as the Northwest Territory and included all of what is now Ohio, Indiana, Illinois, Michigan, and Wisconsin. During the next few decades increasing numbers of white settlers moved into the Wabash Valley and their effects on the Wabash River ecosystem escalated dramatically.

Figure 2: The Wabash River near Logansport, Indiana, 1848. George Winter, *Scene on the Wabash,* **oil on canvas, 29 x 36 inches. Photo courtesy of the Gerald Peters Gallery, Sante Fe, New Mexico.**

HISTORIC INFLUENCES
THE SETTLEMENT ERA (1780 - 1880)

In 1800 the white population of Indiana was only 5,641 (Melish, 1822 in Lindley, 1916), most in scattered small communities near rivers. In 1810 it was 24,520; in 1815, 68,784; and in 1820 it was 147,178 including whites, free blacks, and slaves, but not counting Indians. These early residents found a ready market for their surplus agricultural products in New Orleans and floated them on flatboats down the Wabash and White Rivers to the Ohio and then on to New Orleans during the spring floods.

However, by 1827 the New Orleans market became saturated and ways were sought to transport the goods to and from the eastern markets (Esarey, 1912). Steamboats were already plying Wabash waters as far upriver as Terre Haute and even to Ouiatanon south of Lafayette.

Until 1815 few written accounts of the rivers and streams are available, although French fur-traders had lived in harmony with the Indians for at least a century. In 1792 Heckewelder (McCord 1970, pg. 35) wrote of the river near Vincennes ..."the Wabash as clear as the Monocasy, full of fish..."

Indian opposition to settlers moving north of the Ohio River had been dealt with by 1815 and settlers entered in increasing numbers because of the rich agricultural lands. Caleb Lownes in a letter to Oliver Wolcott (Sec. Treas. under Presidents Washington and Adams) wrote in 1815 (McCord, 1970): ..."The first rate lands lie on the Wabash all the way to the lakes on the most beautiful stream in my recollection-it is about 250 yards wide at this place (Vincennes) and preserves its width very nearly for 400 miles...It is a beautiful and valuable stream--the water generally perfectly clear and transparent--exhibiting a clean gravelly bottom--It abounds with fish of various kinds--Bass-Pickerel, Pike-Perch-Catfish &c. The Catfish are of every size up to 122 1/2 lb. one of this size was caught (at Fort Harrison 80 miles above this)--The perch (probably smallmouth bass) are from 12 to 20 in length-this appears to be dealing in the marvelous but it is nevertheless correct--a large White fish about 2 1/2 feet long with very little bone was yesterday caught by a gentlemen on a party said to be excellent."

Most early accounts focused on the fertility of the land and the rivers were examined mostly as potential sites for mills and water for floating boats loaded with future produce. There was no recognition that blocking small tributaries of the Wabash River might have some effect on the reproductive success of some species of fish. Enoch Honeywell wrote in his diary (McCord, 1970): "Apr (May) 1816. saw Ft. Harrison. There is about 12 families living in and near the fort and 10 to 12 more at 6 miles distance southeast 2 at 8 miles south, which are all of 25 miles. The river here is 50 to 60 rods wide, very deep, clayey banks, always navigable for keel-boats except over the grand rapid below Vincennes in low water. In freshes it inundates its banks very

Figure 3: Fort Harrison located on the east bank of the Wabash River near Lost Creek in Vigo County. (Vigo County Historical Museum, Terre Haute, Indiana)

bad; on the west side the river here it floods about a mile, but the soil being light the water soon drains off or soaks in."

By far the finest early account of the area is that of David Thomas (1819). Had his interest in rivers and lakes been half that of his interest in vegetation and wild life, we would have a very detailed, clear account of aquatic life, indeed. Nevertheless, some illuminating comments about the Wabash River as far north as Otter Creek were made.

"The water of the Wabash forms a good lather with soap, At Pittsburgh, for washing, the river water was good, but it becomes harder in its descent. At Cincinnati an increase of lime was evident; and near the mouth of the Wabash, the water of the Ohio was hard"...

"The Wabash is four hundred yards wide at its mouth, three hundred at Vincennes, and two hundred at Fort Harrison. It is fordable in many places."

"The Wabash has a gentle current, except at the Rapids, twenty-three miles below Vincennes"...

"Whenever a high piece of land appears on one side of the River, the opposite shore is low and sunken; and from Raccoon Creek, fifteen miles above Fort Harrison to the mouth of the river, I believe there is no exception to this remark."

"There is one inconvenience attending this country, exclusive of the overflowing of the Wabash. All its tributary streams after a heavy shower of rain, rise above the banks; and overflow the low land adjoining, which on all, is of considerable extent. In time of high water, it is one of the most difficult countries to travel through, I ever saw. I have known it for more than four weeks at one time, that no person could get away from Union Prairie without swimming his horse, or going in a boat."

"The Wabash abounds with fish of many kinds; which, in the months of April, May and June, may be readily caught with the hook and line."

"The Gar or Bill fish is more than two feet in length. It is quite slim. The bill is about six inches long, tapering to a point. Its scales are very close, thick, and hard." "The strength of this fish is great. In a small creek which flows into the Wabash, I discovered a considerable number, and caught several in my hands; but was absolutely unable to hold one."

"There are three kinds of Cat-fish: the Mississippi cat, the mud cat, and the bull head. Some of the first have weighed one hundred and twenty pounds. The mud cat is covered with clouded spots and is a very homely fish. The head is very wide and flat. Some have weighed one hundred pounds."

"The real sturgeon is found in the Wabash though the size is not large. These have been taken from twenty to sixty pounds weight."

"The shovel fish or flat nose is another species of sturgeon. It weighs about twenty pounds."

"The pond pike is taken in ponds from one to three feet long, but very slim. It is an excellent fish" (Northern pike). "The river pike is large and highly esteemed, but scarce" (Muskellunge).

"The drum or white perch weighs from one to thirty pounds. It is shaped like the sunfish."

"The black perch or bass is excellent, and weighs from one to seven pounds" (Smallmouth bass). "The streaked bass is scarce" (White bass).

"The Buffalo fish is of the sucker kind, and very common. Weight from two to thirty pounds."

"The rock Mullett is sometimes seen three feet long. It is slim and weighs from 10 to 15 pounds." (Smallmouth buffalo?)

"The red horse is also of the sucker kind. It is large and bony, weighing from five to fifteen pounds."

"The Jack pike or pickerel is an excellent fish, and weighs from six to twenty pounds." (Walleye pike)

"The eel is frequently taken in the Wabash, and weighs from one to three pounds. I was told that no fish was found in these waters of a good quality for pickling; and the facts, that mackerel are brought over the mountains from Philadelphia, and white fish from Detroit, tend to confirm that statement."

"The fresh water clam or muscle is so plenty, as to be gathered and burnt for lime. Twenty years ago, I am told, no other kind of lime was procured."

Figure 4: Darwin Ferry located between Merom, IN and Terre Haute, IN. circa 1850. (Vigo County Historical Museum, Terre Haute, Indiana)

Figure 5: Darwin Ferry today with Darwin, IL across the river.

"Craw fish, which resembles the lobster, is very common in the low lands of this country. It is a size larger than the common crab. It works in the ground, and throws up heaps of earth about six inches high, and hollow within. These little mounds are very numerous, and the surface of the ground resembles a honey comb."

Thomas was also alarmed about the number of sick people living along the river and offered some recommendations to the traveler and new settler.

"Let no temptation prevail on the emigrant to go fishing in warm weather. Of the smell of the shores I have spoken. To be wet is imprudent; and to be exposed to the chilling damps of the night, greatly increases the danger. But fresh fish are unwholesome, except for a slight change of diet. We know of no new settlement that has been healthy, where the inhabitants live chiefly on fresh fish. If, however, fish must be eaten, buy them; any price is cheaper than health; and if fishing must be done, do it in cloudy weather; but at night be comfortably sheltered."

Figure 6: View of the lower Wabash River, by the noted Swiss painter Karl Bodmer, who accompanied Maximilian on his visit to New Harmony.

More than one observer remarked upon the great clarity of water. Mrs. Lydia Bacon (McCord, 1970) wrote of 1811 Vincennes "the local situation of the place is very pleasant, lying on a clean stream of Water which affords them a variety of fish & facilitates their intercourse with the Neighbouring States & Territories."

"It was in the month of April (1825) when I first saw the Wabash River...Schools of fishes--salmon, bass, redhorse, and pike--swam close along the shore, catching at the blossoms of the red-bud and plum that floated on the surface of the water, which was so clear that myriads of the finny tribe could be seen darting hither and thither amidst the limpid element, turning up their silvery sides as they sped out into deeper water." (Cox, 1860).

Hugh McCulloch (McCord 1970, pg.147) visited the Wabash River at Logansport in late May, 1833 and wrote ". . I followed an Indian trail that led along the banks of the Wabash, which had not then been deprived of any of their natural beauty by either freshets or the axe of the settler. The river was bank-full. Its water was clear, and as it sparkled in the sunlight or reflected the branches of the trees which hung over it, I thought it was more beautiful than even the Ohio..."

Fish were plentiful and easy to obtain at this time. Rafinesque (1820) stated "Fishes are very abundant in the Ohio, and are taken sometimes by the thousands with the seines.." "The most usual manners of catching fish...are, with seines or harpoons at night and in shallow water, with boats carrying a light, or with the hooks and line, and even with baskets."

Rafinesque (1836), the first scientist to study the life of the Ohio River and its tributaries, descended the river from Pittsburgh during the summer of 1818 where he first "began to study the fishes which we caught or bought, making drawings, &c". Most of his direct observations were made on Ohio River species and there is no indication that he himself studied fish specimens from the Wabash River. He has been described as brilliant and eccentric with a roving character of mind which was to lead him into mental vagabondage, wandering at will over the entire field of books and nature. Certainly he was the consummate "splitter", finding a multiplicity of species where only one, in actuality, existed. Nevertheless, prior to his studies only about 12 resident species had been properly named and described. By 1819 he had described about 100 species.

One of the people to which Rafinesque's 1820 book was dedicated was Charles Alexander Lesueur, a naturalist who had first described species of fish from Lakes Erie and Ontario. Lesueur arrived in New Harmony on the lower Wabash in January 1826 in the company of many other noted scholars including Thomas Say, the so-called "Father of American Zoology." Here he worked for 10 years, intending to complete a work on the fishes of North America. He was an excellent artist and sketched extensively in his travels up and down the entire Mississippi River, but he was a poor writer and never completed his intended work.

Most of Lesueur's limited writings and extensive sketches returned with him to France in 1837 after the death of Say. In 1845 he became the first curator of the Museum d'Histoire Naturelle at Le Havre

Figure 7: Charles-Alexandre Lesueur drawn by Swiss artist Karl Bodmer, who accompanied Maximilian.

(Pitzer, 1989). Some of his 1200 sketches have been published in three volumes by Bonnemains (1984).

Thomas Say sketched and scientifically described many species of insects and molluscs and completed several excellent publications before his death and burial at New Harmony.

Species of fish which Rafinesque regarded as abundant or common in the Ohio River included: drum, the three species of black bass, bigmouth buffalo, shovelnose sturgeon, channel catfish, yellow bullhead, goldeye, mooneye, emerald shiner, golden shiner, logperch, greenside darter, and fantail darter. Hook and line fishing was too slow for the settler, therefore, they often used seines and speared or gigged fish from the clear waters characteristic of most streams at that time.

On the smaller Wabash tributaries still other means were employed: "At John Stitt's mill below town (Crawfordsville), on Sugar river, there is a fish-trap, and in one night we caught nine hundred fish, the first Spring we were in the country (1825), most of them pike, salmon, bass, and perch. Some of the largest pike and salmon (Walleye) measured from two to four feet in length, and weighed from twelve to twenty-five pounds." (Cox, 1860).

Dunn (1910) records the abundance of fish in the White River according to early residents of Indianapolis as follows: "George W. Pitts commented "There was no end of fish in the streams in those days. I went up to McCormick's dam (just above the Country Club) four miles above town on the river one day and sat down at a chute that had broken out and where fish were running through. There were wagon loads of fish, and I threw out with my hand eighty-seven bass, ranging in size from one pound up to five."

"Amos Hanway says there were 'bass, salmon (walleye and/or sauger), redhorse, ordinary suckers, quillbacks, or as they were sometimes called spearbacks, perch, pike, catfish, etc. The biggest salmon I ever caught weighed sixteen pounds. I once caught a pike (Muskellunge) that measured four feet and two inches; at another time a gar-fish that measured over three feet, and a blue catfish that weighed sixteen and a quarter pounds. The finest rock bass (largemouth bass) I ever took was one which weighed eight and a quarter pounds, and that was near Waverly; while the biggest river bass (smallmouth bass) I ever lifted from the water weighed six and one-fourth pounds.'" He went on to say that once in Morgan County, above the Cox dam, when the fish were running, he and his brother Sam "at one haul seined twelve barrels of fish, and there were thirty fish that averaged, undressed, ten pounds each. They were mostly bass and salmon, but there were also large redhorse, white perch, quillbacks and ordinary suckers."

The importance of an abundance of excellent fish is stressed. Flint (1826, in Lindley, 1916) maintained that "The streams, and especially those that communicate with Lake Michigan, are abundant in fish of the best qualities. The number and excellence of the fish, and the ease, with which they are taken, are circumstances of real importance and advantage to the first settlers, and help to sustain them, until they are enabled to subsist by the vails of cultivation."

Maximilian, Prince of Wied (1843) overwintered from Oct. 19, 1832 to March 16, 1833 at New Harmony, which at that time had a population of 600. "The Wabash, a fine river, as broad as the Moselle, winds between banks which are now cultivated, but were lately covered with thick forests." He wrote about the large size and diversity of trees, and described the understory. He was familiar with many birds and other animals. At that time bison, elk, bear, and beaver were "now entirely extirpated." although in the first decade of the 1800's there were many bear & wolves, and small numbers of elk and beaver. "The Virginia deer is still pretty numerous, but is daily becoming more scarce." "The wolf is still common, . ." Grey and red fox, racoon, opossum, groundhog, muskrat squirrel, polecat, otter, & mink were all common and rabbits and pine martin were sometimes seen.

Snapping, softshell, and other emys "are numerous" "The proteus (Menobranchus lateralis, Harl.) of the Ohio and of the great Canadian lakes, is found in the Wabash."

This may have been either hellbender, mud puppy or both. "There are many kinds of fish in the Wabash, on the whole the same as in the Ohio and the Mississippi. 100 pound catfish, several species of sturgeon and pikes, the horn-fish, the buffalo, . . a large fish resembling the carp, &c., paddlefish" "At places where the flat boats, laden with maize, land, the fish collect and assemble in great numbers, and fall an easy prey to the fisherman."

He spent a winter which was unusually mild and dry and, on a trip with Le Sueur, commented "The water of the river is clear and dark green, and the bottom, which is plainly seen, is covered with bivalve shells (Unio), as well as with several kinds of snails."

"Indians were reported to have lived around Harmony until 1810, but in the year preceding the battle of Tippecanoe they all removed, and did not return." He cryptically says that Indians were "now totally extirpated and expelled from Indiana, and the country enjoys the advantage of being peopled by the backwoodsman."

Of the local residents Maximillian says they are ". .called backwoodsmen because they live in the remote forests. . . a robust, rough race of men, of English-Irish origin." "They dwell very isolated, scattered in the forests, and seldom come to town, only when business calls them."

Great changes were to occur between 1830 and 1845, as Cox (1860) poignantly writes: "I can well recollect when we used to wonder if the youngest of us would ever live to see the day when the whole of the Wea plain would be purchased and cultivated; and our neighbors on the Shawnee, Wild Cat, and

Nine Mile prairies were as shortsighted as we were, for they talked of the everlasting range they would have for their cattle and horses on those prairies--of the wild game and fish that would be sufficient for them, and their sons, and their sons' sons. But those prairies, for more than fifteen years past, have been like so many cultivated gardens, and as for venison, wild turkies and fish, they are now mostly brought from the Kankakee and the lakes."

George Winter, one of the first professional artists to live and work in Indiana at Logansport, recorded his observations in diaries and letters. In 1841 he writes: "The sprightly Wabash was low (July) and its rocky bed was ocassionally visible, yet it flowed wildly on. The river is a clear and rushing stream, dotted by small islands-which threw their images upon the glassy surface. It was a mixed scene that presented itself to the eye, combining the wild with the partial markings of civilization: on the southern side of the Wabash is the great Miami Reservation, known for its unsurpassed excellence of soil and valuable timber. It is a noble but Fated forest, and the sound of the axe had already reverberated in its shady recesses."

A few years later Winter commented "Now A.D. 1845 we witness . . . the effects of the partial clearing up the country . . . has had a striking effect upon the affluents of the Wabash-the beautiful Islands . . . are beginning to wash away under the influence of the greater volume of water that fills the banks and increased rapidity of the current of the river."

The changes, of course, came about as the direct and indirect result of a rapidly increasing population and the establishment of extensive agriculture which, in turn, led to the exportation of surplus agricultural goods. Navigable waters throughout the state were clearly needed, but the physical clearing of the land itself acted against this need in many of the smaller streams.

"Unquestionably White River is not so easily navigable now as it was ninety years ago, though probably as much water passes out through its channel in the course of a year as there did then. The flow is not so steady because of the clearing of the land and improved drainage make the surface water pass off more rapidly. And this has increased the obstructions in the streams, for the soil, sand, and gravel wash much more easily from cleared land. Moreover, in the natural state, most of the timber that got into the river came from the undermining of banks on which it stood, and this usually did not float away but hung by the roots where it fell. But after the axmen got to work, every freshet brought down logs and rails which formed drifts at some places. Some logs stranded as the water went down, decayed, became waterlogged, and made bases for sand and gravel bars." (Dunn, 1910)

HISTORIC INFLUENCES
THE CANAL ERA (1825 - 1875)

Sediment pollution from a rapidly expanding agriculture soon had a powerful ally when the state began an extensive, but unrealistic program of canal building from 1828 to the mid 1850s. Beginning at Fort Wayne in 1828, the Wabash and Erie Canal crept steadily westward toward the mouth of the Tippecanoe River, which was considered to be the head of navigation for the Wabash River at that time. The canal paralleled the Wabash River on the north between Fort Wayne and Delphi and on the south and east from Delphi to Terre Haute (Figure 8). Elevated, waterproof aquaducts were constructed over each tributary and feeder canals channeled water into the Canal (Figure 9). Dams were constructed on several tributaries in order to provide the canal with water during the dry summer periods.

By 1834 about one thousand Irishmen were at work when they weren't drinking or fighting and the entire state was clamoring for canals. An appropriation was asked to open almost every stream in the state large enough to float a canoe (Esarey, 1912).

The upper portion of the Wabash and Erie Canal was in operation by July 4, 1835, but tolls were inadequate to keep it in repair. Many wooden aqueducts had already rotted and were in disrepair. In 1836 construction was begun on the Whitewater canal; it included two large dams when completed in 1839. Several sections of canal were constructed on the White River in 1836 and a 19-mile section of Pigeon Creek near Evansville dried up right after completion.

In August 1838, 859 Potawatomi natives were rounded up by General John Tipton and started on their ill-fated march toward Kansas, a trip during which more than a quarter of the group would perish. "The enforced Indian migration allowed unsatiable whites to engage in the free enterprise of demolishing forests, polluting streams, and building drab towns. Settlements were aesthetically no improvement over tribal villages, but all were puffed with pride" (Fatout, 1972).

Meanwhile, the Wabash and Erie Canal crept southward, reaching Lafayette in 1843 and Terre Haute in 1847. As originally planned, the canal was to connect with the Wabash River at Terre Haute. However, it was decided to extend it on to the Ohio River via the lower portion of the Central Canal which was to be constructed to Indianapolis. Thus, the canal proceeded southeast to the Eel River, on past Worthington and Bloomfield, and to Maysville just west of Washington. It continued on south through Petersburg, crossed the Patoka and finally linked up with the canal at Pigeon Creek.

Throughout the 1840s difficulties were encountered in supplying sufficient water to that part of the canal south of Lafayette and, as a result, so-called "feeders" were developed from tributaries all along the canal. The problem was particularly aggravating below Terre Haute and dams were constructed across Splunge Creek, Adams Creek near Monrovia, and Birch

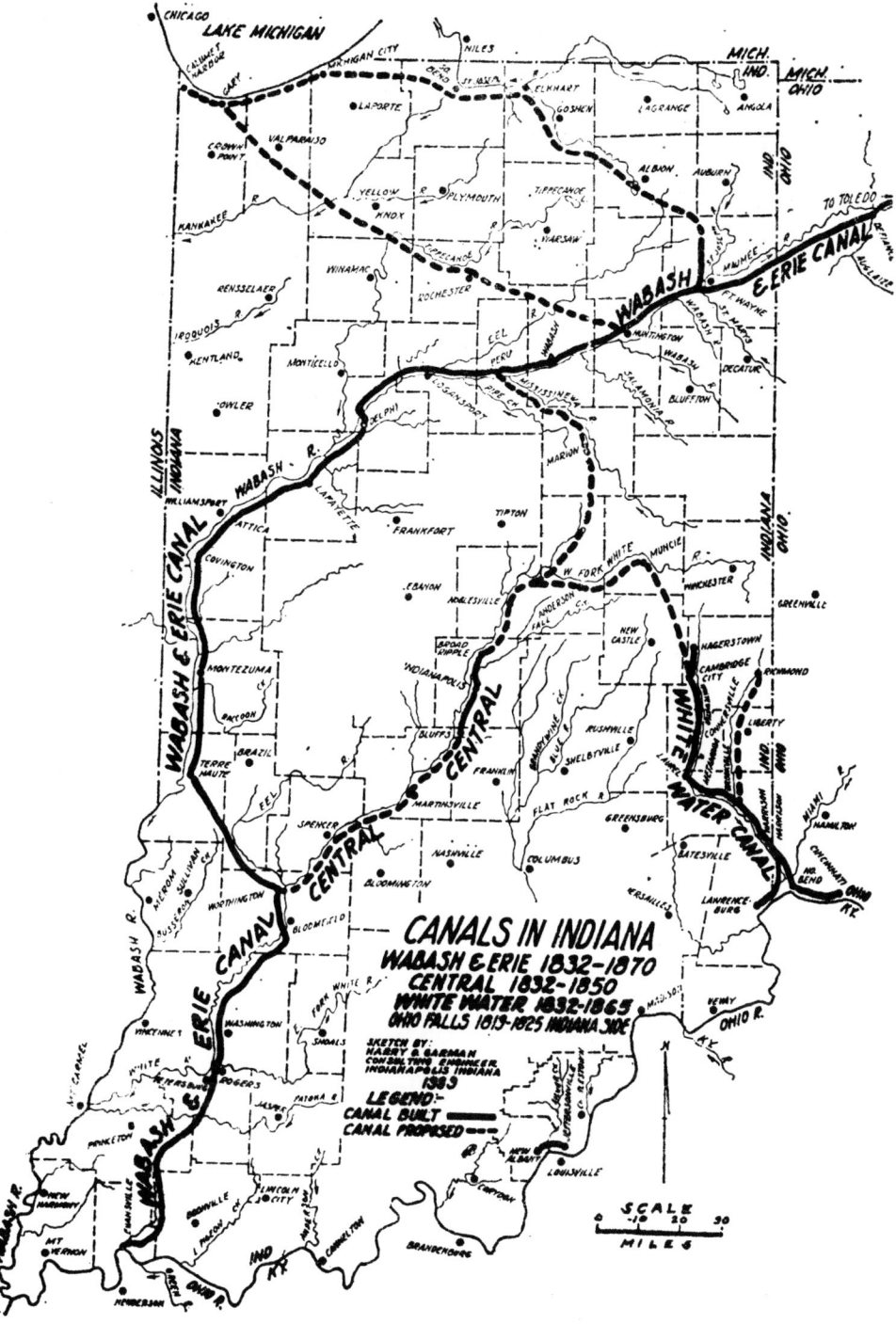

Figure 8: Indiana's canal system - constructed and proposed.

creek at Saline City. The six-square-mile reservoir at the latter site was subject to frequent acts of sabotage because it was believed to be the cause of a malaria outbreak and finally was drained completely.

The Canal was directly instrumental in increasing the human population, enlarging farms by clearing and draining lands which before were not considered worth cultivation (Anonymous 1907). More than 5000 bushels of corn were shipped in flatboats to Toledo in 1844 (Figure 10). This increased 100-fold in 1846, and amounted to 2,775,149 bushels in 1851. In that year also there were 9 flouring mills, 8 saw mills, 3 paper mills, 8 carding and fulling mills, 2 oil mills, and one iron foundry, all of which were operating from water power obtained from the Canal. It at once became the highway for handling firewood and the manufacture and shipping of lumber was begun and maintained for a long time on an enormous scale. In addition, stone quarries arose and lime was manufactured.

The steam boats which made their way up the Wabash River were small compared to those which navigated the Ohio and Mississippi Rivers, but vessels of 40, 80, or 100 tons were common (Cammack 1954). Lafayette became an important commercial center because of the increasing boat traffic on both the Wabash River and the canal. A city wharf was constructed which was already too small by 1853 to accommodate the demand by large and small craft. Steamboats over 100 tons were charged $4, those less than 100 tons $3, and other smaller craft $1 for 48 hours docking. Tippecanoe County also boasted 8 boat builders in 1850.

Many steamboats carried both freight and passengers, and some were showboats.

The "Floating Palace", for example, was a circus with scheduled shows at Independence, Attica, and Covington on April 27, 28, and 29, 1853 (Cammack 1954).

Although the Wabash and Erie Canal was separate from the Wabash River, the aggregate damage to tributary stream habitat and the Wabash River itself during this period must have been significant, but can only be inferred since few descriptive records exist. The entire canal program which caused most of the environmental problems would soon become completely worthless.

The zenith of Canal usage was 1850 when perhaps 500 boats navigated it. By this time repair or replacement was a regular feature of the Canal which had 9 aquaducts, 37 locks, 5 dams, 71 road bridges, and 139 culverts between the State Line and Perrysville alone. Water weeds clogging the Canal had to be removed repeatedly by a specially invented submarine mower. Cholera epidemics caused death and panic among residents and the Irish Canal laborers alike.

Spring floods in 1854 wrecked the Sugar Creek aquaduct and damaged another at Raccoon Creek. By this time, Indiana had about 1300 miles of operating railroads and another 1600 under construction. A worse repetition of floods, breaks, and droughts occurred in 1857 and 1858. A flood on the upper Wabash wrecked the Wildcat dam, carried away aquaducts over Wea and Shawnee Creeks, and breeched banks at a dozen places. Navigation was abandoned south of Terre Haute in 1860 and by 1870 little more than a succession of stagnant pools marked the site of the canal (Esarey, 1912), a casualty of gross fiscal mismanagement and competition from railroads.

Figure 9: Aqueduct over the St. Mary's River at Fort Wayne, Indiana.
(Indiana Historical Society)

Figure 10: Covered flatboat used to convey goods downriver.

THE SCIENTIFIC ERA (1875 - present)

The first reliable scientific records about the nature of the fish populations of the Wabash River date only since the 1870's when David Starr Jordan was beginning in Indianapolis what was to become a distinguished career in higher education. While teaching at Shortridge High School and Butler University, Jordan and Herbert E. Copeland avidly collected fish from the White River and began studies of the life-history of some species of darters. They also collected at the Falls of the Ohio River near Jeffersonville, attempting to make sense out of the much earlier, hasty work of Rafinesque (1820). His efforts expanded when Jordan moved to Indiana University.

Jordan's first paper (1875), subsequently republished in the 1892 Biennial Report of State Fish Commission, described the various species he had personally examined from the White River, the Wabash River, and the Ohio River. Taxonomic uncertainty abounded. The list was updated and clarified by Jordan in 1877 and ultimately appeared with altered terminology in excellent summaries by Eigenmann and Beeson (1894) and Hay (1895).

Between 1875 and 1888 Jordan, his students, and colleagues collected fish from several sites on the Wabash River and its tributaries (Jenkins 1886, Evermann and Jenkins 1888). His assessment of the river appearance during this period includes the following statements: "The upper Wabash and most of its tributaries are clear streams, . . ." and "Towards its junction with the Ohio R. the Wabash becomes a large river with moderate current, the water not very clear, and the bottom covered with gravel and sand in which grow many water plants. The tributary streams are mostly sluggish and yellow with clay and mud" (Jordan 1890). He found the "fish fauna of the Lower Wabash . . to be unexpectedly rich,. . ." especially in the number of species of darters and their abundance. His 1888 collections were accompanied by remarks on abundance; eg. scarce, not rare, common, abundant, and very abundant.

Rolfe (Forbes and Richardson 1920) stated "The waters of the Wabash are, like those of the Illinois and the Kaskaskia, commonly brown and opaque with suspended silt, never clearing even at the lowest stages; and the same is true of most of its tributary streams, especially those of the lower Illinoisan glaciation."

Culbertson (1908) pointed out the association of deforestation in Southern Indiana with the erosion of soil and the pattern of flood and drought, and noted that "Streams that thirty years ago furnished abundant power for mills during ten months of the twelve now are even without flowing water for almost half the time." He went on to state that they have also "had a serious effect .. upon the animal life of these streams."

This early scientific period was dominated by scientists who relentlessly sought to collect and catalogue new species of fish wherever they could be found. It was evident that rivers at that time were not pristine, for Jordan noted in an address to the State Fish and Game Convention on December 19, 1899 "...That there never were such (smallmouth) bass streams as in Indiana, and that White River is the best bass stream they have ever known. I think probably nothing better could be done--if we could devise a way--than to bring the bass back, and where there are now a dozen scattering fish put two or three thousand."

Thus, the seeds of an extensive stocking program were already sown prior to 1900. A few years earlier, the First Annual Report of the Commission of Fisheries of Indiana (1883) carried Jordan's "Catalogue of the Fishes of Indiana" in an Appendix, but chiefly discussed the coming availability of a great new species which would put fish back into Indiana lakes and streams--the carp. The Commissioners also discussed a recent fish-kill extending 20 miles downstream from Kokomo on Wildcat Creek, the result of pollution from a strawboard factory. If a new strawboard factory being constructed then at Anderson were to run its refuse into White River, they continued, "then it is "goodby fish" from Anderson to the Ohio River."

It is clear that extensive changes in the native fish populations had already occurred by the time Jordan arrived in Indiana in 1874, alterations at first resulting from clearing land for fields and building dams for grist mills and a bit later the wholesale destruction of habitat in a gigantic state effort to provide better commercial links between the fledgling state of Indiana and the eastern seaboard. Jordan was in time, however, to witness the first side effects of industrial development in Indiana, a development which would superimpose an additional heavy burden on the waters of Indiana.

During the first decade of the 1900's Forbes and Richardson (1920) collected fish at about five sites on the lower Wabash River between Illinois' Embarrass River and Little Wabash River. All of these early investigators employed the seine as their primary means of collecting fish and, therefore, were restricted to the more shallow, hard-bottomed sections of streams.

Shortly after these studies had been completed the problem of gross organic pollution and the human health aspects of pollution began to be perceived. "Before our population was so concentrated, sewage disposal by dilution was satisfactory from a physical standpoint, but now the condition of many of our streams has become such that for a part of the year at least, the odors from them are quite obnoxious and a nuisance to the cities and to the population living along the banks, as well as a menace to their health." (Craven 1912).

Following these early investigations there was a gap of nearly 40 years during which few studies were made. Blatchley (1938), a geologist by profession and a former student of Jordan, mostly summarized the earlier work. Not until the early 1940s did Gerking (1945) systematically collect fish at 412 sites scattered throughout Indiana in a comprehensive program, again using a quarter-inch mesh seine as the principal collecting technique. He seined 14 sites on the Wabash River downriver from Delphi and included commercial gill-net captures. His comprehensive work established a firm basis for future comparative studies.

Speaking mainly of streams smaller than

the Wabash River, Gerking wrote "Streams in the northern third of the state often run clear.." "Creeks of the central and southern part of Indiana are usually turbid and warm." "Many of the southwestern streams are slow-moving and usually carry a heavy load of suspended material."

Gerking (1945) found that the number of species of darters at sites sampled by the earlier investigators had diminished greatly since Jordan's time. He found only 3 species compared to 13 species earlier at Delphi, no species compared to 4 species at Mt. Vernon, and 3 species compared to 12 species at New Harmony. He attributed this reduction in diversity to increased siltation from soil erosion, but also had some observations on other possible influences from "city sewage, cannery waste, coal mine drainage, paper mill waste, and dairy-products factory waste." "The establishment of treatment methods of these wastes before their deposition in streams has done much to alleviate the problem, but much work remains to be done before pollution control is fully realized."

Visher (1944), addressing the causes for the increasing frequency of floods in Indiana, cited four main reasons: (1) abundant rainfall, (2) concentration of rainfall, (3) inadequate size and numbers of runoff channels, and (4) changes produced by man. "Deforestation has greatly increased runoff as have extensive drainage operations by open ditches and by tiles. Indiana has many miles of drainage ditches, 20,787 miles according to the 1930 Census of Drainage. Moreover,..there are..many thousand miles of small tile in addition to the 10,439 miles of large tile such as storm sewers ... the extensive erosion of cultivated hillsides have carried large amounts of soil and other materials from the higher levels into the stream channels.."

It is safe to say that the investigative effort exerted on the Wabash River during the past 25 years far exceeds the combined efforts of the past. The research effort during this period has been directed primarily toward assessing man's activities as they influence elements of the river ecosystem. State environmental organizations such as the Department of Conservation and Environmental Protection Agency in Illinois and the Department of Natural Resources and Department of Environmental Management in Indiana have implemented extensive chemical and biological sampling programs. Governmental and industrial sponsorship of research programs have reached new heights. Immediate answers were sought for environmental questions which were often inadequately understood.

The fish populations of the Wabash River have been studied by our group since 1967, initially concentrating on thermal effects near the Wabash River Electric Generating Station (EGS) north of Terre Haute, Ind. and Cayuga EGS near Cayuga, Indiana. In 1973 we extended the studies to longer stretches of the river because we realized that heat was but one of many environmental influences on the biota of the river. Since 1978 we have studied nearly 170 miles of river extending from Delphi, Ind. on the north (RM 330) to Merom, Ind. on the south (RM 161). Most of the following sections summarize this research through 1997.

PHYSICAL ATTRIBUTES OF THE WABASH RIVER

The Wabash River originates in the agricultural drains and ditches in northwestern Ohio near Fort Recovery at an altitude of about 267 m and flows southwesterly 764 km to enter the Ohio River at an altitude of about 97 m near Mount Carmel, Illinois. It flows freely throughout its length except for Huntington Reservoir at river kilometer (rkm) 662 (RM 411). Its average rate of descent is about 0.22 m/km. However, the river is best divided into two sections: a relatively steep upper section with a rate of fall of 0.454 m/km and the longer, lower section with a rate of fall of only 0.123 m/km (Figure 11). The transition between these dissimilar segments is abrupt and occurs at about rkm 570 (RM 356) near Logansport.

Except for the Tennessee River basin, the Wabash River basin is the largest Ohio River tributary basin with a total land area of 85,500 km². It receives water from 62,000 km² of Indiana, 22,540 km² of Illinois, and 740 km² of Ohio. The Wabash River drains 65.6% of Indiana's area (Clark 1980). Excluding Lake Michigan, it is the largest body of water in Indiana, with more than 100 km² of surface area.

About two-thirds of the watershed is devoted to agricultural cropland (Figure 12), the highest proportion within the Ohio River basin (ORSANCO 1990). An additional 8.2% is in pasture or grassland. Forests or woodland constitutes only 13.5% of the basin's land area, the second smallest proportion within the Ohio River basin. Agricultural development is considerably

greater in the upper half of the basin. The lower half of the basin has a greater percentage of forests, as well as most of the surface coal mines. The upper half of the Wabash River lies in the **Eastern Corn Belt Plains** Ecoregion, while the lower half is contained within the **Interior River Lowland** Ecoregion (Omernik and Gallant 1988).

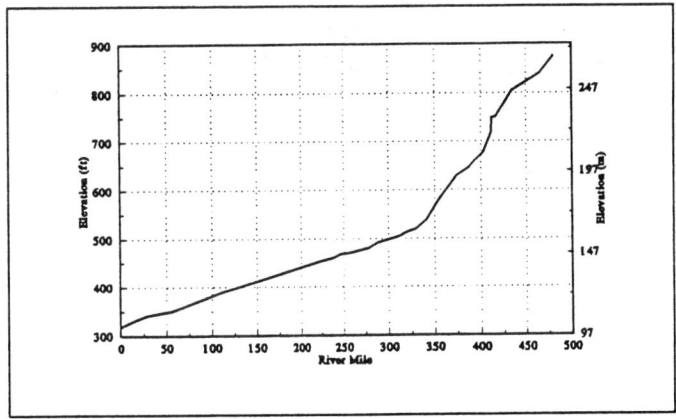

Figure 11: Elevation profile of the Wabash River.

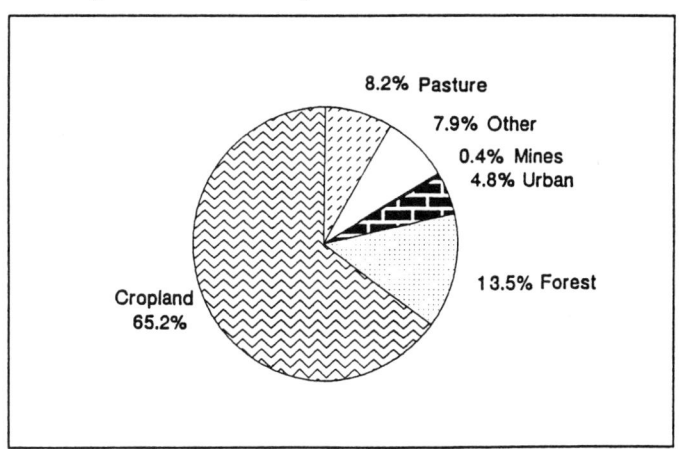

Figure 12: Primary landuse in the Wabash R. basin.

During the Pleistocene Epoch, glaciers moved into Indiana at least three times. The Illinoian boundary advanced nearly to the Ohio River, resulting in deposits of glacial drift ranging in thickness from less than 15 m in the south to more than 90 m in the north. Extensive glacial till covers the northern basin with sand and gravel outwash deposits occurring along major rivers and streams.

The middle Wabash River flows through an extensive deposit of loess (Figure 13). Here, the major underlying material consists of Pennsylvanian and Mississippian bedrock.

Homoya et al. (1986) delineated 12 natural regions of Indiana, 11 of which are based importantly on the dominant natural vegetation composition. Their **Central Till Plain** natural region, which roughly corresponds with the **Eastern Corn Belt**

Plains Ecoregion, is subdivided into three Sections: **Bluffton Till Plain, Tipton Till Plain,** and **Entrenched Valley.**

The Wabash River flows within the Bluffton Till Plain from its origin at rkm 764 (RM 475) to rkm 570 (RM 354) near Logansport, Indiana. From there to rkm 370 (RM 230) near Clinton, Indiana, it is contained within the **Entrenched Valley,** and for the remainder of its journey it flows through the **Southern Bottom-lands** natural region.

Homoya et al. (1986) also distinguished a separate **Big Rivers** natural region that includes the lower Wabash from about rkm 467 (RM 290) near Attica, Indiana to its mouth and also the White River from the confluence of the East and West forks to its mouth.

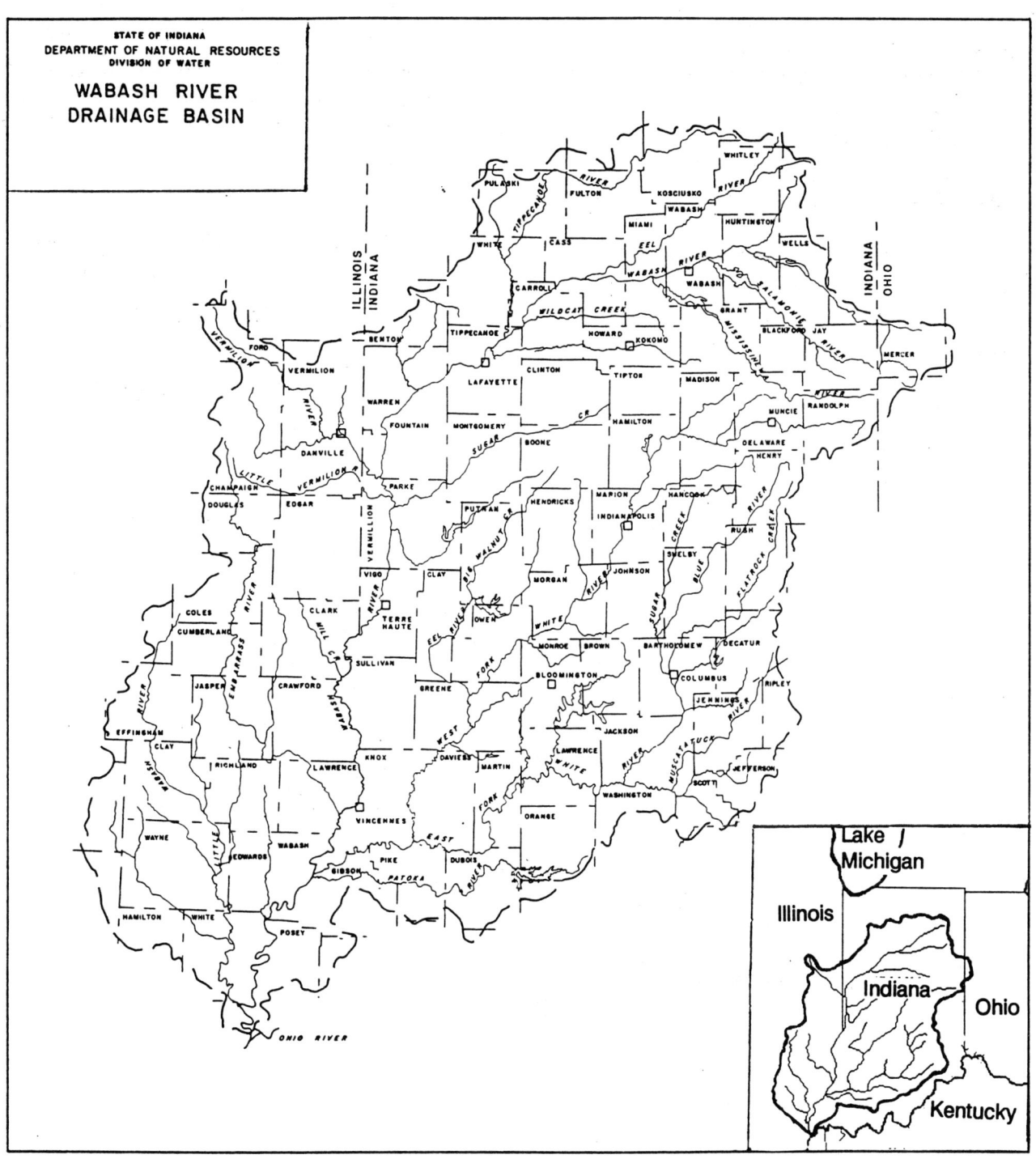

Figure 13: The Wabash River drainage basin.
(Indiana Department of Natural Resources, Division of Water)

24

Flow Regime of the Wabash River

In terms of average discharge, the Wabash River is about one-fourth the size of the Ohio River where the two rivers meet. On the average nearly 850 m³/s is discharged into the Ohio River, making the Wabash River the twelfth largest river in the United States (Todd 1970). The White River (drainage basin area = 29,394 km²), a major tributary, is 69.1% as large as the Wabash River basin (42,538 km²) where they join at rkm 154 (RM 96). In terms of discharge, however, the two rivers produce nearly the same average volume of water, 335.6 m³/s for the White River and 341.5 m³/s for the Wabash River (Arvin 1989).

Although Huntington Reservoir is the only mainstem reservoir, water discharge in the upper Wabash River is influenced by the Salamonie and Mississinewa Reservoirs. Annual discharge in the upper river (Peru) is lower and more stable than more southerly stations (Figure 14).

Significant contributions of water between Peru and Lafayette are made by the Eel and Tippecanoe rivers and Wildcat Creek. Lakes Shaffer and Freeman, located on the lower Tippecanoe River, not only influence the flow of the Wabash River, but also provide a source of clarified water.

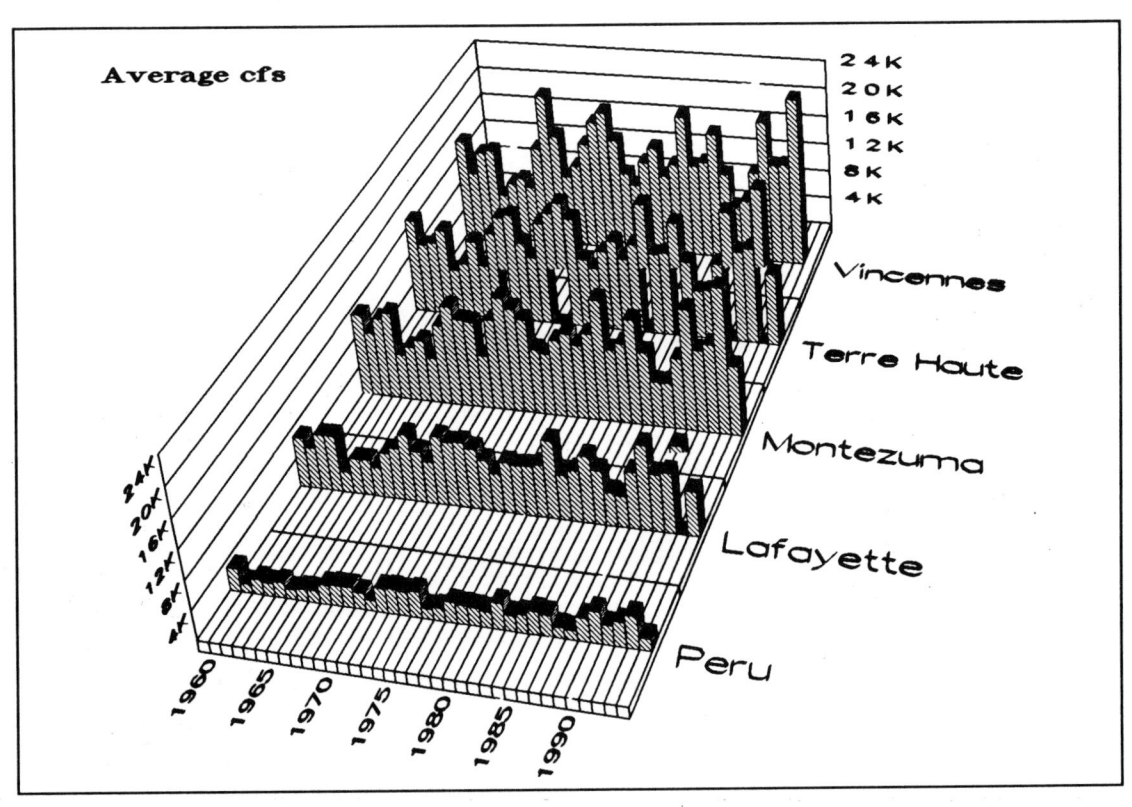

Figure 14: Mean annual discharge of the Wabash River at Peru, Lafayette, Montezuma, Terre haute and Vincennes from 1959 through 1994.

Big Pine Creek, Big Vermilion River, Sugar Creek, and Big Raccoon Creek enter the Wabash River between Lafayette and Terre Haute. Reservoirs are found only on the Middle Fork of Big Vermilion River and Big Raccoon Creek (Mansfield Reservoir).

Seasonal variations in discharge are substantial (Figure 15). The highest average discharge in March or April (18,000 to 20,000 cfs) is about four times larger than the flow in August, September, or October (5,000 cfs).

River discharge based on seasonal or annual periods of time is valuable information for some purposes, for example, in determining the yield of water from a particular watershed. It may even be useful in characterizing rivers in a general way. However, it is limited in terms of its ecological value.

Rivers rarely flow at a steady, average rate. Indeed, the variability of discharge or flow in most rivers is one of their most distinguishing and vexing characteristics. Rivers are among the least stable environmental systems on our planet and, because of this, they are among the most difficult systems to study ecologically.

Today's Wabash River is unusual in that it flows unconstrained for the last 662 km (411.4 miles) of its journey south to the Ohio River. Few rivers in the United States can make that claim, especially rivers east of the Mississippi River.

The presence of reservoirs on some tributaries, in addition to the single mainstem reservoir (Huntington Lake) moderate flow variability to a limited extent. However, most of these reservoirs are small because of

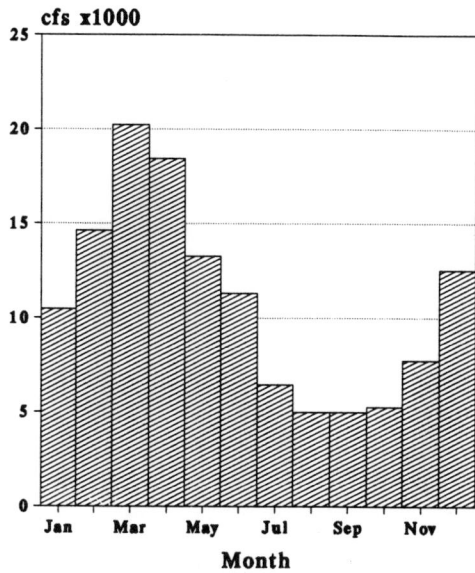

Figure 15: Mean monthly discharge of the Wabash River at Montezuma 1970-1989

the gently sloping topography. For example, Huntington Reservoir was constructed in 1969 and is approximately 905 acres in surface area (Clark 1980). Salamonie Lake was completed in 1966 and is approximately 2,800 acres in surface area. Mississinewa Lake, completed a year later, is approximately 3,180 acres. Lakes Shafer and Freeman on lower Tippecanoe River are less than 3,000 acres combined.

The variability of the discharge affects all of the biotic components of lotic ecosystems. Horowitz (1978) has shown that sections of rivers having more constant discharge rates tend to have greater fish species richness than other more changeable sections (Horwitz 1978).

The daily flow of the Wabash River measured at Montezuma during the summer is shown in Figure 15. The superimposed horizontal bars indicate electrofishing periods. Electrofishing success under different flow caused us to develop a different flow caused us to develop a collecting criteria based on river level. We decided to collect only when the river dropped to certain levels as determined by U.S.G.S. gauging stations at Lafayette and Terre Haute, Indiana. In the following dis-charge figures the dates we electrofished are shown as horizontal lines. Long lines usually indicate that the entire length of the middle Wabash River was electrofished. Short lines usually indicate that only smaller segments were electrofished.

Jim Thayer, Rick Wright, and Bob Gammon motoring downriver
to the next collecting station.

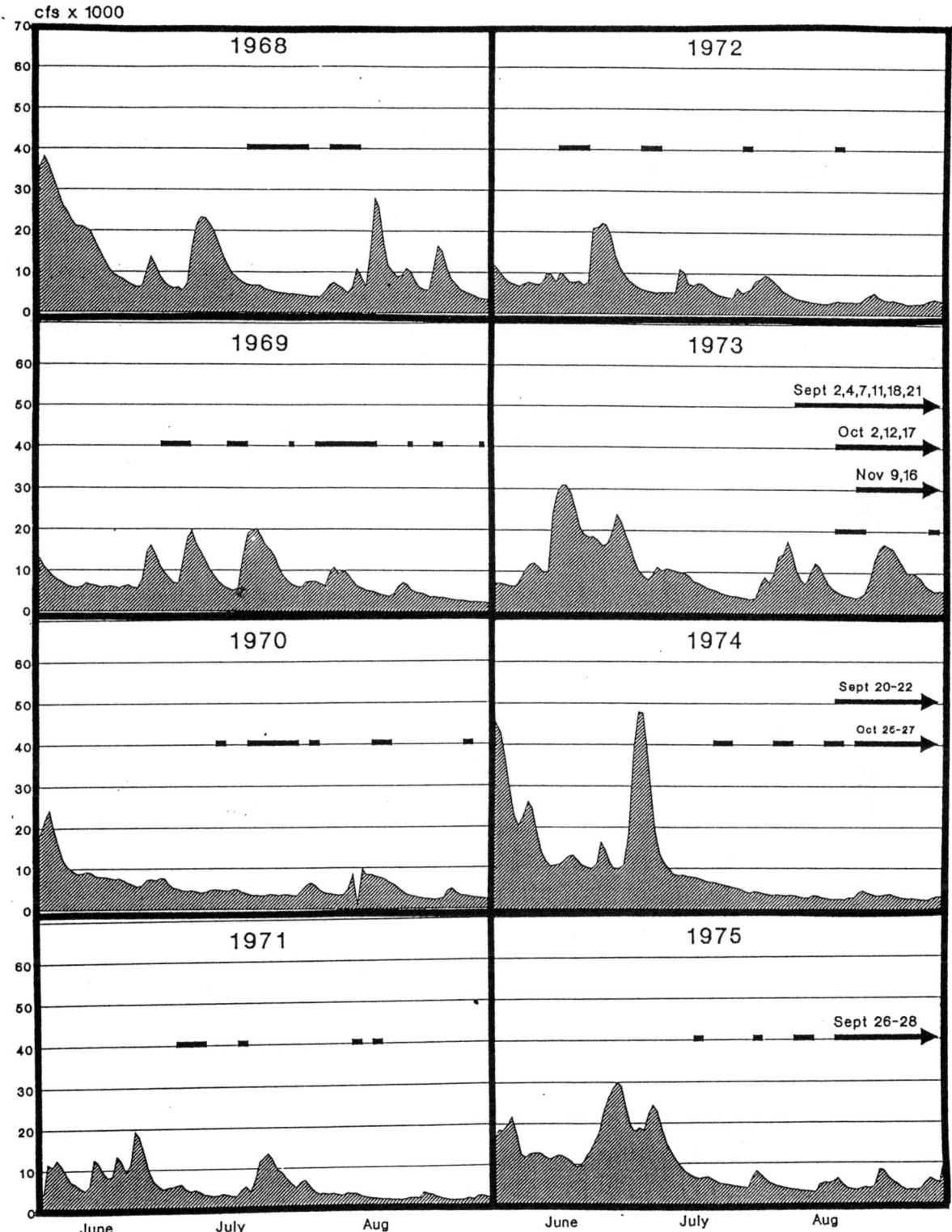

Figure 16: Daily discharge (cfs x 10^3) of the Wabash River as measured by the U.S.G.S. gauging station at Montezuma, Indiana from May 23 to August 31.

28

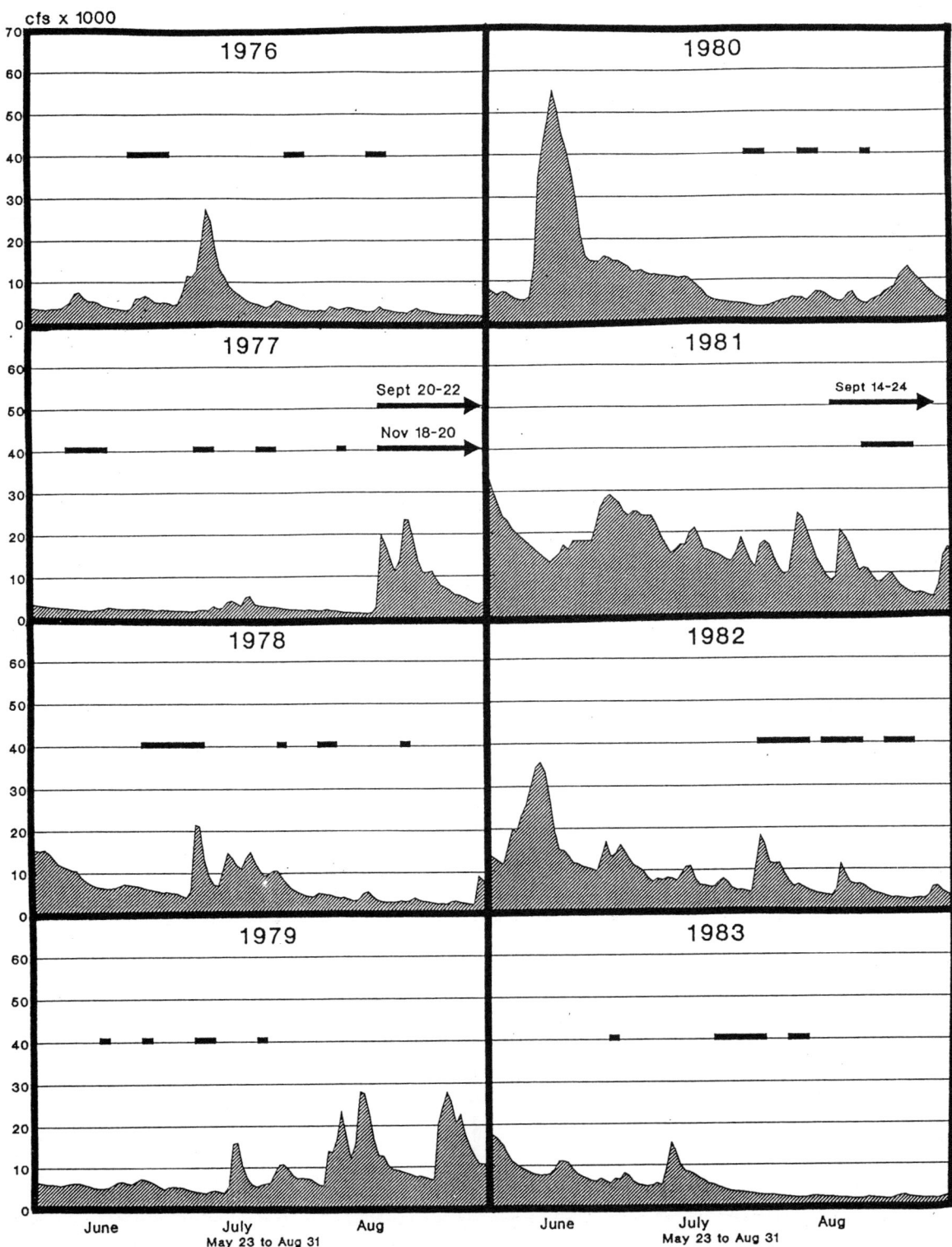

Figure 16: (continued)

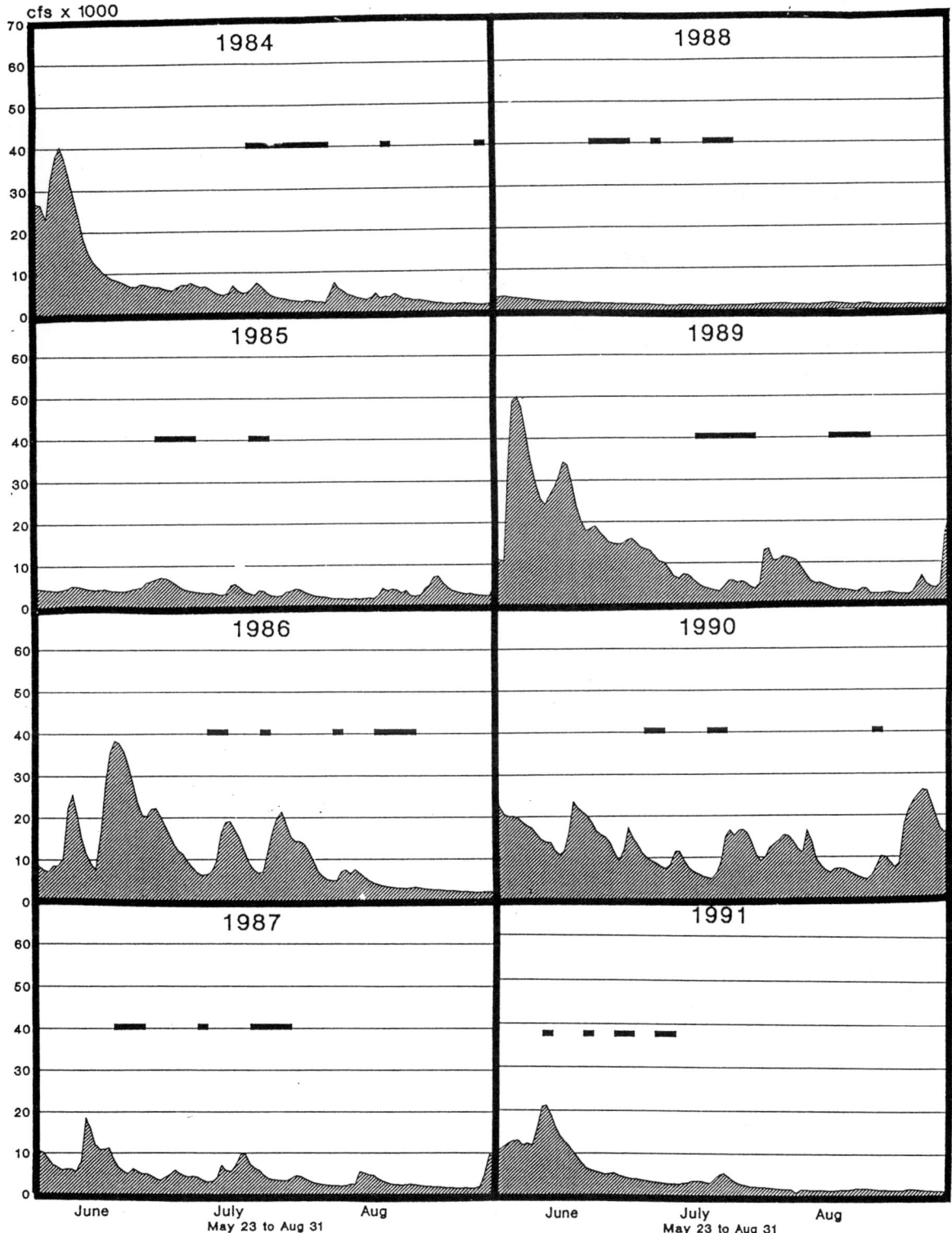

Figure 16: (continued)

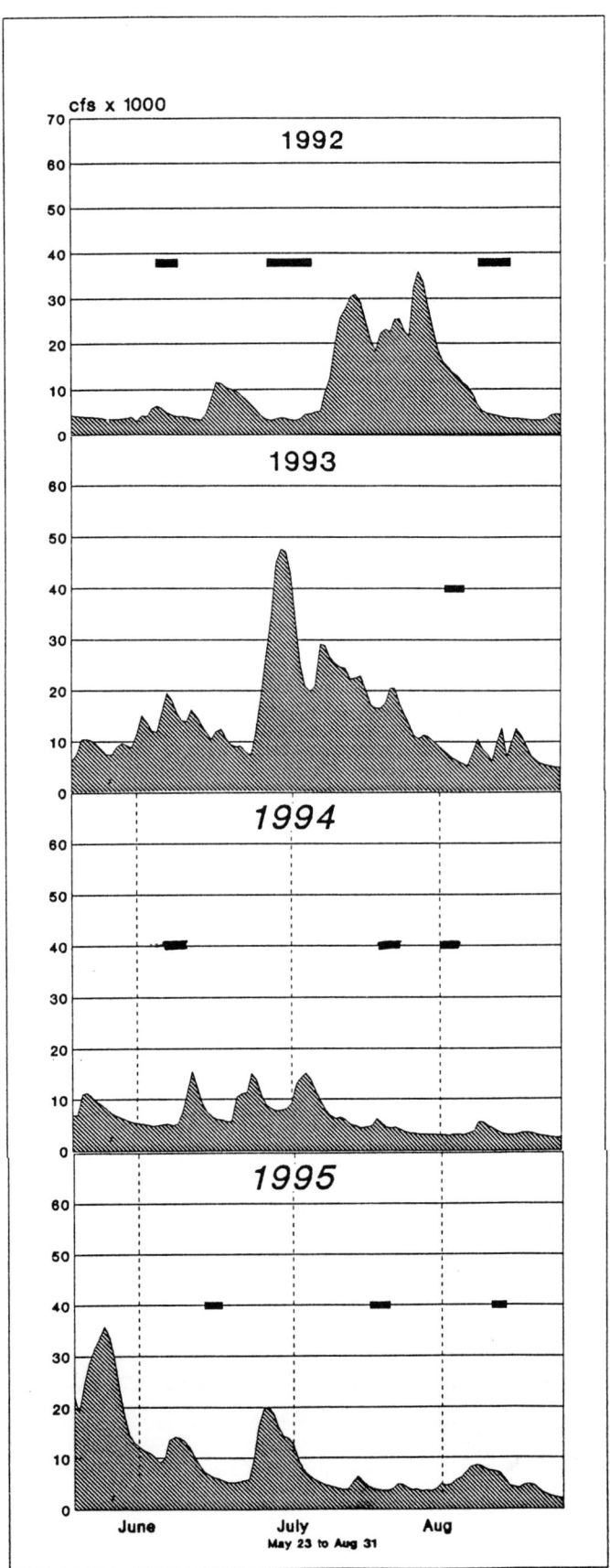

cfs x 1000

Figure 16: (continued)

Temperature

Water temperatures during the winter months are normally near 0°C and then rise steadily from March through June (Figure 17). In years of cold winters the river is ice-covered north of the Cayuga EGS. Temperatures often exceed 25°C during the low-flow months of July, August, and early September and then decline sharply in October.

The diurnal changes in water temperature of flowing waters are normally less extreme than for air temperatures. The temperature differential between air and water may exceed 10-15°C although the daily range in rivers is usually less than 4°C (9°F). A typical annual temperature pattern is shown in Figure 17 for data collected by USGS at Lafayette in 1969. The weekly mean, high, and low temperature is shown by the vertical line.

Reservoirs, lakes, and ponds usually stratify thermally during the warm summer months. However, this rarely occurs in flowing waters because of mixing, even in large rivers which have low dams such as the Ohio River. Temperatures in unshaded shallows such as near shore and riffles may increase substantially. Most streams provide the biotic residents with a diversity of thermal habitats. Cooler water from shaded tributaries together with springs and areas of upwelling ground water provide valuable refugia during periods of thermal stress.

Many research reports have been written on the effects of electric generating stations (EGS) on the biota of the Wabash River. Examples appear later in this report. A complete list of detailed reports may be found at the last section.

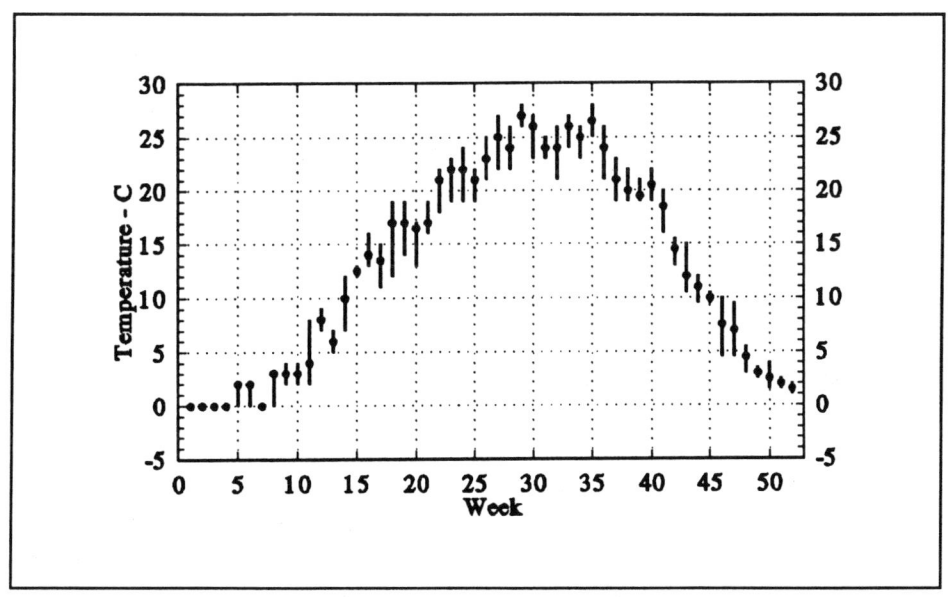

Figure 17: Pattern of thermal change in the Wabash River at Lafayette, IN in 1969.

Suspended Solids Concentration

Most rivers are murky to some degree because of a heterogeneous mixture of suspended particles. Some of this is inorganic matter such as sand, silt, and clay. Another part consists of microscopically small fragments of decomposing organic matter including bits of leaves and dead insects. In larger rivers living phyto- and zooplankton are carried along by the current as part of the suspended solids load. Rivers originating from melting glaciers carry large loads of mostly inorganic glacial "flour". Rivers draining swamps or bogs are often rootbeer brown in color because of colloidal humic organic substances in suspension.

The suspended solids concentration (SSC) is of considerable importance both to the river's inhabitants and to humans who use the river as a resource.

From an ecological standpoint the deposition of part of the suspended solids load may negatively influence reproductive success for some species of fishes. For some fish and most benthic organisms at least part of the SS load is food in the form of detritus, bits of dead organic matter, and phytoplankton. For other species of fishes the turbidity which accompanies the SS load severely limits visual capabilities and interferes with the location of food. Most of the so-called "sport" fishes such as smallmouth bass (*Micropterus dolomieu*), spotted bass (*Micropterus punctulatus*), sauger (*Stizostedion canadense*), white bass (*Morone chrysops*), and crappie (*Pomoxis spp.*) are included in this category. Additional non-game predaceous fishes which locate food visually include goldeye (*Hiodon alosoides*), mooneye (*Hiodon tergisus*), and skipjack herring (*Alosa chrysochloris*).

That part of the SS load consisting of decomposing organic matter and phytoplankton may influence the biotic community through its effect on the dissolved oxygen concentration, an aspect which will be addressed in the next section.

The SS load is also important to all human communities which utilize surface waters for human consumption. Suspended solids have to be removed in treating the water for human consumption. High concentrations of suspended solids in rivers or lakes are also aesthetically undesirable from a recreational perspective. Few people care to swim, canoe, or fish in turbid water which is perceived as being "dirty" water.

The average annual suspended solids concentration (SSC) in different sections of the Wabash River during the period 1977-1987 is shown in Figure 18. The average SSC doubles from Peru to Vincennes and increases still further by the time the river reaches the Ohio River. The observed pattern of increasing SSC with increasing river size is probably typical of most free-flowing rivers.

Crawford and Mansue (1988) examined partial-records for the Wabash River at Lafayette for the years 1964 through 1982 (N = 59). A statistically significant increase occurred in discharge ($P < 0.05$), but no change over time was indicated for SSC, flow-adjusted SSC, or suspended-sediment discharge. The median SSC was 66 mg/L (mean = 87 mg/L; range = 6 to 980 mg/L). The median suspended-sediment discharge was 714 tons/day (range = 13 - 81,500 tons/day).

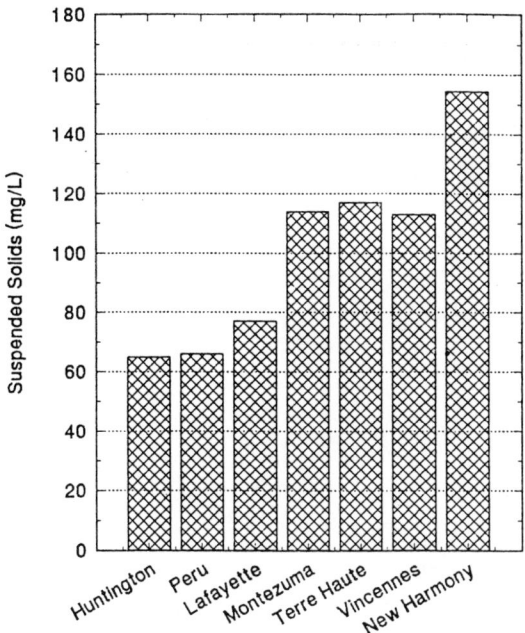

Figure 18: Mean suspended solids concentration (mg/l) 1977 - 1987.

The daily SS load at Lafayette was examined for a shorter period during 1978-80 (N = 784). The median suspended-sediment concentration at this time was 45 mg/L (mean = 61 mg/L; range = 6 to 677 mg/L) and the median suspended-sediment discharge was 1,341 tons/day (mean = 1,882 tons/day; range = 21 to 79,600 tons/day).

The SSC is strongly influenced by river discharge and both parameters are related to rainfall events in complicated fashion. Crawford and Mansue (1988) used partial-records for the period 1951 to 1980 to mathematically describe the relation between water discharge and suspended-solids discharge for the Wabash River at Lafayette as a first-order polynomial equation:

$$Ln \ Q_s = -6.5993 + 1.5832(Ln \ Q_w)$$

where LnQ_s is the natural logarithm of the suspended-solids discharge in tons per day; and

LnQ_w is the natural logarithm of the water discharge in cubic feet per second.

HydroQual (1984) examined the SSC and SS load as a function of Wabash River flow for the summer period only, using ISBH data from 1968 to 1981. Figure 19 indicates the SS load as a function of river flow and gives the equation describing SSC as a function of river flow for three flow regimes (Q = flow in cfs):

(a) for flows less than 2500 cfs
$$SSC(mg/L) = 5800 \ x \ (1.00132^Q/Q)$$

(b) for flows between 2500 and 6300 cfs
$$SSC(mg/L) = 5600 \ x \ (1.000303^Q/Q)$$

(c) for flows exceeding 6300 cfs
$$SSC(mg/L) = 131,000 \ x \ (1.000185^Q/Q)$$

An analysis of the particle sizes of suspended-sediment was performed on 6 samples taken when the river discharge was between 5,220 and 23,600 cfs (Crawford and Mansue 1988) when the SSC ranged from 98 to 185 mg/L. Particles less than 0.004 mm in diameter (clay) constituted 68% of the SS load by weight; particles measuring between 0.004 mm and 0.062 mm in diameter (silt) constituted an additional 27% by weight. Only 5% of the SS load consisted of particles greater than 0.062 mm in diameter. No estimates of bed-load were made.

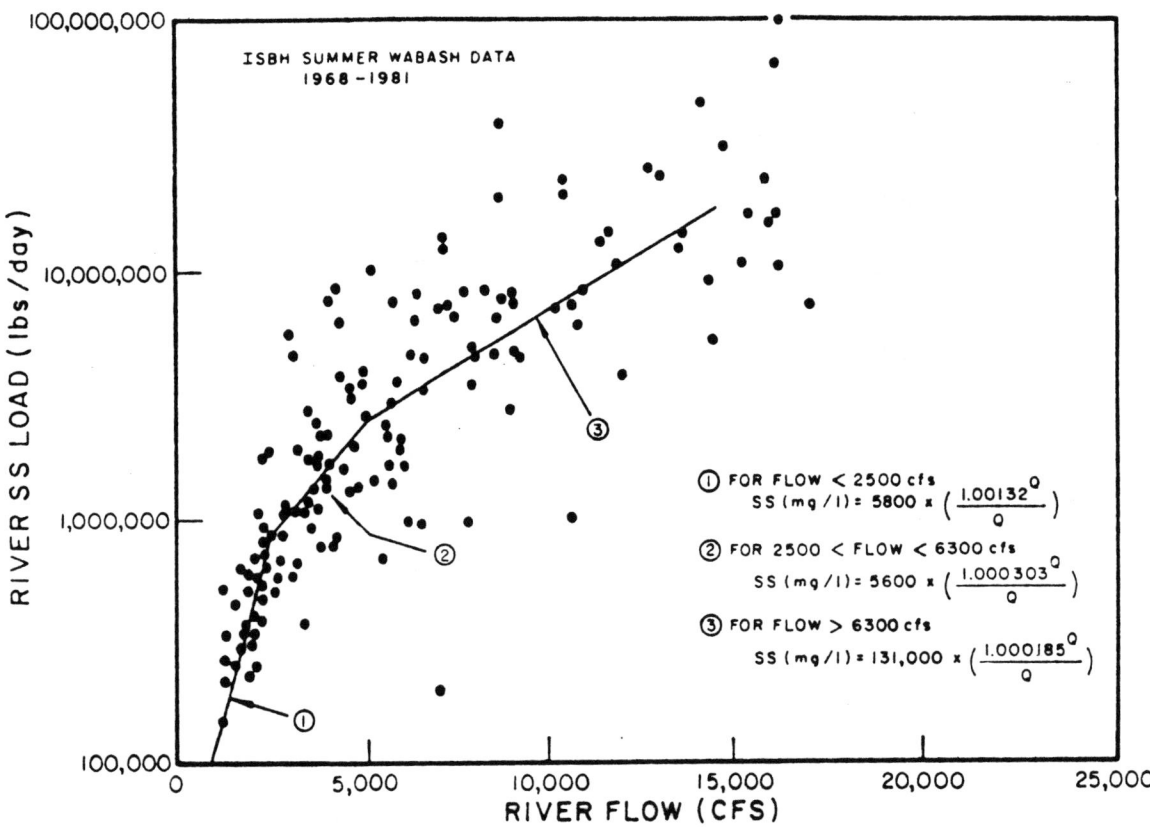

**Figure 19: Suspended solids loads (lbs/day) as a function of river discharge (cfs)
based on Indiana State Board of Health data 1968 - 1981.**

Partial-records (N=67) for the Wabash River at New Harmony, Indiana for the years 1975 through 1984 indicated a median SSC of 116 mg/L (mean = 150 mg/L; range: 27 - 601 mg/L) and a median SS discharge of 5.66×10^6 kilograms per day (6,240 tons/day with a range of 391 to 76,600 tons/day).

Smith et al. (1987) analyzed water-quality trends in major U.S. rivers for the period 1974 - 1981 and state that an increase in SSC for the Ohio River occurred over this period of time. Sediment increases occurred wherever cropland erosion rates were high (more than 5,600 kg\ha\year). Crawford and Mansue (1987) estimated that for the Wabash River above Lafayette the mean annual suspended-sediment yield was 474 kg/ha (135 tons/mi^2)

The SS load strongly influences the depth to which light penetrates. The relationship of extinction coefficient values of Wabash River water to SS loads characteristic of summer flows in 1981 and 1982 was determined by HydroQual (1984) and is shown in Figure 20.

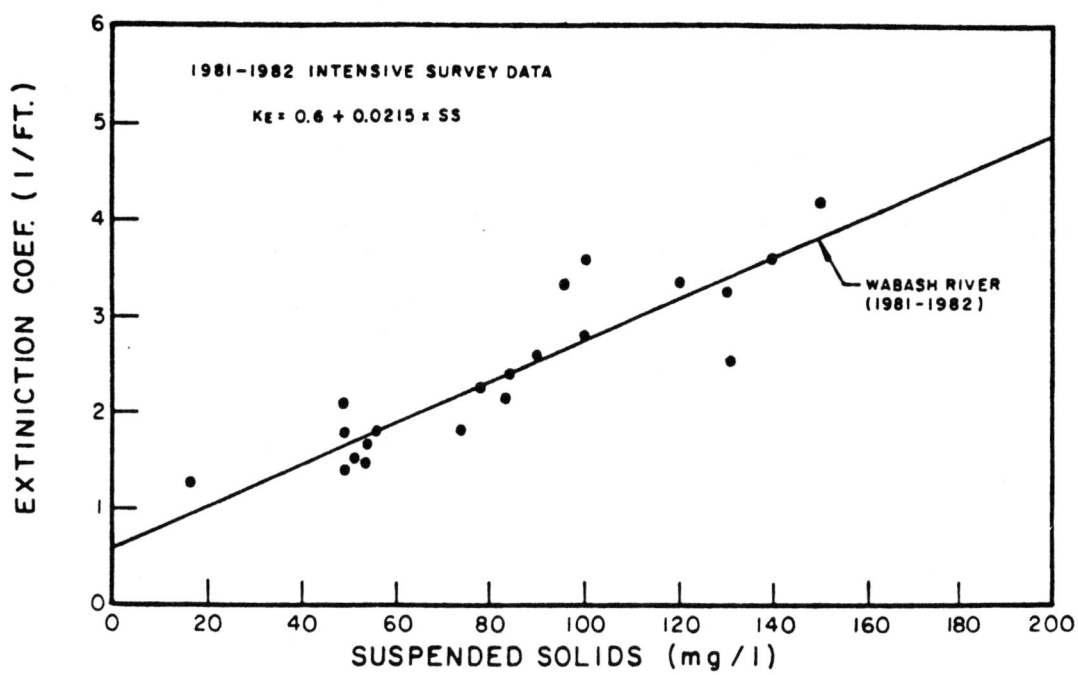

Figure 20: Extinction coefficients as a function of suspended solids concentration (mg/l) in the Wabash River, summers 1981 - 1982.

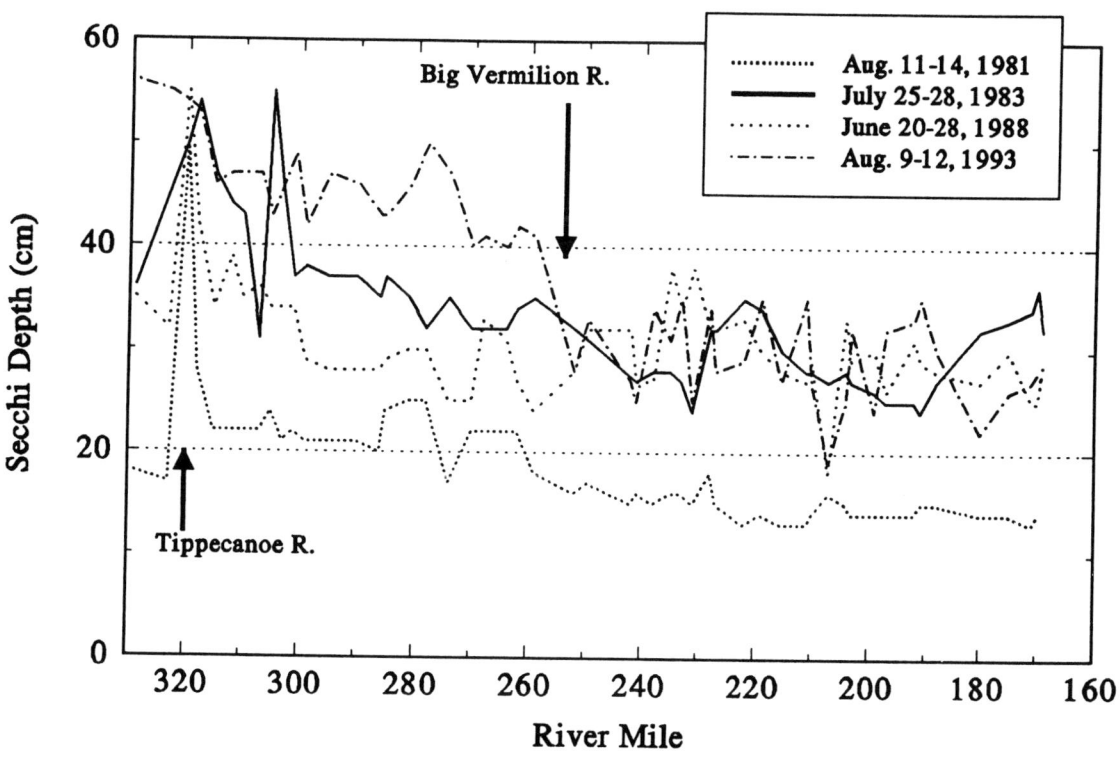

Figure 21: Profiles of Secchi depth (cm) of the Wabash River.

The secchi disk, a simple plate with alternating black and white quarters, is an eminently practical device for determining water clarity. The Secchi depth (Z_{SD}) is the depth below the surface at which the black and white quarters disappear from view. Whatever its limitations, the Secchi disk is portable, durable, easy to use, and well adapted to survive the environmental rigors of the fish collecting boat, compared to more complicated instruments. Secchi depths were routinely determined by the fish collecting crew at each fish collecting stations since 1981.

Profiles of the Secchi depth along the middle Wabash River during the low-flow summers of 1983 and 1988 are presented in Figure 21, together with two other periods. All but one of these profiles were taken during very low and stable discharge conditions. The exception was the profile obtained during the period August 11 to 14, 1981 was taken when the discharge rate was greater than normal.

The 1993 profile shows the effect of the infamous 1993 flood which had scoured away fine sediments and benthic algae and reduced the suspended solids concentration for a period after it subsided. This unusually clear water, however, was obliterated by the entry of Big Vermilion River water with its heavy load of suspended solids. This secchi profile was obtained in early August. Had another profile followed two to three weeks later it may have reverted to a more normal level because of additional phytoplankton densities.

The increased clarity of water at about RM 320 marks the entry of the Tippecanoe River into the Wabash River. Two reservoirs in tandem on the lower Tippecanoe River, Lake Schafer and Lake Freeman, act as settling basins to remove suspended solids before the clarified water enters the Wabash River. There are secondary increases in clarity where Wildcat Creek and Sugar Creek join the Wabash River, but only during some years.

River Channel Morphology

A thorough analysis of river channel morphology for the entire middle Wabash River has not yet been attempted. However, studies were initiated in more limited segments because of fish kills which occurred in 1977 and 1983 between the Cayuga Electric Generating Station and Montezuma, Indiana (Gammon and Reidy 1981).

High rates of sediment oxygen demand (SOD) were found in this part of the river as compared to reaches further upriver or downriver (Bell 1983). Additionally, it was determined that when discharge was low algal densities and particulate sedimentation rates increased in this same section of river (Parke 1985; Parke and Gammon 1986). The deposition of suspended materials, especially algae, and their subsequent decomposition apparently led to a depression of dissolved oxygen (DO) concentrations in the water column.

Although one possible reason for the problem was entrainment mortality of phytoplankton by the Cayuga EGS and its subsequent deposition and decomposition, another potential contributor was suspected to be river morphology. The Wabash River is partially obstructed at the mouth of Sugar Creek during low-flow periods by a gravel bar which creates a more lentic environment, thereby promoting the deposition of suspended particulates.

Width, mean depth, and cross-sectional areas of the Wabash River were determined annually at 0.5 km intervals from 1985 through 1988. The first series of measurements was taken from the mouth of Big Vermilion River to Montezuma, Indiana

on July 29 and 30, 1985 when the river discharge averaged 51 m^3/sec (1880 cfs) at Montezuma. The studies were extended from Montezuma, Indiana to I-70 bridge south of Terre Haute, Indiana on August 11, 12, and 13, 1986 when the river was gradually rising. River discharge at that time averaged 86 m^3/sec (3170 cfs).

The measurements were repeated on August 19 - 24, 1987 using the same transects between the State Road 234 bridge east of Cayuga, Indiana and Clinton, Indiana. The discharge of the river at Montezuma, Indiana averaged 43.8 m^3/sec (1615 cfs) and was gradually decreasing during this period.

In 1988 the Wabash River did not flood during spring and by mid-May the river had already stabilized at levels low enough for fish studies to be initiated. During the ensuing drought the river discharge continued to decline throughout the summer and early fall. Shoal areas normally covered by at least a few inches of water became exposed early in the summer and were soon colonized by terrestrial vegetation which grew two to three feet high by August. Even the normally bare and steep mud and clay banks turned green with weedy growth.

While the low 1988 summer discharge was far from ideal for collecting fish by electrofishing, conditions were nearly perfect for measuring river morphology during extremely low flows. River discharge was estimated to be 34.1 m^3/sec (1258 cfs) on July 19, August 1, and August 2, 1988, when breadth, depth, and cross-sectional area was measured from Big Vermilion River to Clinton, Indiana.

Contours of the river bottom were determined for transects of the channel using a Lowrance Model X-155B sonar transported by boat which motored at a constant speed from one shore to the other. The depth of the river was measured directly in a few shallow places. River width was measured optically using a Leitz rangefinder.

The cross-sectional areas were estimated from the transects using a K. and E. Compensating Polar Planimeter. Mean depth was computed by dividing cross-sectional area by channel width. The profile of the cross-sectional areas determined in 1988 are shown in Figure 22. For comparative purposes this profile was divided into six segments. The segments upstream from Sugar Creek and Clinton, Indiana both prominently exhibit the relatively large cross-sectional areas which are characteristic of "pool" morphology.

Average cross-sectional areas, depths, and widths for each segment in 1988 are shown in Figure 23. The same general pattern existed in 1985, 1986, and 1987, but with slightly higher values. The river was narrowest and had the smallest cross-sectional area from Big Vermilion River to the Cayuga Electric Generating Station discharge. The section of river between Montezuma, Indiana and the Eli Lilly-Clinton plant was widest and most shallow.

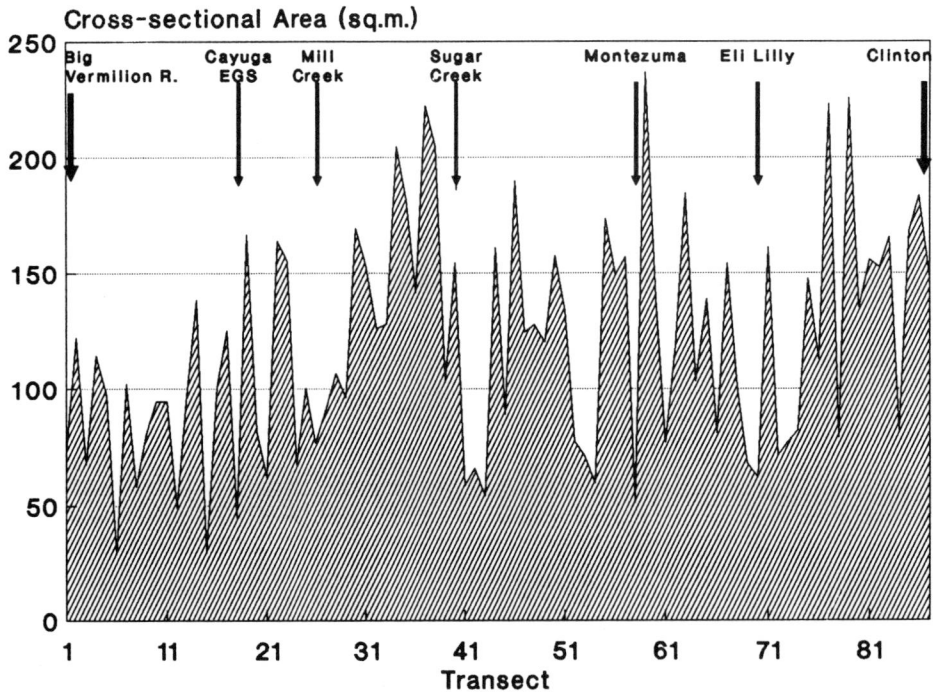

Figure 22: Cross-sectional area of the Wabash River from Big Vermilion River to Clinton, Indiana in 1988.

39

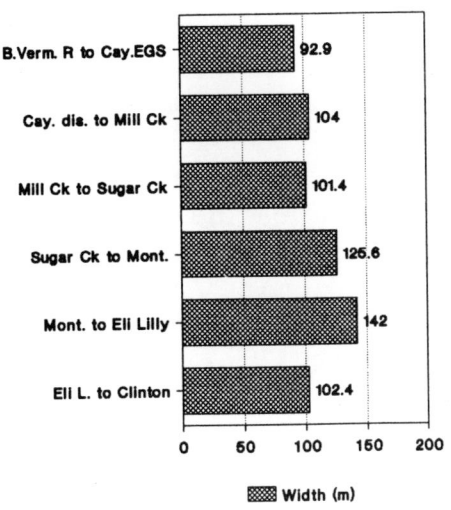

Mean Width - 1988

Segment	Width (m)
B.Verm. R to Cay.EGS	92.9
Cay. dis. to Mill Ck	104
Mill Ck to Sugar Ck	101.4
Sugar Ck to Mont.	125.6
Mont. to Eli Lilly	142
Eli L. to Clinton	102.4

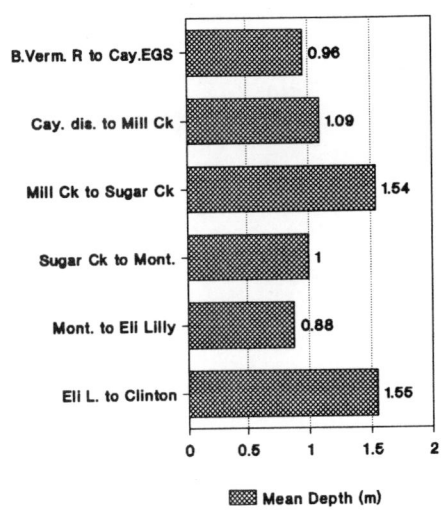

Mean Depth - 1988

Segment	Mean Depth (m)
B.Verm. R to Cay.EGS	0.96
Cay. dis. to Mill Ck	1.09
Mill Ck to Sugar Ck	1.54
Sugar Ck to Mont.	1
Mont. to Eli Lilly	0.88
Eli L. to Clinton	1.55

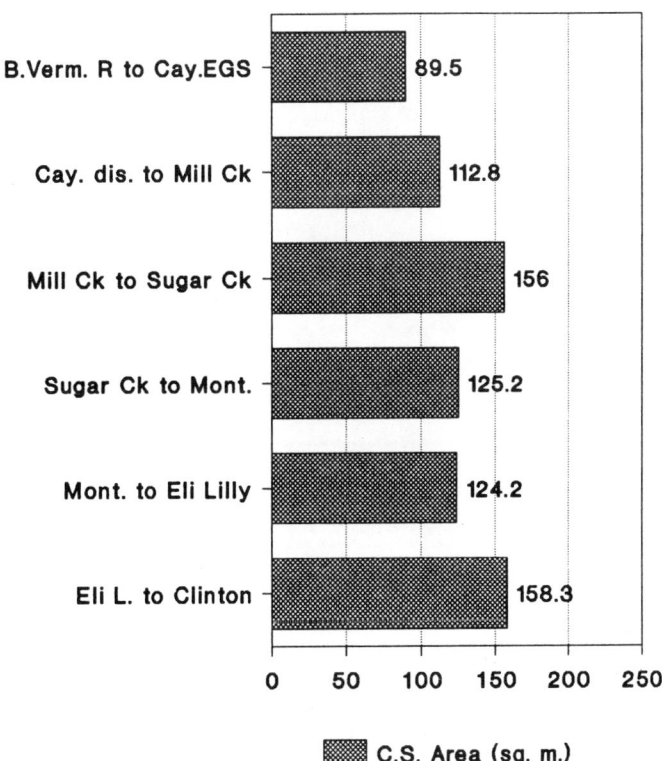

Cross-sectional Area - 1988

Segment	C.S. Area (sq. m.)
B.Verm. R to Cay.EGS	89.5
Cay. dis. to Mill Ck	112.8
Mill Ck to Sugar Ck	156
Sugar Ck to Mont.	125.2
Mont. to Eli Lilly	124.2
Eli L. to Clinton	158.3

Figure 23: Mean width, depth, and cross-sectional areas of six segments of the Wabash River between Big Vermillion River and Clinton, Indiana.

These measurements, which are representative of varying degrees of low-flow conditions, suggest that river morphology changes when the river is very low. Some anomalies are apparent when the channel morphology which existed in 1987 (low-flow) is compared to that of 1988 (very low-flow). The mean width decreased at very low-flows most places except between Montezuma, Indiana and Eli Lilly plant at Clinton where it changed little. The mean depth of the "pooled" segments actually increased in 1988 compared to 1987, although it decreased in the other segments.

In 1988 the segment from Mill Creek to Sugar Creek was about 50% greater in cross-sectional area than the segments from Big Vermillion River to Mill Creek. The lower segment from Eli Lilly plant at Clinton to Clinton, Indiana was also about 30% larger in cross-sectional area than the segment from Montezuma, Indiana to Eli Lilly plant. This relationship was also present in 1987, although it was less evident.

The mean width of the river increased steadily as it flowed south and, as mentioned previously, was particularly wide and shallow in the segment extending from Montezuma, Indiana to Eli Lilly plant at Clinton. The river channel downriver from Clinton, Indiana was examined only in 1986. At that time there were no "pooled" segments. However, the river at Terre Haute, Indiana was unusually wide in places.

Neil Parke determines the width of the river with a rangefinder.

THE RIPARIAN WETLAND CORRIDOR

Riparian wetlands found throughout Indiana, including the middle Wabash River, consist primarily of the bottomland forests which flank the river. The original riparian community was probably dominated by tree species such as those found by Lindsey (1962) at Beall Woods (now Beall Woods State Park) near Keensburg, Illinois. He found that a high proportion of trees exceeded 76 cm (30 inches) in basal diameter. A year earlier none of the stands which he examined along the Wabash River from Logansport, Indiana to Vincennes, Indiana represented the original, presettlement type (Lindsey et al. 1961) and ". .it appeared that every example of such superb timber had been sacrificed. .".

Lindsey et al (1961) found that the chief species on first bottoms or lowest terrace were black willow (*Salix nigra*), silver maple (*Acer saccharinum*), American elm (*Ulmus americana*), and cottonwood (*Populus deltoides*) with important contributions from sycamore (*Plantanus occidentalis*), red elm (*Ulmus rubra*), cork elm (*Ulmus thomasi*), box-elder (*Acer negundo*), white ash (*Fraxinus americana*), and hackberry (*Celtis occidentalis*). On the less frequently flooded second terraces there was a shift to species which were less water tolerant, but more shade tolerant.

Lindsey et al. (1961) also reported that on the cut banks of the Wabash and Tippecanoe rivers, rapid erosion undermined and toppled trees of any size, but that erosion proceeded slowly where some old elms sheathed the steeply sloping bank with a complex network of exposed living roots, mostly several centimeters in diameter which effectively protected both trees and banks.

Representatives of most of these same species still border Indiana's streams and rivers today. However, ". . . the floodplains were rapidly cleared" after settlers discovered that rich bottomlands would grow the best corn after the trees had been cleared" (Petty and Jackson, 1966). As a result, little of the original ecotone remains to buffer water courses from agricultural activities.

Our first riparian survey was undertaken in 1983 because we observed an accelerated rate of tree cutting along the river corridor during the previous several years. The middle Wabash River includes approximately 265 km (165 miles) of river between Delphi, Indiana and Merom, Indiana. For comparative purposes this section of river was subdivided into 12 Reaches as indicated in Table 1.

The 1983 evaluation was conducted from the river's surface entirely by boat. A boat was also used in 1994 for assessing riparian wetlands between Delphi, Indiana and Montezuma, Indiana, but the survey from Montezuma to Merom, Indiana was made using a light plane.

Three classes of riparian condition were determined: (1) banks which were devoid of trees (although some brushy growth might be present), (2) banks with one or two trees growing between the river edge and fields beyond, and (3) banks with more than two trees.

Table 1: Location of riparian Reaches of the middle Wabash River.

Reach	River Mile (Rkm)	Description of Location
1	330 to 313 (531-502)	Delphi to north Lafayette, Ind.
2	312 to 302 (501-486)	Lafayette & W. Lafayette area
3	301 to 286 (485-460)	Lafayette to Attica, Ind.
4	285 to 269 (459-433)	Attica to Covington, Ind.
5	268 to 251 (432-404)	Covington to Coal Creek
6	251 to 246 (404-396)	Cayuga Elec. Generating Station
7	247 to 233 (396-375)	Cayuga EGS to Eli Lilly, Clinton
8	232 to 218 (374-351)	E.Lilly, Clinton to Otter Creek
9	217 to 212 (349-341)	Wabash Elec. Generating Station
10	213 to 203 (343-327)	Terre Haute, Ind. area
11	202 to 186 (325-299)	Terre Haute STP to Darwin, Ill.
12	185 to 160 (298-257)	Darwin, IL to Merom, Ind.

Distinguishing between categories (2) and (3) was sometimes difficult and created most judgmental problems. The middle Wabash flood-plain is excellent for growing corn and has few pastures or grasslands and the tilled fields abut directly on riparian forests. We did not attempt to determine the width of forested riparian borders where it was substantial because of constraints of time and funding.

The extent of bare banks and banks with a vegetated border consisting of only 1-2 trees is summarized in Table 2 and Figure 24. The location of these areas and the willow groves is shown in greater detail in Figures 25A-Q.

In 1994, 28.82 km (5.56%) of bare banks and 27.21 km (5.25%) of banks with only 1-2 trees were found within the 518 km of the middle Wabash River. The 1994 estimates were considerably lower than the 1983 figures which were 37.92 km (7.32%) bare banks and 35.0 km (6.76%) of areas with 1-2 trees.

The proportion of bare banks along the Wabash River in 1994 was highest in Reach 6 at the Cayuga Electric Generating Station (EGS) where virtually the entire outer bank of the large oxbow is bare. More than 9% of banks in Reach 12, the longest Reach extending from Darwin, Illinois to Merom, Indiana, were barren of woody growth (Figure 24). Other sections with relatively high percentages of bare banks included Reaches 2, 6, 7, 8, and 10.

The extent of banks protected minimally with only 1-2 trees in 1994 was greatest (9.77%) in Reach 11 from Terre Haute, Indiana to Darwin, Illinois. Reach 1 (Delphi, Indiana to Lafayette, Indiana) also contained many short sections with limited bank protection. In most other Reaches the proportion of banks with only 1-2 trees was less than the proportion of bare banks.

In 1994 we found many groves of small willows growing on banks which were bare or only thinly vegetated in 1983. Most were small patches, but some fairly extensive groves occurred in the following areas: (1) Clinton, Indiana, the mouth of Otter Creek, (2) the oxbow river bend south of I-70 at Terre Haute, and (3) the section upriver from York, Illinois. Some portion of the lower values found in 1994 compared to the 1983 data may be attributable to this pio-

neering invasion of willows. Another portion may be attributable to interpretive differences.

Riparian wetlands are important buffers of streams from sediment generated by activities such as road-building, logging, mining, building construction, and agriculture. The relationship is complex because of differences in topography, geologic material, soils, and climate, but

Table 2: Results of 1994 survey of eroding banks and banks with only 1-2 trees of the Wabash River from Delphi, Indiana to Merom, Indiana.

Reach	Total Bank Length (km)	Bare Bank (km)	Banks with 1-2 trees (km)	Bare + Banks w\ 1-2 trees (km) (%)
1	58	0.89	4.29	5.18 8.93%
2	30	1.90	1.59	3.50 11.65%
3	50	1.04	1.81	2.84 5.69%
4	54	2.96	1.52	4.48 8.30%
5	56	1.54	1.78	3.33 5.94%
6	16	2.46	0.75	3.20 20.02%
7	42	2.94	1.81	4.75 11.30%
8	46	3.04	1.66	4.70 10.22%
9	18	1.74	0.82	2.56 14.22%
10	32	2.17	0.72	2.89 9.04%
11	54	2.24	5.28	7.52 12.92%
12	81	7.33	5.18	12.51 15.44%
Totals	518 km	28.82 km	27.21 km	57.46 km 11.09%

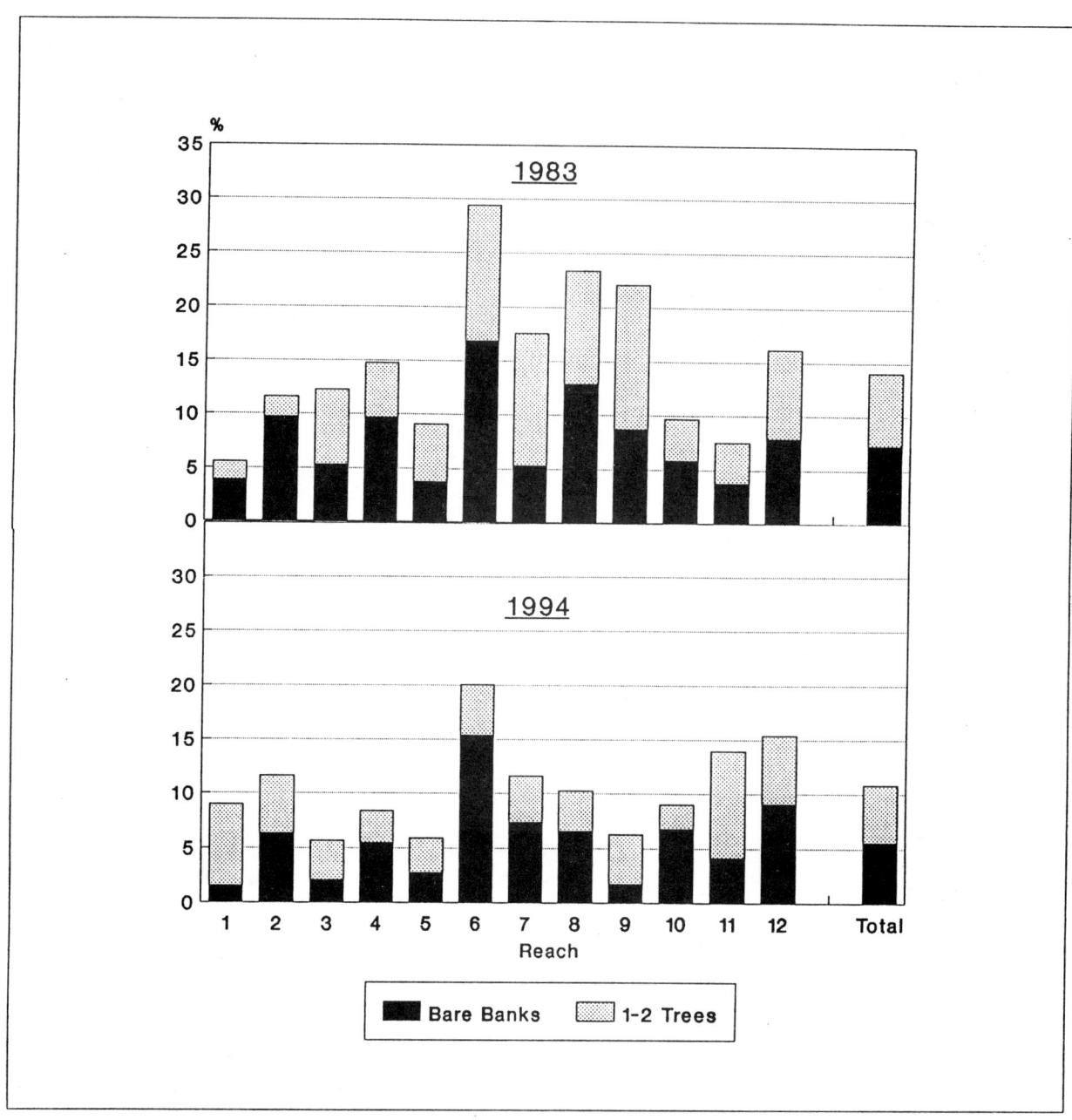

**Figure 24: Percent bare banks and banks with only 1-2 trees
between the edge of the Wabash River and agricultural fields in 1983 and 1994.**

relatively narrow riparian buffer strips act effectively to remove a substantial portion of sediment carried by sheet erosion (Peterson and Correll, 1984; Kovacic et al., 1990; Osborne and Kovacic, 1993). The use of grassy waterways in field depressions to reduce soil erosion is a familiar rural feature in the cornbelt states. Without an adequate riparian buffer sediment and nutrients move directly into rivers and their tributaries. Kovacic et al. (1990) and Osborne and Kovacic (1993) examined the effectiveness of such riparian wetlands in agricultural settings in Illinois. They found that a 16 m wide wooded corridor and a 39 m wide grassland absorbed 97% and 90% of nitrate-N, respectively, nutrients which would otherwise have entered a stream.

Riparian wetlands also trap sediments during flood events which would otherwise be deposited on agricultural fields. When floodwaters exceed bankful capacity most of the coarse suspended sediment is deposited within the strip of riparian forest close to the stream bank. In sections with no riparian forest, piles of coarse sand may be deposited on agricultural fields rendering them unsuitable for the growth of typical crops.

Riparian wetlands also function to preserve bank stability. One characteristic of unregulated rivers and streams is a high degree of variability in discharge. During periods of flooding the roots and trunks of streambank trees absorb the grinding force of floating ice sheets, woody debris, and suspended sand and gravel which would otherwise erode or even relocate the stream channel. Condit and Roseboom (1989) concluded that stream channel erosion on floodplain rowcrop fields was the primary form of flood damage to landowners in the Court Creek, Illinois watershed.

In the lower 19 km (12 miles) of Sugar Creek more than 30 acres of cropland were eliminated from 1955 to 1978 through accelerated lateral erosion assisted by removal of riparian wetlands. Furthermore, suspended sediment from actively eroding banks were responsible for depressing the fish community during a wet summer (Gammon and Riggs, 1983).

The elimination of riparian wetlands along the middle Wabash River is clearly of major concern since approximately 5-7% of its banks are devoid of woody vegetation and another 5-7% provide a minimal buffer consisting of 1-2 trees between croplands and the river's edge. Furthermore, these minimal wetlands consist mostly of many short segments which could easily be undermined and subsequently induce a surge in the rate of lateral erosion. The ecological effects that eroding banks may exert on the Wabash River ecosystem is not known at the present time.

Inadequate riparian buffering of tributaries to the Wabash River also undoubtedly contributes to the oversupply of nutrients in the Wabash River mainstem. A recent study of the riparian condition of three Wabash River tributaries concluded that they were incapable of adequately protecting them from agricultural nonpoint source pollution (Gammon 1996).

As will be discussed in the next section, nitrate-N, ammonia-N, and phosphate concentrations are all higher in the upper Wabash River than farther downriver, the result of extensive tiling of agricultural fields and channelization of tributaries in this area. The influence of this eutrophication on the entire Wabash River ecosystem will be discussed later.

46

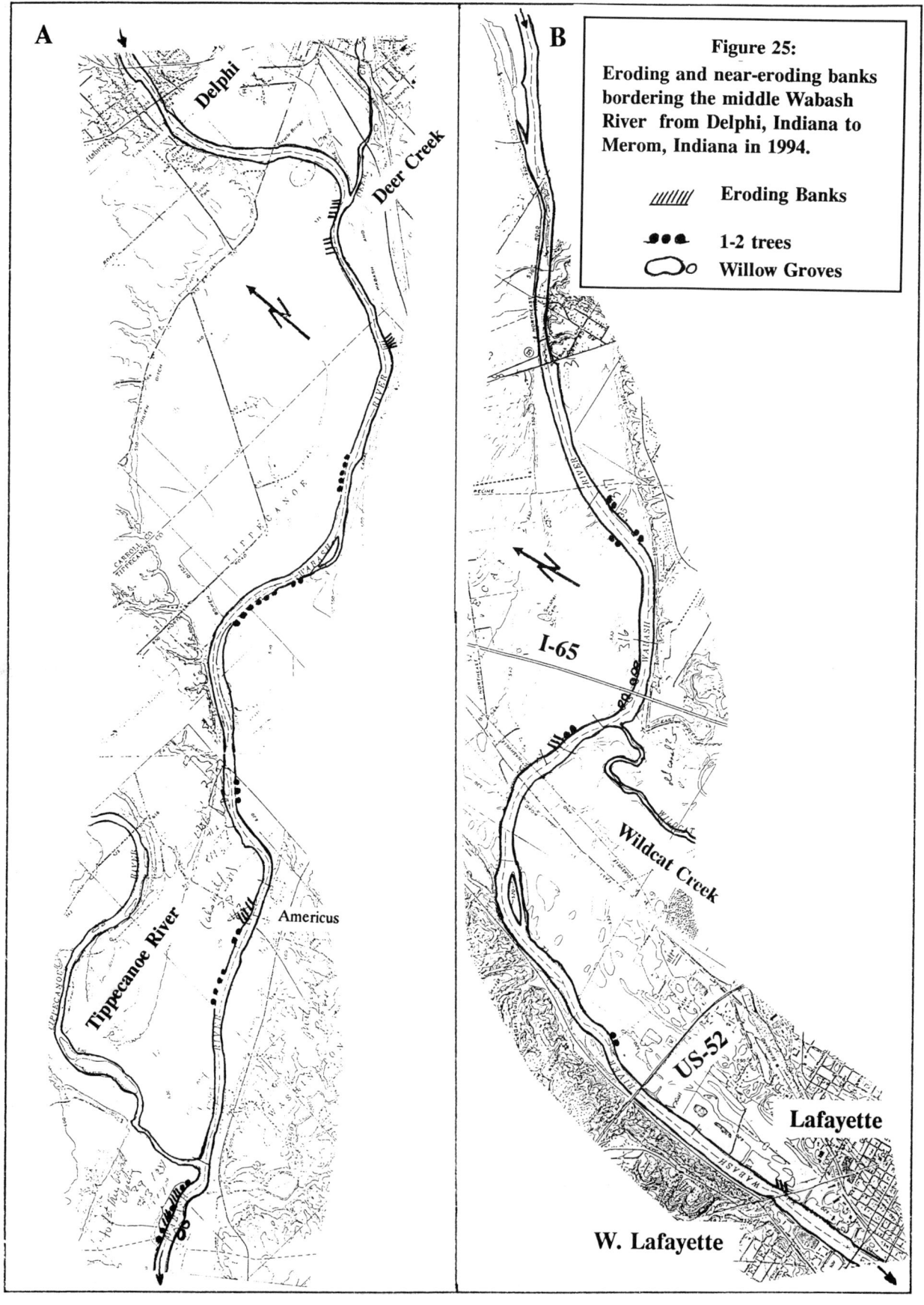

Figure 25:
Eroding and near-eroding banks bordering the middle Wabash River from Delphi, Indiana to Merom, Indiana in 1994.

///// Eroding Banks

••• 1-2 trees

Willow Groves

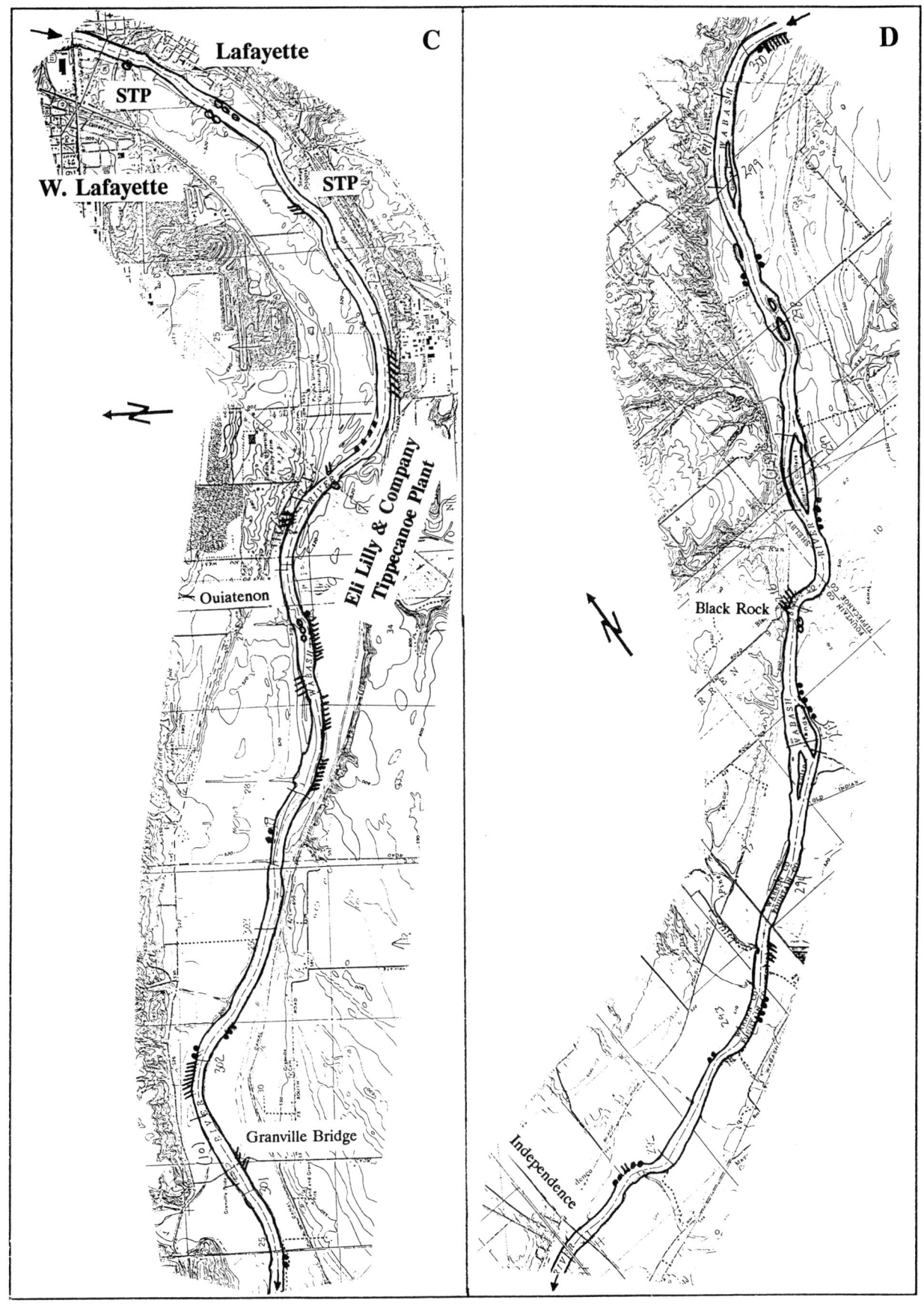

C

Lafayette

STP

W. Lafayette

STP

Eli Lilly & Company
Tippecanoe Plant

Ouiatenon

Granville Bridge

D

Black Rock

Independence

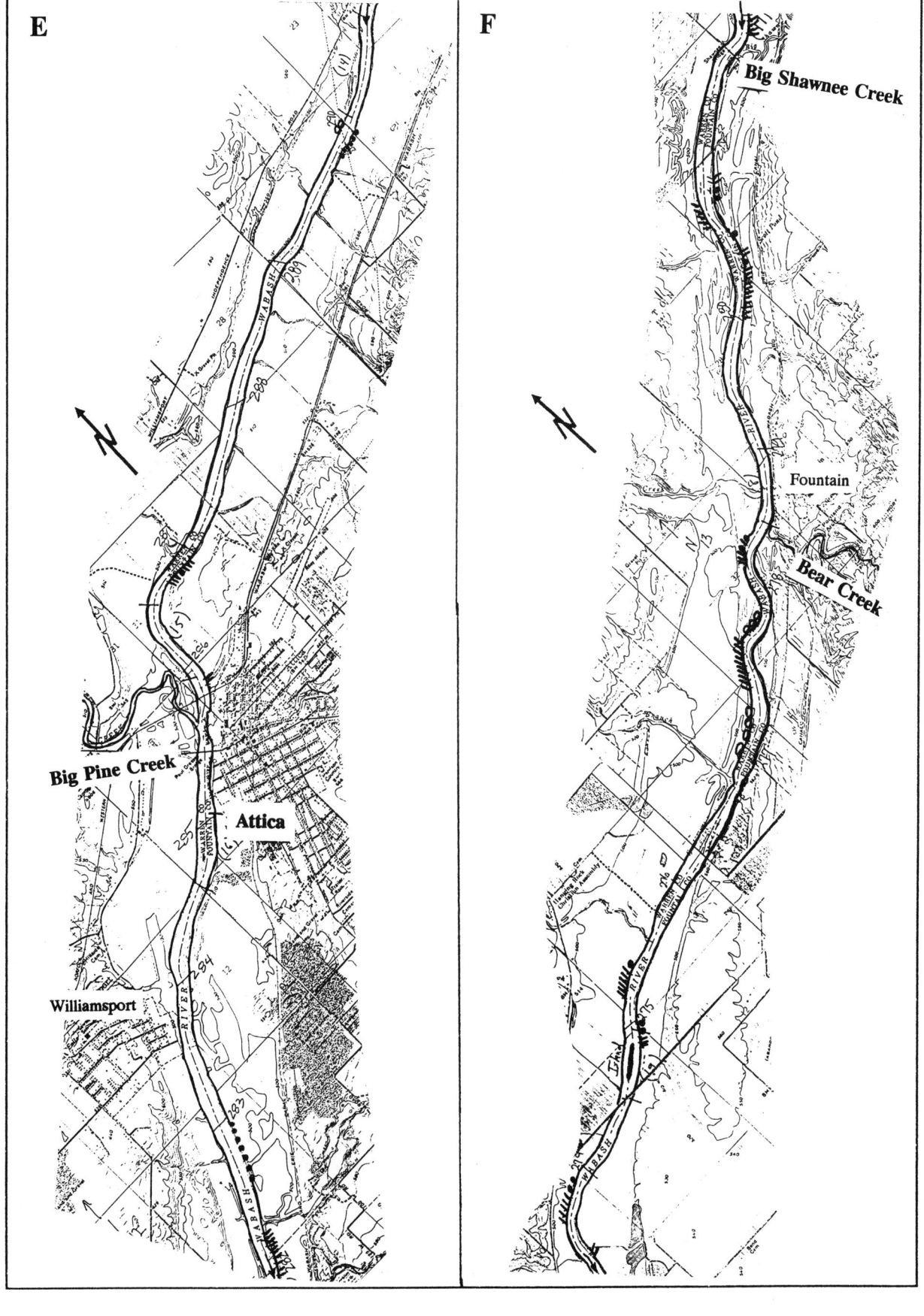

E

Big Pine Creek

Attica

Williamsport

F

Big Shawnee Creek

Fountain

Bear Creek

49

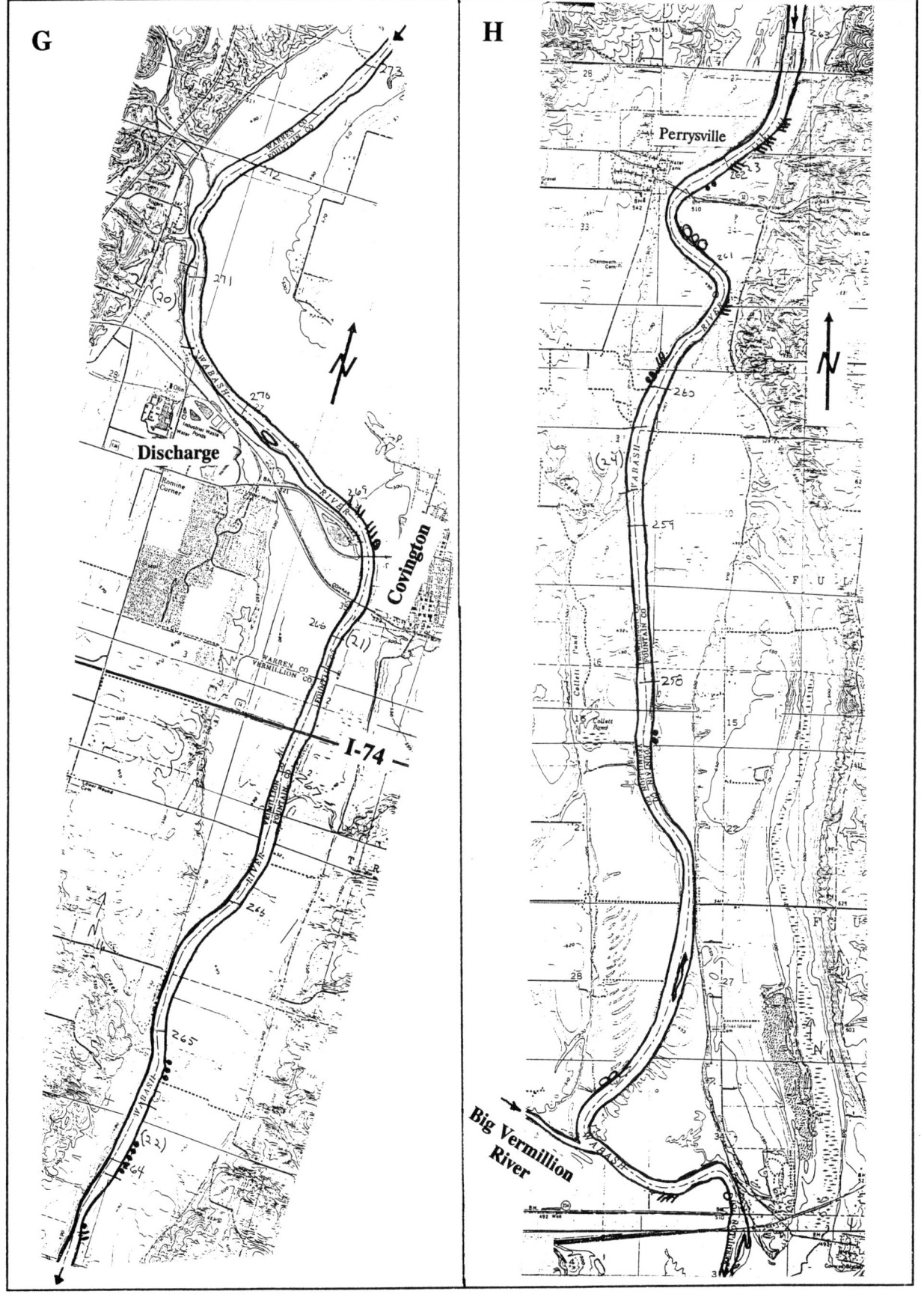

50

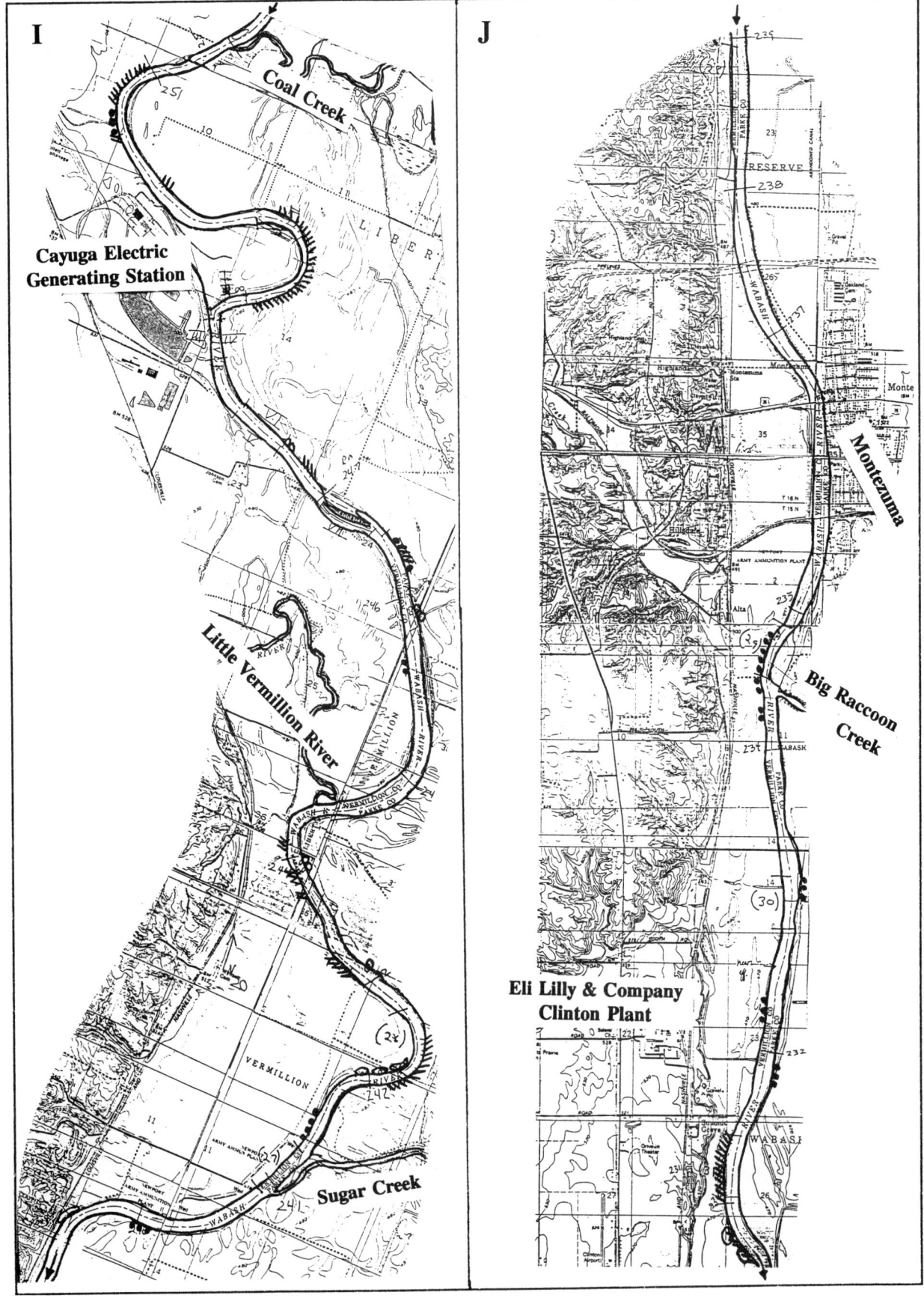

I

Coal Creek

Cayuga Electric
Generating Station

Little Vermillion River

Sugar Creek

J

Montezuma

Big Raccoon
Creek

Eli Lilly & Company
Clinton Plant

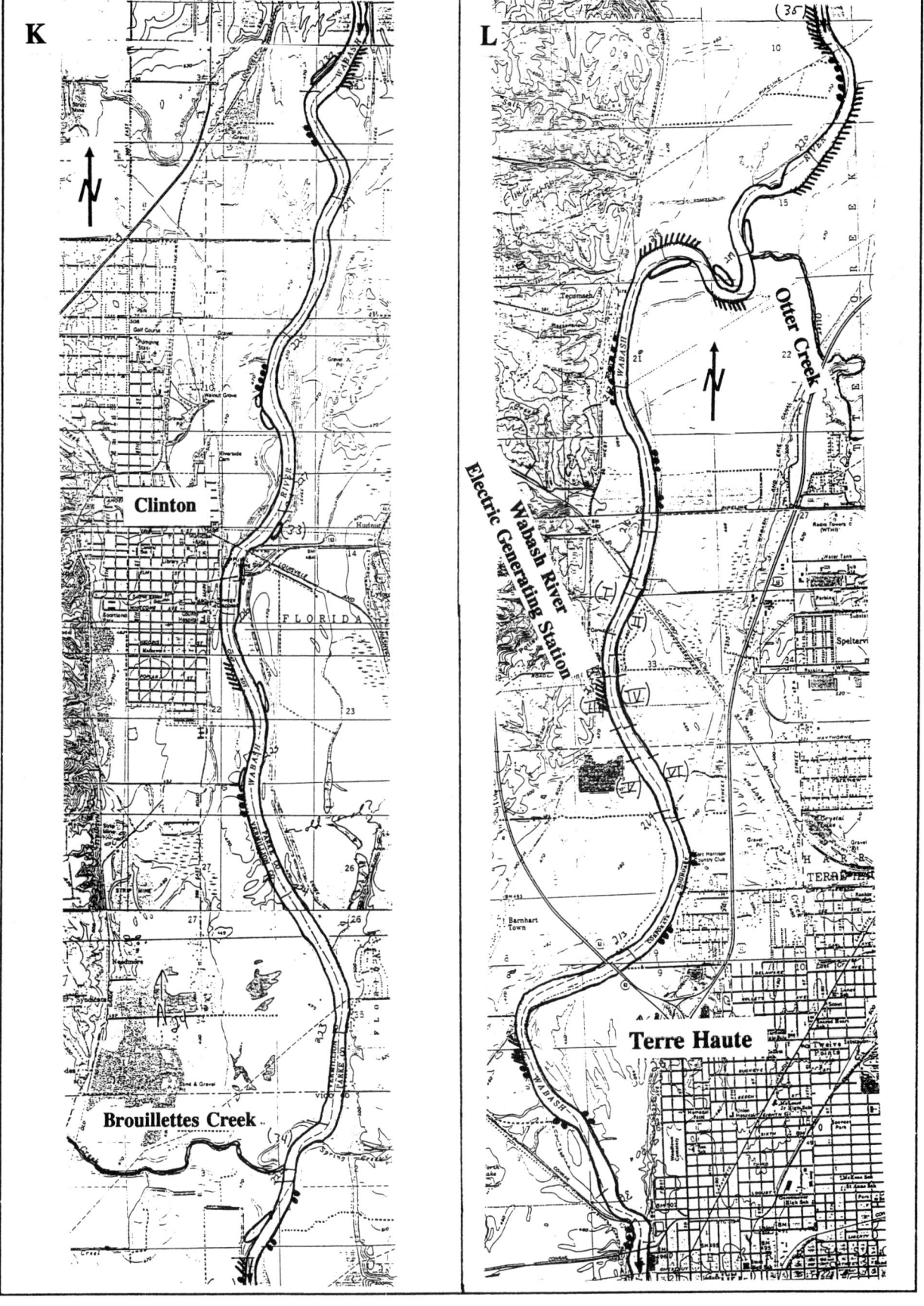

K

Clinton

Brouillettes Creek

L

Otter Creek

Wabash River
Electric Generating Station

Terre Haute

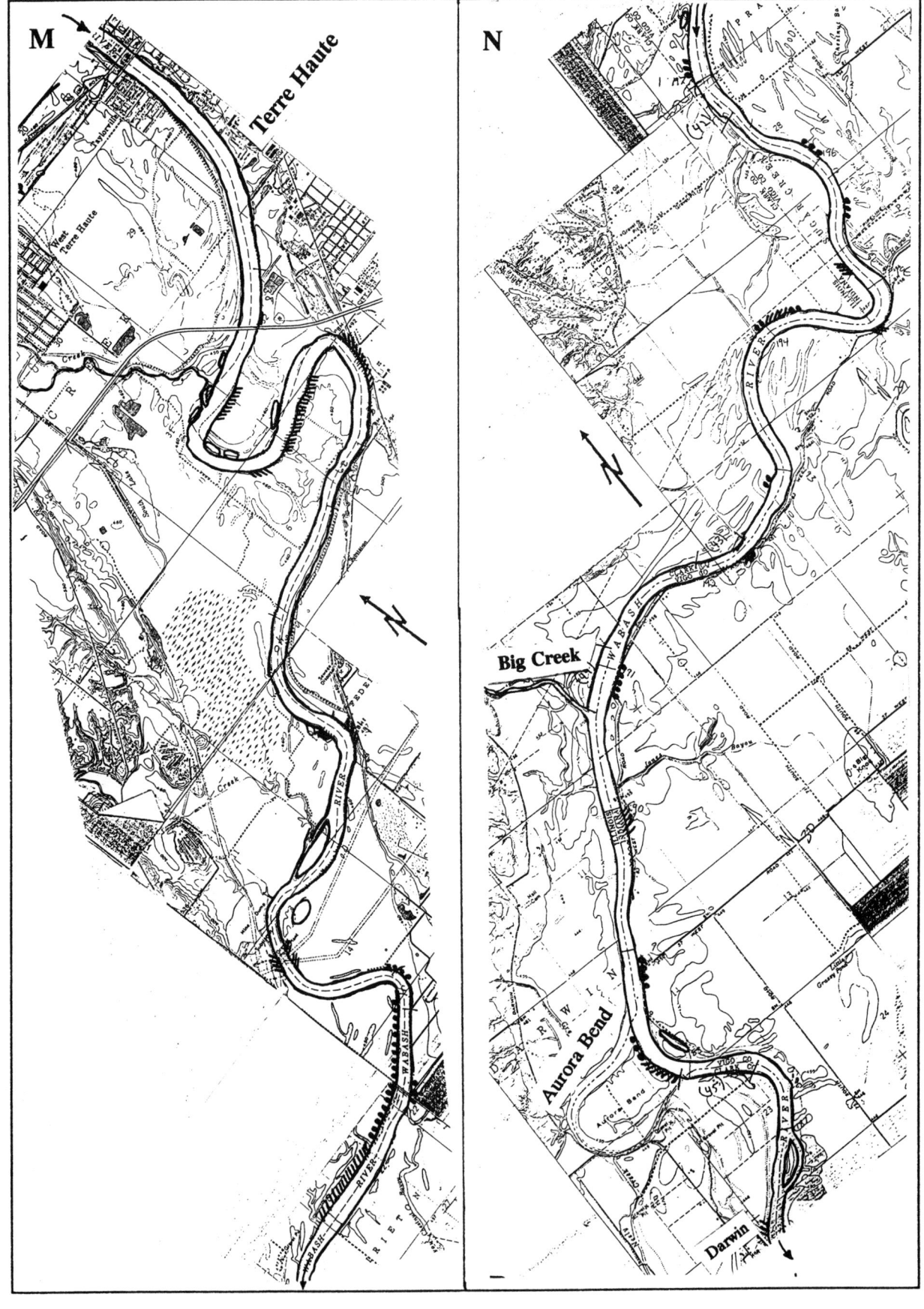

M

Terre Haute

West Terre Haute

N

Big Creek

Darwin

Aurora Bend

Darwin

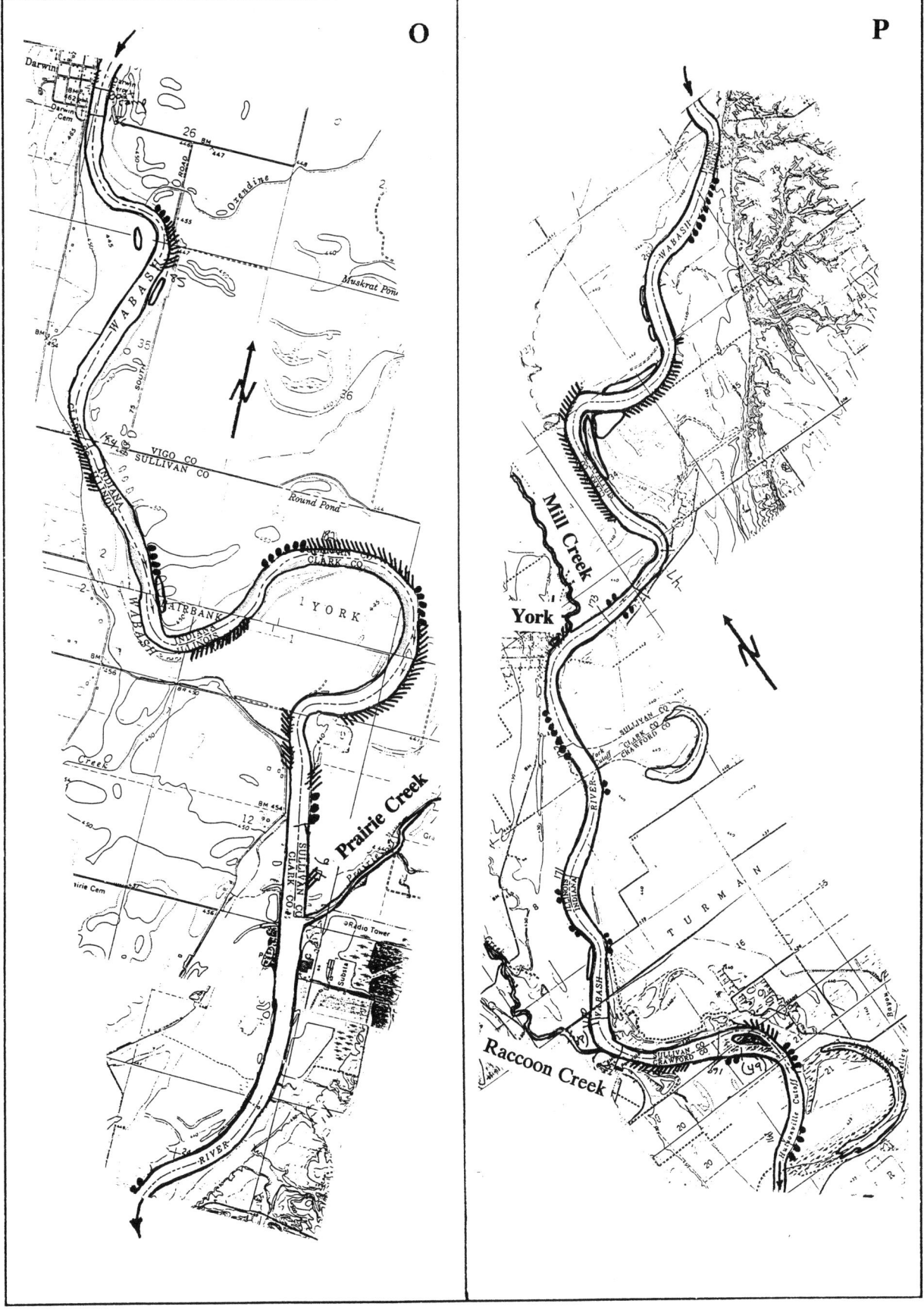

O

P

Mill Creek

York

Prairie Creek

Raccoon Creek

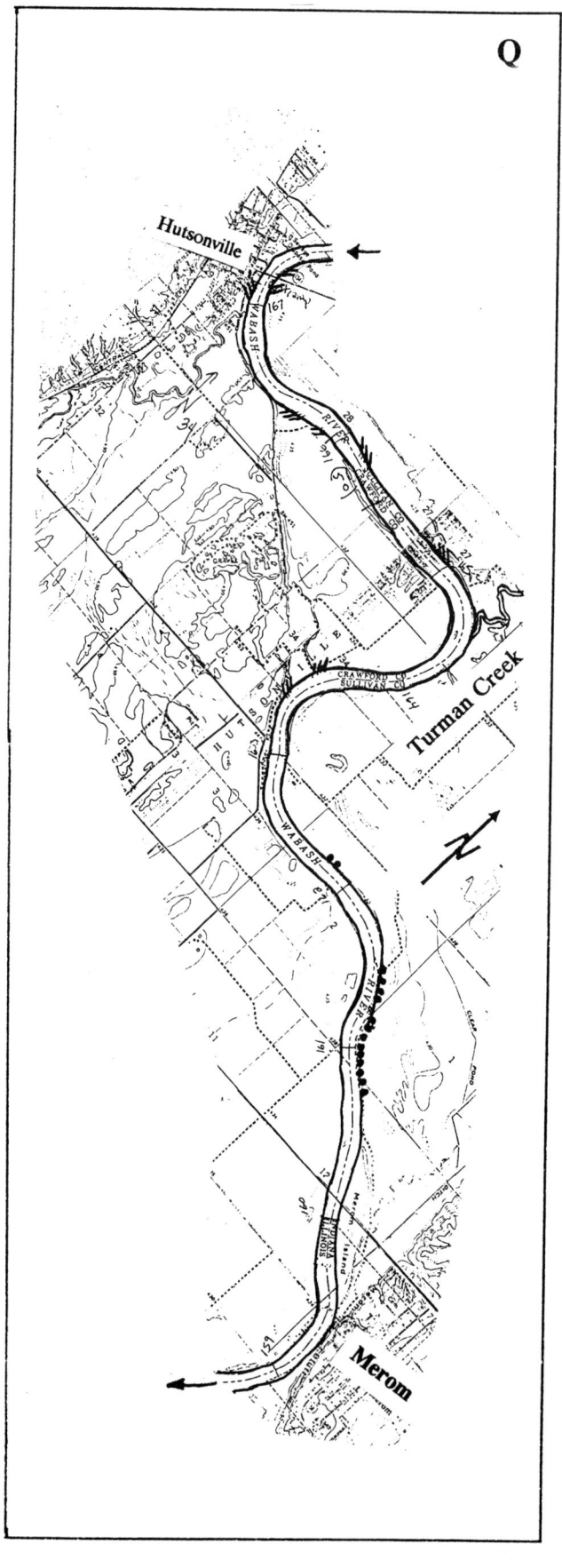

CHEMICAL ATTRIBUTES - NUTRIENTS

In addition to the water which flows between their banks and the multitude of organisms living there, rivers also contain a large variety of dissolved inorganic and organic materials. It is not the intent here to thoroughly analyze these chemical attributes of the Wabash River. However, it is necessary to examine existing information for changes in water chemistry over time and differences in important chemical environmental factors from place to place.

The major modifications which have been observed over the past decade in the fish community are probably correlated with long-term changes in the chemical\physical environment. The character of the fish communities of the Wabash River differs from place to place. These differences might be associated with spatial variations in the physical/chemical environment.

The chemical environment of the river has been examined for a relatively long period of time. In April 1957 the Division of Sanitary Engineering, Indiana State Board of Health established 49 sites from which biweekly samples were collected for physical, chemical, and bacteriological analyses. Changes have been made over time in the location of collecting sites, in chemical procedures, and in frequency of sampling. In 1986 the Indiana Department of Environmental Management was formed and the Office of Water Management took over the program. The most recent report (1991) summarizes analyses from 103 stations throughout Indiana.

ORSANCO (1990) analyzed trends in data from 1977 to 1987 at multiple stations within the Ohio River basin from using the Seasonal Kendall Test. Records at 21 stations on the Ohio River mainstem and 12 stations on tributaries were examined, including the Wabash River at New Harmony, Indiana. A decrease in copper was the only statistically significant change noted at this station. Inadequate data existed for testing changes in total phosphorus, ammonia$=$N, NO_3/NO_2-N, total Kjeldahl-N, and total N. No statistical change occurred for total suspended solids, total dissolved solids, hardness, sulfate, phenolics, iron, lead, mercury, and zinc.

Martin and Crawford (1987), in addition to other records, examined data from eight water-quality stations of IDEM located on the Wabash River between Lafayette and Vincennes. They found that sulfate and suspended solids concentrations increased downriver. However, no spatial differences were noted for specific conductance, pH, or total alkalinity.

Table 3 summarizes water-quality data from the Ohio River Sanitation Commission (ORSANCO) (1990) study together with Indiana Department of Environmental Management (IDEM) data over the same time period. In addition to the IDEM and ORSANCO data, only a few other published chemical records are available for the Wabash River (Lesniak et al, 1973; Siefker and McCleary, 1979).

Table 3: Mean concentration and range of some physicochemical water-quality parameters for the period 1977-87[a] in different sections of the Wabash River.

Parameter	Huntington (rkm 658)	Peru (rkm 595)	Lafayette (rkm 502)	Montezuma (rkm 386)	Terre Haute (rkm 351)	Vincennes (rkm 209)	New Harmony[b] (rkm 83)
Total suspended solids (mg/L)	64.9 (5-1000)	66.2 (3-1260)	77.0 (3-1300)	113.7 (4-2430)	116.7 (7-2110)	112.8 (6-980)	157.2
Conductivity (umhos/cm)	597.6 (200-1160)	562.3 (280-1600)	533.4 (230-870)	595.7 (341-920)	562.2 (226-900)	552.0 (240-880)	488.7
Alkalinity[c] (mg/L as $CaCO_3$)	173.0 (90-275)	177.4 (116-258)	197.2 (117-270)	198.4 (122-276)	192.8 (90-260)	187.8 (112-252)	
Phosphate (mg/L)	0.217 (0.06-1.4)	0.182 (0.05-1.0)	0.170 (0.03-0.74)	0.202 (0.06-1.30)	0.207 (0.05-1.40)	0.204 (0.05-0.51)	0.300
Sulfate[d] (mg/L)	89.1 (21-230)	64.1 (21-100)	63.1 (28-90)	66.9 (30-90)	68.1 (31-100)	69.1 (28-120)	60.1
Ammonia (mg/L)	0.260 (0.1-0.2)	0.195 (0.1-1.8)	0.166 (0.1-1.3)	0.153 (0.1-1.0)	0.145 (0.1-1.8)	0.157 (0.1-1.0)	0.145
Nitrate (mg/L)	4.09 (0.1-15.0)	3.50 (0.5-8.9)	3.29 (0.2-10.1)	3.67 (0.1-10.6)	3.43 (0.1-9.0)	3.19 (0.1-7.4)	2.17
Organic Nitrogen (mg/L)			0.996		0.993	0.900	1.183
pH	7.65	7.72	7.80	7.81	7.77	7.86	

[a] Except where noted, data from Indiana Department of Environmental Management, Office of Water Management, Monitoring Station Records - Rivers and Streams. Published annually.
[b] ORSANCO. 1990. Long-term trends assessment of fifteen water quality parameters in the Ohio River. Toxic Substances Control Program, Ohio River Valley Water Sanitation Commission, Cincinnati, Ohio. 26 pp.
[c] 1986-90 only
[d] 1977-85 only

Conductivity

Conductivity is the ability of water to conduct an electric current. It is measured physically as the reciprocal of the electrical resistance of water expressed as umhos/cm. The electrical conductance of water depends upon the amount of dissolved anions and cations. Distilled water has a very low electrical conductance. The addition of charged particles makes the solution conductive although the variable mixture of chemicals in natural waters makes the relationship far from simple.

For any specific natural river there is generally a good linear relationship between electrical conductance and the dissolved solids concentration. The range of the slope constant for most rivers is generally between 0.55 and 0.75, although it may range up to 0.96 in some extreme cases (Hem, 1985). For rivers in the Ohio River basin including the Wabash River this relationship is described as:

Tot. Diss. Sol. = 0.625 x Cond.(umhos/cm)

Crawford & Mansue (1988) found that the specific conductance of the Wabash River averaged lower during summer (507-528 umhos/cm) than at other seasons (annual average 550-560 uS/cm). They noted no differences over space for either annual or summer conductance. They also found that total alkalinity was slightly lower during summer than it was for the entire year (annual alkalinity = 190-200; summer alkalinity = 180-190) and that there were no noticeable spatial differences.

In recent years the average conductivity or conductance varies only slightly along the Wabash River (Figure 26). The Ohio River Sanitation Commission (ORSANCO) (1990) found lower conductance values at New Harmony than at most other stations and that changes over time were statistically insignificant.

Figure 27 illustrates annual averages for conductance from 1959 through 1989 at five stations sampled by the Indiana Department of Environmental Management (IDEM). It appears that conductance may have decreased at all stations during the 1980's and that over the period of record the conductance at the Peru station may have decreased most. However, this observation needs statistical verification.

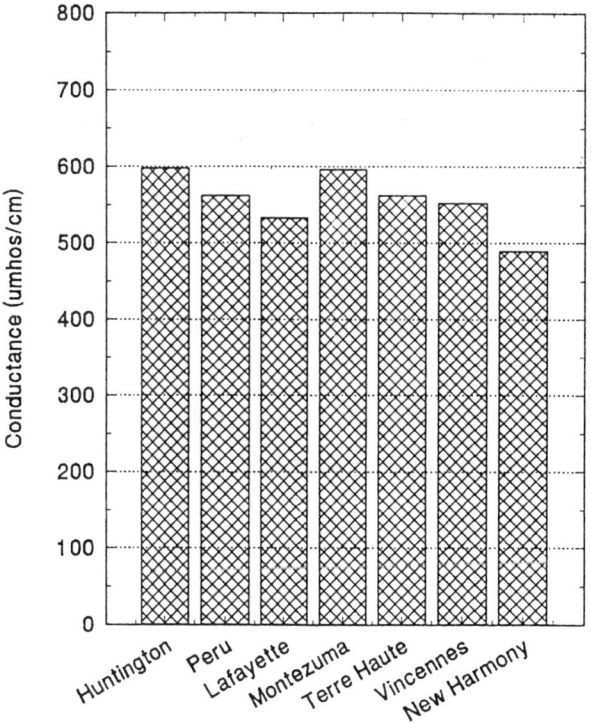

Figure 26: Mean conductivity of the Wabash River 1977 - 1987.

58

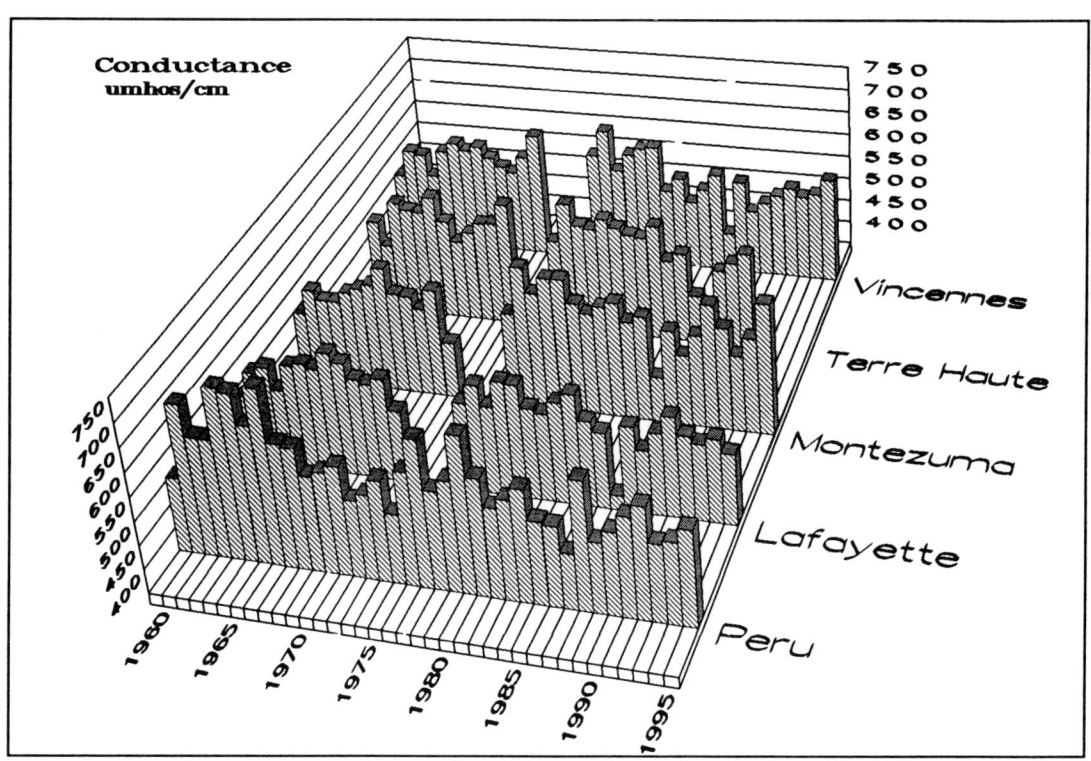

**Figure 27: Mean annual specific conductance (umhos/cm)
of the Wabash River at five locations.**

Nitrogen

The various chemical forms of nitrogen include elemental nitrogen, nitrates, nitrites, ammonia, and organic nitrogen. Elemental nitrogen (N_2) is not very soluble in water. Maximum saturated concentrations of N_2 are about 15-20 ml/L during winter and 10-15 ml/L during summer (Wetzel, 1975). In the hypolimnion of reservoirs, however, it may become more concentrated and may become supersaturated when this water is released into spillways (Trefethen, 1972). Fish exposed to supersaturated N_2 may develop gas bubble disease and elevated mortality. Free-flowing waters such as the Wabash River would be expected to be saturated with N_2 at all times.

Nitrogen is an essential requirement for living organisms since it is a constituent of proteins and nucleoproteins. Atmospheric nitrogen may be fixed directly by blue-green algae, but most aquatic algae utilize inorganic nitrogen in the form of nitrates, nitrites, or ammonia. All of these forms of nitrogen are highly soluble in water.

Elemental nitrogen is added to water during denitrification when NO_3 is reduced to N_2. However, Owens and Nelson (1973) found denitrification rates in the Wabash River were very low and limited primarily by both high concentrations of dissolved oxygen and low temperatures during the winter months.

Ammonia (NH_4^+) and nitrites (NO_2^-) are seldom present in significant concentrations in surface waters except where industrial or organic sources of pollution enter. Ammonia nitrogen is important because it is potentially toxic to aquatic fauna at relatively low levels of concentration at high pH. It is

produced mostly by bacterial deamination of organic compounds and by hydrolysis of urea and, hence, is contained in effluents from wastewater treatment and various industrial plants. All aquatic fauna also generate ammonia, but usually diffusely and in negligible amounts.

The average concentration of ammonia ranges from 0.15 to 0.25 mg/L (Table 3 and Figure 28). The higher values in the upper Wabash River are probably the result of the more intensive agriculture and channelization found in the upper basin.

Ammonia is extensively applied as a fertilizer to agricultural lands during the spring and early summer, but it does not easily leach from soils because it adsorbs

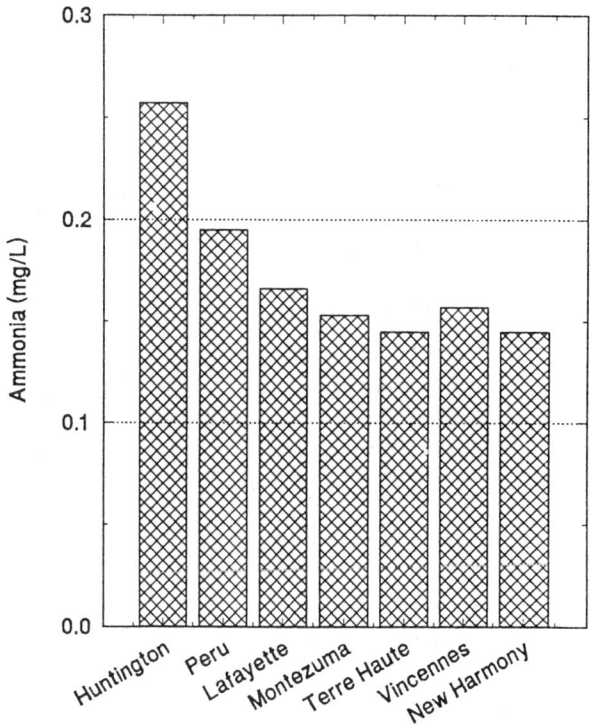

Figure 28: Mean ammonia concentrations (mg/L) of the Wabash River 1977 - 1987.

tightly to clay and other soil particles. Algae and some aquatic macrophytes can directly utilize ammonia as a source of nitrogen, but most ammonia is oxidized by specific bacteria to nitrite and nitrate. Active bacterial oxidation of ammonia probably occurs at specific sites within biofilms on and in bottom sediments or cooling towers. In most of the Wabash River these bacteria would probably be in a quiescent state (Lewis and Gattie, 1991).

Nitrite and nitrate concentrations together constitute total oxidized nitrogen. Nitrates are the primary essential nitrogen source for most aquatic flora and, together with phosphorus, are of great significance in eutrophication of lakes and streams. In most of the Wabash River the concentration of nitrites and nitrates averages 3.0 to 3.5 mg/L

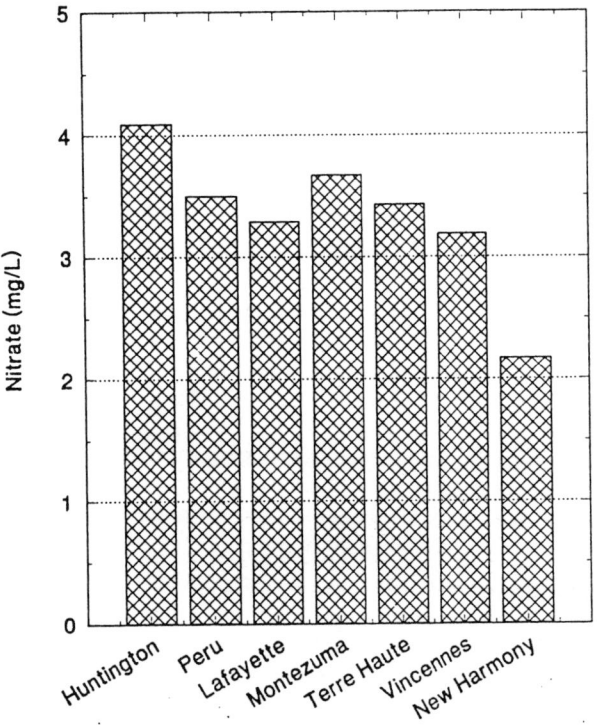

Figure 29: Mean nitrite + nitrate concentrations (mg/L) of the Wabash River 1977 - 1987.

(Figure 29). However, at New Harmony the concentration was only slightly greater than 2.0 mg/L (ORSANCO, 1990). As with ammonia, average nitrate-nitrite concentrations were highest in the upper basin at Huntington where it was more than 4.0 mg/L.

In addition to originating from the oxidation of ammonia, nitrites are generated by reduction of nitrates in wastewater treatment plants and by their direct use in industrial process water as a corrosion inhibitor. Nitrite is the actual etiologic agent of methemoglobinemia or "blue-baby syndrome" in infants and may also participate in the formation of carcinogenic compounds. Fortunately, nitrites usually occur only in low concentrations in natural river environments.

Nitrates, however, are seasonally abundant in agricultural areas where many small and medium-sized rivers may have nitrate concentrations exceeding 10 mg/L NO_3 (Hem 1985). Omernik (1977) found that nutrient concentrations were directly proportional to the percentage of land in agriculture and inversely proportional to the percentage of land in forest. With conventional farming practices an average of 140 pounds of nitrate fertilizer per acre is applied. Approximately 80 pounds will be taken up by corn while the remaining 60 pounds leaches into the soil water and eventually enters streams.

Nitrate concentrations in the lower Ohio River averaged between 3.2 and 3.9 mg/L between 1954 and 1961, but then increased during the next decade to as much as 7.7 mg/L. Average concentrations then declined to less than 5.0 mg/L between 1975 and 1979. These increases parallel the rise in nitrogen fertilizer applications.

Organic nitrogen is defined as organically bound nitrogen. It includes such natural materials as proteins and peptides, nucleic acids and urea, and a variety of synthetic organic materials. It does not, however, include all organic nitrogen compounds. In practice, organic nitrogen and ammonia are determined together as "kjeldahl nitrogen" in which analysis such forms of nitrogen as azide, azine, azo, hydrazone, nitrate, nitrite, nitroso and others are not accounted for. Organic nitrogen of the Wabash River averages approximately 1.0 mg/L.

Flow-adjusted concentrations of ammonia-nitrogen, total Kjeldahl nitrogen, and total nitrogen throughout the Ohio River basin decreased from 1977 to 1987 (ORSANCO, 1990) possibly because of improved waste-water treatment. In 1978, for example, it was estimated that only 47% of waste-water treatment plants provided adequate treatment. By 1988 the proportion of plants which treated wastes adequately increased to 90%. Levels of nitrate/nitrite nitrogen, however, remained unchanged over this same period of time.

The Wabash River and surrounding terraine at Clinton, Indiana

Phosphorus

One of the most important chemical constituents of river water for aquatic organisms is phosphorus. It is required by every living cell (DNA, RNA, ATP, etc.). However, the various forms of inorganic phosphate generally are quite insoluble in water and, therefore, are limited in availability to those organisms. Phosphorus, among other required nutrients, is especially limiting to aquatic algae which absorb it avidly when it is present.

Phosphorus enters rivers through rainfall, dry fallout, erosional sediments, and anthropogenic sources such as sewage and industrial outfalls and, especially in agricultural areas, erosion from fertilized fields. Important chemical fractions of phosphorus include particulate-P, soluble-P, soluble "ortho" phosphate or PO_4-P, and total-P.

Most of the phosphorus in rivers is attached to particulate matter.

In 1971 the Indiana General Assembly enacted a Phosphate Detergent Law (IC 1971, 12-1-5.5) which banned the sale of laundry detergents containing more than 8.7% phosphorus by weight. This law was amended in 1972 to ban all phosphorus from detergents. A second amendment was passed in 1973 allowing the addition of trace amounts of phosphorus, less that 0.5% by weight. This legislation was prompted by recommendations of the International Joint Commission as part of an effort to reduce the eutrophication rate in the lower Great Lakes, especially Lake Erie. Since the ban on phosphate based detergents, the phosphorus concentration in Wabash River has decreased significantly (Figure 30).

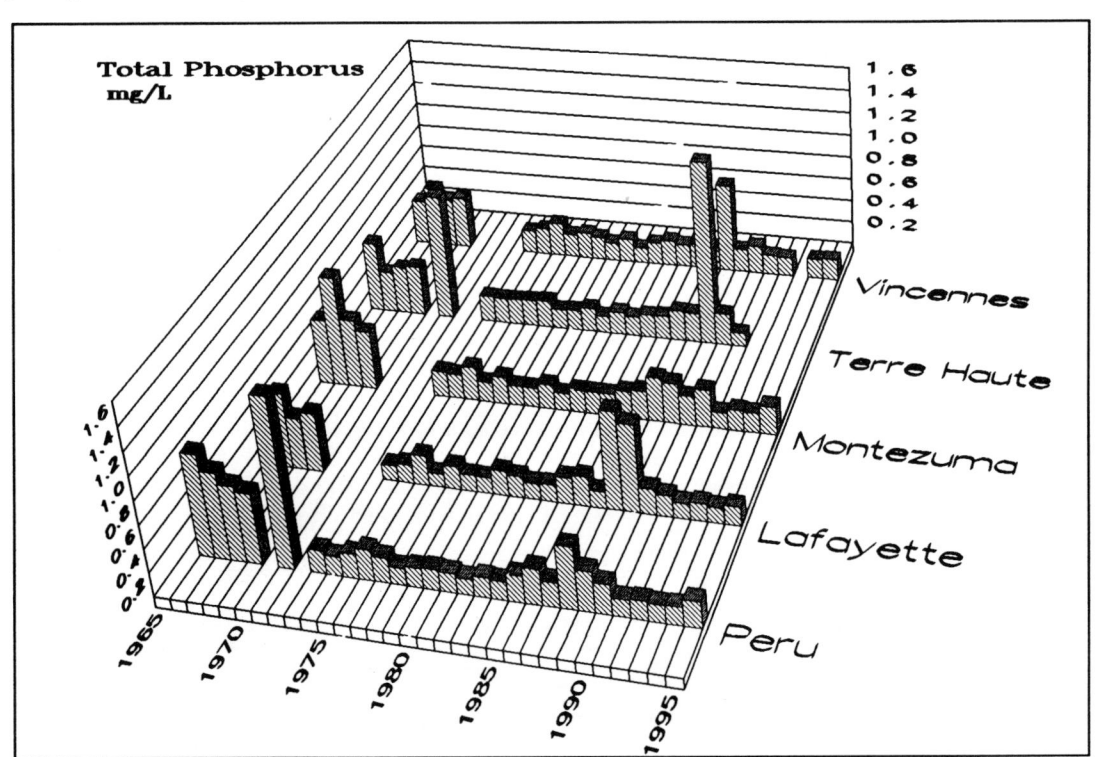

Figure 30: Mean phosphorus concentration (mg/L) of the Wabash River.

The actual magnitude of the decrease is somewhat difficult to assess because of changes in analytic procedures by the Indiana Department of Environmental Management (IDEM). Prior to 1971 only the dissolved phosphate concentration was measured, while after that date the analytical methodology determined total-P.

Information about the changes in phosphorus loadings to rivers is included in the Indiana Stream Pollution Control Board's 305(b) Report (1977). A state-wide survey of phosphorus content of raw sewage indicated that overall concentrations of 10 -12 mg/l in 1971-72 declined to 7 mg/l or less by 1973-76. It was concluded that phosphorus loadings to streams were reduced by 25%-30% overall. Their Figure 50 indicates average phosphorus concentrations in the Wabash River of approximately 0 in 1973, 0.25 mg/L in 1974 and 1975, and 0.32 mg/L in 1976.

The 1977 305(b) Report also cited the results of an NSF-funded study by the Department of Bionucleonics of Purdue University, which examined the status of phosphorus for the Wabash River at Lafayette. This study indicated phosphorus levels in the river were lower after the phosphate detergent ban (0.26 mg/L) than before (0.82 mg/L). It also concluded that agricultural runoff due to improper fertilizer application and heavy rainfall accounted for a large portion of total phosphorus present in the river.

The average concentration of phosphorus from 1977 to 1987 was similar from Peru to Vincennes, but considerably higher at New Harmony (Figure 31). This is somewhat surprising considering the close association of phosphates with suspended sediment which increased steadily downriver.

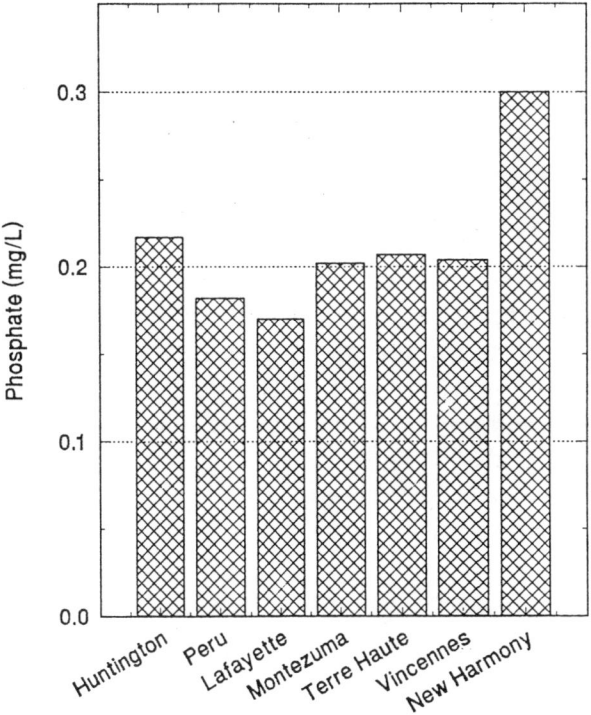

Figure 31: Mean phosphorus concentration (mg/L) of the Wabash River 1977 - 1987.

Smith et al. (1987) found a strong statistical association ($p < 0.001$) between phosphorus trends and trends in suspended sediment. Increased phosphorus concentrations were associated statistically with high values of fertilized cropland ($p = 0.103$) and density of cattle ($p = 0.002$).

Phosphorus concentrations in Indiana rivers are far higher than in "natural" waters despite recent actions. Meybeck (1982) estimated that naturally occurring dissolved inorganic phosphate in river water should average only about 10 ug/L as P with perhaps an additional 15 ug/L as dissolved organic phosphate. He also noted that about 95% of the phosphorus carried in river water is in particulate form. His estimates indicate that human activities increase phosphorus concentrations in European and North American rivers ten- to one-hundred-fold.

CHEMICAL ATTRIBUTES - DISSOLVED OXYGEN

The amount of dissolved oxygen (DO) in any freshwater ecosystem is arguably the single most important chemical determinant of the character of that ecosystem. Most microorganisms and all animals require an external source of oxygen for life processes. Plants also require oxygen, but they create their own supply if light is available and need an external source of oxygen only in the dark.

Oxygen solubility in pure water is determined primarily by temperature and secondarily by atmospheric pressure and other factors (Figure 32). Over the normal range of temperatures experienced by organisms in our rivers twice as much oxygen is available at 0 °C as at 34 °C. Regardless of temperature, however, the amount of dissolved oxygen in water is miniscule compared to the amount available to air-breathers such as ourselves. Whereas 20.95% of the air we breath consists of oxygen, most aquatic organisms thrive in water with dissolved oxygen concentrations of only 5 to 15 mg/L or only 0.005 to 0.015%. Their morphological and physiological adaptations to this scarcity of oxygen are marvelously efficient and the product of millions of years of evolution.

The actual dissolved oxygen concentration in streams and rivers is dependent upon the physical characteristics of the stream, the nature and intensity of in-stream chemical reactions, and the biological activities of the aquatic community. The influence of adjacent terrestrial activities such as agriculture is of much greater importance for streams and rivers than it is for lakes.

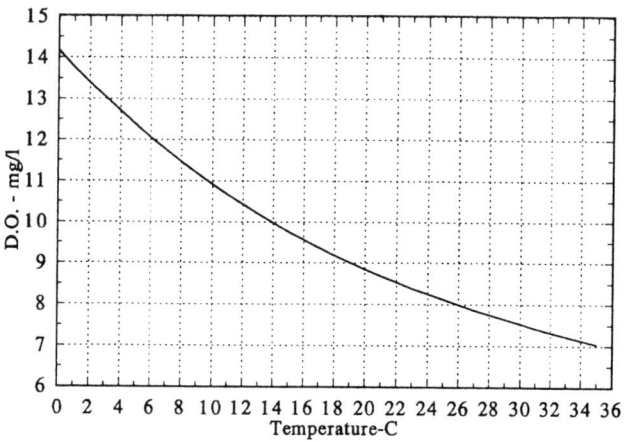

Figure 32: The relationship of D.O. concentration (mg/L) to temperature (°C).

A general model of important determinants of dissolved oxygen in the Wabash River is shown in Figure 33. Major contributions of oxygen are made by incorporating atmospheric oxygen through the water surface - reaeration - and by the generation of excess oxygen resulting from photosynthesis of aquatic plants. Major routes of oxygen removal include, in addition to nonphotosynthesizing aquatic plants, the oxidation of reduced compounds such as ammonia and ferrous iron, aerobic decomposition of dissolved and particulate organic compounds (carbonaceous BOD), aerobic decomposition of reduced compounds, organic compounds, and the benthic fauna comprising the bottom substrate (benthic demand), and uptake to satisfy the metabolic needs of all animals in the water column. The removal of oxygen by fish and zooplankton is relatively insignificant compared to other avenues of uptake and is usually not even considered.

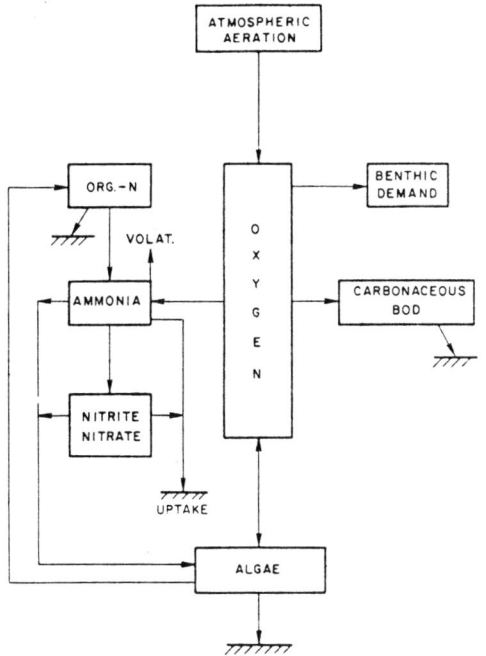

Figure 33: Dissolved oxygen model for the middle Wabash River. (after HydroQual, 1984)

The annual pattern of the average dissolved oxygen concentration in the Wabash River is the reverse of the temperature curve, ie. it is higher during the winter and lower during the summer.

There are two major reasons for this. Firstly, there is an inverse relationship between water temperature and dissolved oxygen concentration at saturation which has just been discussed. Secondly, the relatively low average dissolved oxygen concentration during summer and early fall is caused by the increased metabolism of a larger biomass of aquatic biota, especially phytoplankton and benthic macroinvertebrates which are most abundant and more concentrated during this period of the year. Furthermore, all of these organisms are at their metabolic maxima because of the relatively high temperatures. The dissolved oxygen concentration often

exhibits great diurnal variability especially during the summer (Figure 34).

Much has been made of the ability of rivers and streams to "self-purify" when they receive a load of organic matter. Most of the organic material entering smaller midwestern streams shaded by natural riparian vegetation is in the form of leaves, bark, bud scales, etc. either by direct leaf-fall or as "leaf wads" washed in during rain storm events (Bird and Kaushik, 1981). The stream organisms, benthic macroinvertebrates in particular, are well adapted to take advantage of this annual gift of energy, known ecologically as coarse particulate organic matter (CPOM). After repeated processing larger rivers such as the Wabash River eventually receive the residue in the form of fine particulate organic matter (FPOM) and any nutrients released during decomposition (Cummins et al. 1983).

Streams influenced minimally by man's activities may at times be overwhelmed by the autumnal load of leaves which reduces dissolved oxygen concentration and turns water black (Slack 1955, Schneller 1955).

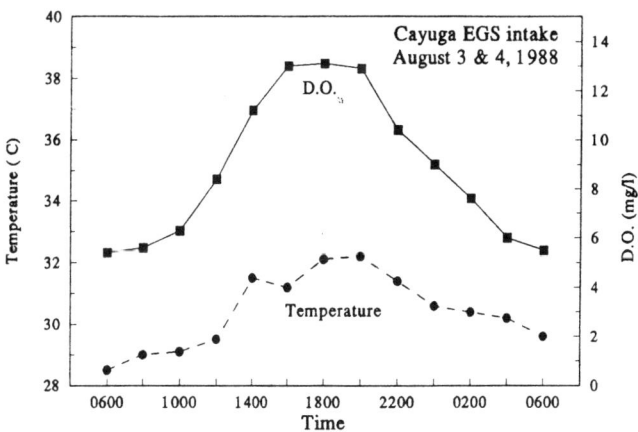

Figure 34: Typical diurnal change in dissolved oxygen and temperature in the Wabash River

With most of the world's expanding human population located near freshwater, anthropogenic sources of CPOM and FPOM have become more important than natural sources. The combined organic contributions of industry, agriculture, and treated or untreated human wastes are now of far greater importance than natural sources. Furthermore, they are released into streams more or less continuously throughout the year. Regardless of origin, all of this material together with microorganisms and phytoplankton, imposes on a river a Biochemical or Biological Oxygen Demand (BOD).

BOD is a heterogeneous collection of large and small molecules and particulates together with attached microorganisms and living phytoplankton. Some of the simpler compounds are quickly oxidized to a stable chemical form, while others require the cooperative work of microorganisms and the aquatic fauna over a long period of time. In this country the BOD is determined by measuring the loss of DO in a water sample incubated at 20 °C in the dark over a 5-day period and expressing the result in milligrams of oxygen per liter. In some countries a 7-day incubation period is used. In addition, an "ultimate" BOD value is often estimated.

A major effort to reduce point sources of BOD was undertaken during the past 20 years. In the decade following passage of the Clean Water Act of 1972 it was estimated that municipal BOD loads had decreased by 46% and industrial BOD loads had decreased at least 71% (U.S. Environmental Protection Agency, 1982). Significant progress in reducing industrial sources of BOD was made in the mid- to late-1970's. Funding for upgrading municipal waste treatment facilities through the EPA's Construction Grants Program also increased during this period and reached their highest levels in 1979 and 1980 (Congressional Budget Office, 1984).

The sources of carbonaceous biological oxygen demand (CBOD) entering the middle Wabash River were determined by HydroQual (1984) during August of 1981 and 1982 (Figure 35). Direct contributions from

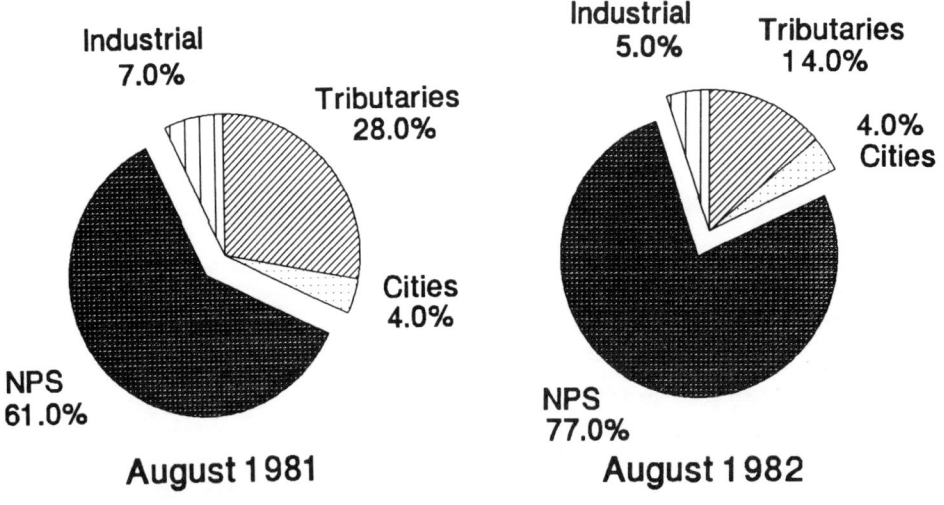

Figure 35: Sources of carbonaceous BOD for the middle Wabash River.

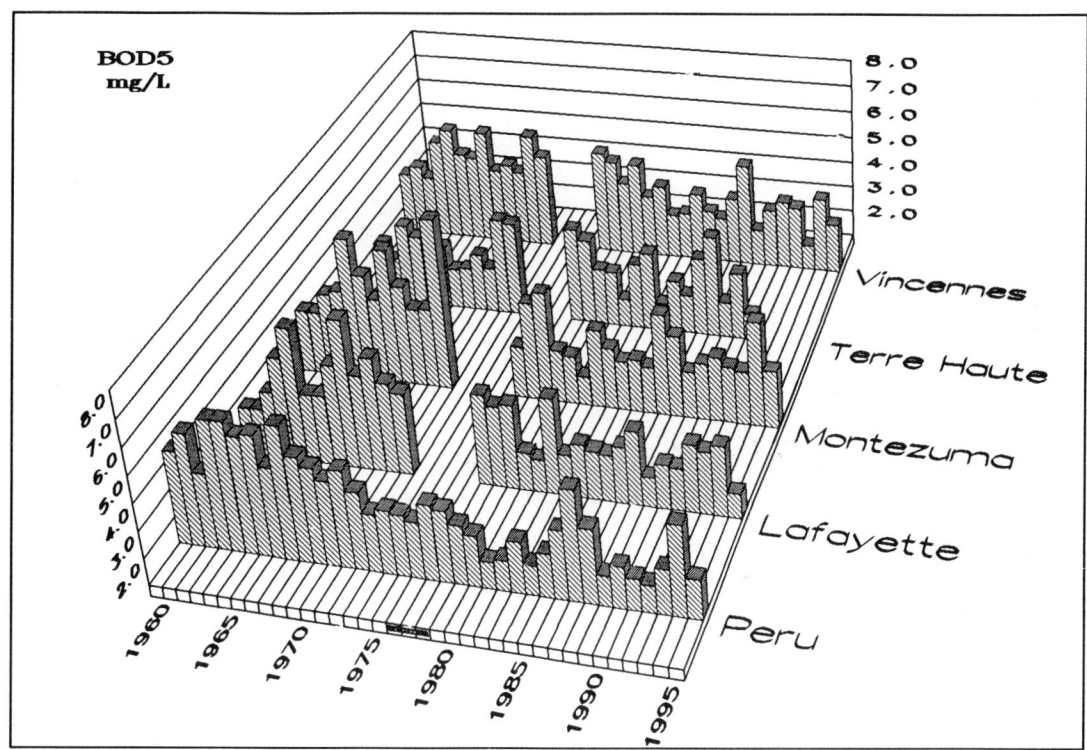

Figure 36: Mean annual BOD (mg/L) of the Wabash River from 1961 through 1995.

industries and municipalities were estimated individually at 100 to 2000 lb/d (220 to 4400 kg/d) and collectively constituted approximately 10% of the total load. The major tributaries were estimated to contribute an additional 15-30%, although it should be pointed out that this CBOD originated from varying sources. The Big Vermilion River, alone, contributed approximately 16,500 lb/d (7500 kg/d), a loading roughly equivalent to all of the other major tributaries combined.

That fraction of carbonaceous biological oxygen demand which entered the river diffusely was determined by subtracting the point-source loadings from the estimated total CBOD. This was found to comprise 60% to 77% of the total CBOD and was estimated to be 136,000 to over 200,000 lb/d (62,000 - 92,000 kg/d). It is likely that most of this was of nonpoint agricultural derivation,

although part of it may have originated *in situ* as phytoplankton.

In addition, it was estimated that 100,000 - 130,000 lb/d CBOD (47,000 - 58,000 kg/d) were contributed by the upper Wabash River. No sources of origin were identified and HydroQual categorized this as a "point" source contribution for modelling purposes. However, it is likely that two-thirds to three-quarters of this amount was derived from agriculture.

In the decade of the 1970's there was a sharp decline in the mean annual BOD of the river (Figure 36). Since then levels appear to have stabilized at about 2.5 to 3.0 mg/L in the 1980's. BOD levels averaged 4.5 to 5.0 mg/L over most of the length of the river during the 1960's (Figure 37). There-

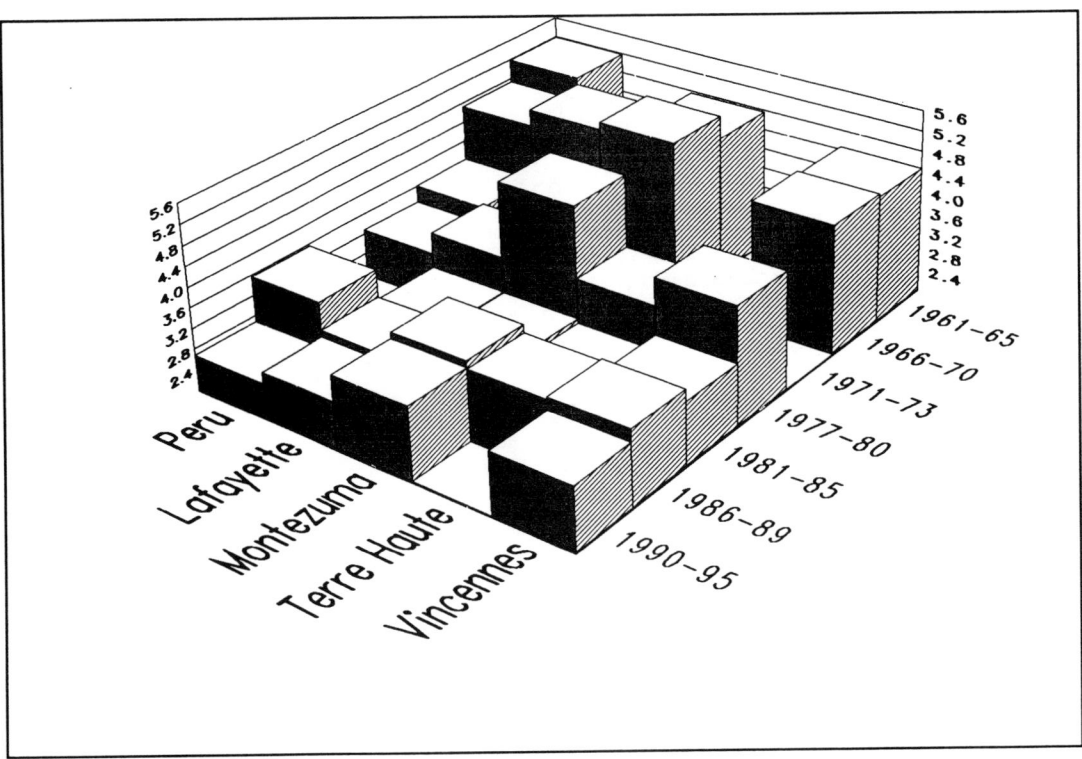

Figure 37: Five-year mean BOD (mg/L) at five locations on the Wabash River.

fore, the national effort to reduce CBOD no doubt included the Wabash River Valley and that may be the major cause for observed reductions of BOD in the Wabash River. Steady improvements in waste treatment were made at Terre Haute, Lafayette, Covington, Kokomo, Frankfort, and most other sizeable communities in the watershed (Anonymous, 1983).

These declining values of the average annual BOD provide encouraging evidence for the efficacy of improving treatment of industrial and municipal wastes. However, when the annual pattern of BOD is examined (Table 4 and Figure 38) it is apparent that today's BOD levels are lower only during part of the year. Biological oxygen demand remains quite high from July to September,

the months of lowest flow. Although there is reduced BOD from industries and municipalities, the favorable temperatures and ample nutrient levels during summer and early fall permit the development of large densities of phytoplankton, an important fraction of the total BOD load. As will be shown later, densities of phytoplankton are particularly high in July, August, and September.

The importance of BOD to both aquatic organisms and the aquatic ecosystem as a whole is no doubt significant, but it is difficult to categorize and evaluate. On the one hand, part of the BOD is food and energy for aquatic biota. On the other hand, particulate BOD contributes toward increasing turbidity to the detriment of species of

69

Table 4: Mean and standard error of monthly BOD of the Wabash River at Lafayette.

Month	1961 - 1969	1970 - 1979	1980 - 1989	1990 - 1995
January	4.01 (0.579)	5.15 (2.958)	1.64 (0.341)	1.20 (-)
February	3.48 (0.515)	4.00 (0.819)	1.70 (0.363)	3.65 (0.950)
March	4.23 (0.949)	5.72 (1.996)	2.12 (0.434)	2.00 (-)
April	4.47 (0.647)	4.02 (0.676)	2.61 (0.472)	2.24 (0.286)
May	4.98 (0.322)	4.01 (0.443)	2.80 (0.554)	-
June	5.71 (1.392)	4.64 (0.628)	2.58 (0.374)	3.83 (1.020)
July	5.56 (0.750)	4.56 (1.044)	4.74 (1.705)	-
August	5.77 (0.534)	4.20 (0.651)	6.19 (1.254)	3.78 (0.668)
September	5.42 (0.474)	4.44 (1.000)	4.06 (1.001)	2.00 (-)
October	4.92 (1.089)	3.16 (0.450)	1.91 (0.356)	2.17 (0.145)
November	3.22 (0.473)	2.58 (0.542)	1.76 (0.334)	1.00 (-)
December	3.62 (0.916)	2.86 (0.558)	2.14 (0.232)	3.15 (0.350)

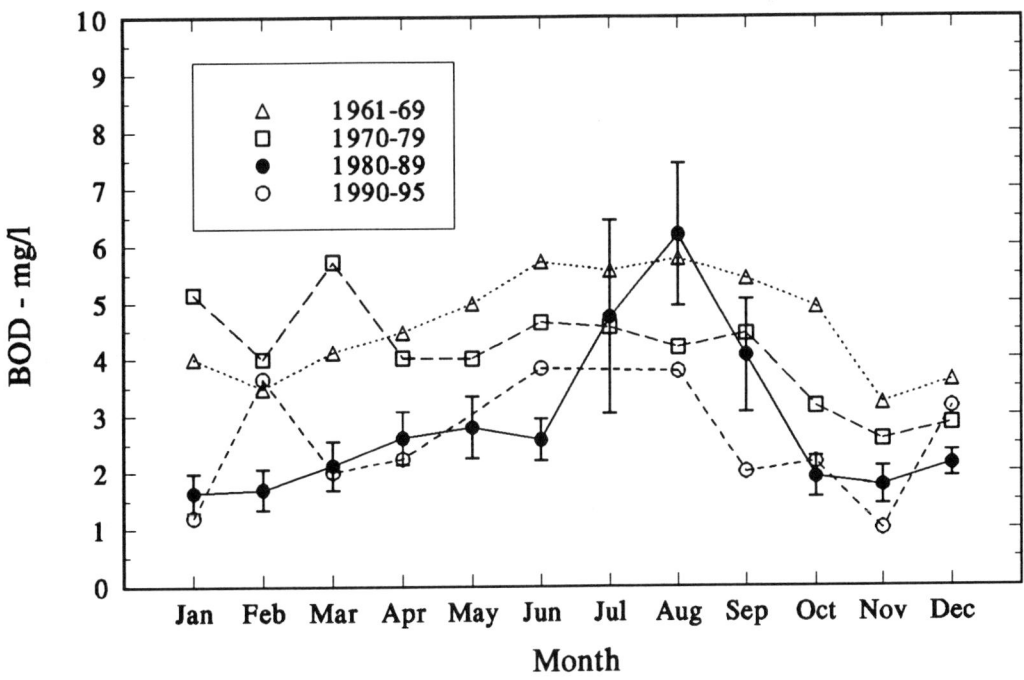

Figure 38: Annual pattern of BOD (mg/L) at Lafayette, Indiana
during the decades of the 1960's, 1970's, 1980's, and 1990's.

fish which locate food visually. It may also physically smother spawning and nursery areas. Its greatest impact, however, probably lies in its influence on the dissolved oxygen concentration of the water which, in turn, affects the aquatic fauna.

One way of characterizing the DO status of a river is to estimate the dissolved oxygen deficit (DOD) at different rates of river discharge by computer modeling. The DOD is the saturation DO concentration at the prevailing temperature minus the actual DO concentration (Smith et al. 1987). A DOD model for the Wabash River was developed by HydroQual (1984) using a version of the DIURNAL computer model and data collected in 1981 and 1982. At low discharge rates this model projected values of 2.0 to 2.5 mg/L from Delphi to Montezuma (Figure 39). Dissolved oxygen deficit levels then increased sharply to reach a value of about 4.0 mg/L from Terre Haute to Merom, Indiana.

Phytoplankton respiration comprised the largest single uptake of DO, an estimated 50-60% of the DOD in the upper reaches and about 70% downriver from Terre Haute. The second largest source was carbonaceous biological oxygen demand (CBOD) which entered the river from multiple point sources and accounted for about 10% of the DOD in the upper reaches and over 15% downriver from Terre Haute. Sediment oxygen demand (SOD) was estimated to be the third largest oxygen sink. Nitrogenous BOD and other sinks also contributed to a minor degree.

Smith et al. (1987) found no significant relationship between the change in BOD loads from municipal treatment plants within 160 km upstream from U.S.G.S. NASQAN stations and the decline in the DO deficit. Most NASQAN stations are located on large rivers. If DOD is mainly caused by respiring algae, as is the case in the Wabash River, then improvements in DOD due to decreased point loadings of BOD might have been obscured.

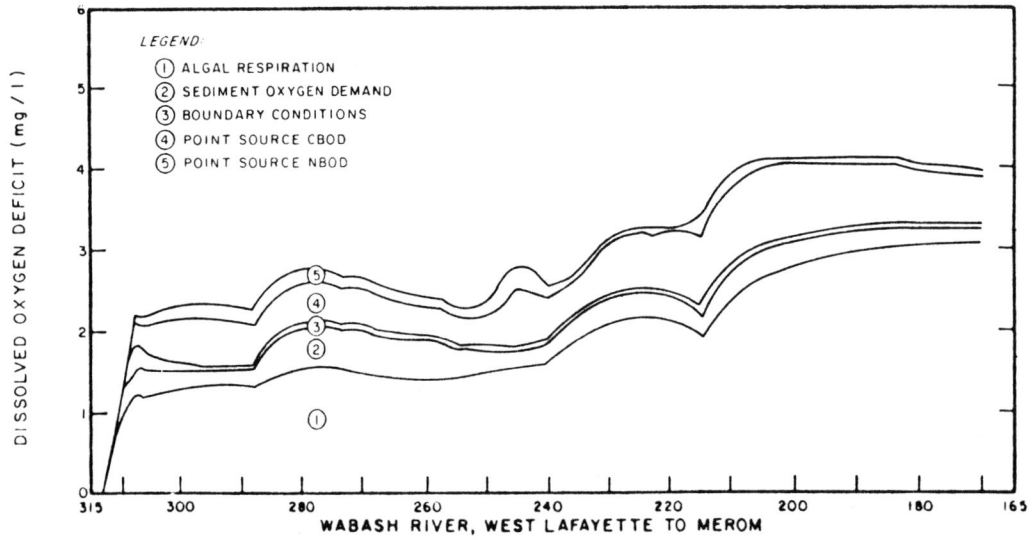

Figure 39: Estimated mean dissolved oxygen deficit (DOD) for the middle Wabash River during low-flow conditions. (After Figure 6.17(A), HydroQual, 1984)

CHEMICAL ATTRIBUTES - HEAVY METALS

It is difficult to assess the temporal and spatial differences of many heavy metals because the sampling effort and locations have changed over time. As mentioned before, copper concentrations decreased statistically over the period 1977 - 1987 at New Harmony (ORSANCO 1990), as it also did elsewhere within the Ohio River basin.

Table 5 summarizes data from the ORSANCO study (1990) and unpublished data from the Indiana Department of Environmental Management. Concentrations of iron, lead, manganese, nickel, arsenic, and copper were higher downriver than upriver.

Table 5: Mean concentration (ug/L) of metallic elements in the Wabash River. Concentrations are total and for the period 1981 - 1987, except where noted.

Metals	Laf.	T.H.	Breed	Vinc.	N.Har.[a]
Arsenic	2.13	2.49	2.59	2.47	-
Berylium	2.00	2.00	-	-	-
Cadmium	-	2.13	2.00	-	-
Chromium (Hex)	10.00	10.00	9.85	10.00	-
Chromium	11.67	12.04	11.62	10.89	-
Copper	6.78[b]	7.07[b]	7.90	-	17.77
Iron	2,195[b]	3,754[b]	4,328	-	4,613
Lead	12.03[b]	13.36[b]	9.18	-	28.54
Manganese	122.0	209.3	223.4	-	-
Thallium	20.00	20.00	-	-	-
Nickel	6.47	8.12	8.23	-	-
Zinc	23.98[b]	23.01[b]	-	-	-
Antimony	0.80	0.60	-	-	-
Selenium	1.15	1.24	-	-	-
Mercury	0.12[b]	0.10[b]	-	-	0.32

[a] 1977 - 1987   [b] 1981 - 1987

The concentration of metals in the Wabash River is related not only to industrial sources, but also to the potential influence on water quality of coal mining. Coal deposits are distributed throughout 20 counties in south-western Indiana. The influence of coal mining, past and present, extends northward only to the level of the Big Vermilion River at Rkm 414 (RM 257).

Underground mines produced most of the coal from the late 1800's until about 1950 after which strip mining predominated. Not until the passage of the Indiana Strip Mine Law in 1968 was there any serious attempt to reduce mining effects. This act required that all acidic material be buried and that all stripped lands be graded and seeded. However, the law applied only to active mined areas and not to older abandoned mines which continue to leach acids and heavy metals into surface waters (Thomas 1978).

Several studies have examined the effects of coal mining on water quality in streams of southwestern Indiana (Corbett and Agnew 1968; Corbett 1969; Wilber et al. 1980; Peters 1981; Wangsness 1982; Wangsness et al. 1981a, 1981b, 1983; Zogorski et al. 1981; and Martin and Crawford 1987). Generally these investigations indicate that specific conductance, dissolved solids, acidity, sulfate, iron, manganese, and aluminum were higher in streams draining mined areas than in streams in unmined areas. Total alkalinity and pH were generally lower in waters in mined areas than in unmined areas.

A few older abandoned mining areas are located directly adjacent to the Wabash River mainstem, but most active and abandoned mines are located near tributaries. Seepage from these mines enter tributary waters which empty into the Wabash River. Big Vermilion River is the largest tributary affected by mining. Other tributaries affected by mine wastes proceeding sequentially in a downriver direction include Coal Creek, Brouillettes Creek, Otter Creek, Coal Creek, Little Sugar Creek, and Honey Creek.

Thomas (1978) assessed impacts of abandoned mines in west-central Indiana and recognized three priority classes. Six mine sites were included as Priority I, the most serious. Thirteen more mines were classified as Priority II, somewhat less serious. Ten more mines which contributed only small amounts of pollutants were designated as Priority III. An additional category, IV, represented areas which were not polluting streams or lakes.

The remains of the so-called "Green Valley Mine" near Little Sugar Creek illustrates the problems presented by old mines. This deep mine is located immediately east of the Illinois state line and about eight miles northwest of Terre Haute. It began production of coal in 1947 and was abandoned in 1963. Gob piles and slurry covered over 100 acres, an ugly pile of barrens devoid of vegetation and capped by a small acidic pond.

Acid runoff entered West Little Sugar Creek and thence to the Wabash River. Frequently the upper stream was milky white and the water near the Wabash River red-orange in color. Thomas (1978) examined water quality from September through mid-November, 1977 and found that the pH was frequently less than 3.0 in the upper stream with conductivity which was often greater than 2000 umhos/cm (Table 4). The delivery of heavy metals and their concentrations were substantial.

The Indiana DNR Abandoned Mine Program has been working to correct some of the water quality problems caused by abandoned mines. John Allen (1991) of the Division of Reclamation estimates that rehabilitating the Green Valley Mine will require $3 million. There has been no biological assessment of Little Sugar Creek. The total contributions of all the active and abandoned mines within the Wabash River basin has not yet been assessed. As will be shown later, the fish communities of the middle Wabash River change quite suddenly very close to the northern extent of mining areas. With the limited information currently available it is not possible to assess chronic biological effects attributable to this major activity.

Table 6: Water quality of West Little Sugar Creek near the Green Valley mine, September to mid-November, 1977.

Water Quality Parameter	Number of Samples	Median Concentration*	Range*
Conductivity (umhos)	8	2,289	937 - 6,254
pH	8	4.31	2.87 - 5.96
Arsenic	13	50.9	29.3 - 641.0
Barium	13	85.9	32.5 - 122.0
Cadmium	12	2.68	1.88 - 23.70
Chromium	13	94.6	7.28 - 517.0
Iron	15	387,000	120,000 - 6,070,000
Lead	6	7.63	5.13 - 30.40
Manganese	15	4,650	1,100 - 31,000
Nickel	6	255	201 - 1,030

* ug/l except for conductivity

WABASH RIVER FISHES
INTRODUCTION AND METHODS

The Wabash River mainstem may be divided arbitrarily into four regions: 1) the headwaters which lie upriver from Huntington Reservoir (rkm 660) (RM 410), 2) the upper mainstem extending from the Huntington dam south to Logansport, Indiana, 3) the middle mainstem from Logansport to the mouth of the White River (rkm 206 to 660) (RM 130 to 410), and 4) the lower mainstem from White River to the Ohio River.

The fish community of the Wabash River has received the attention which it deserves only in recent decades. Earlier this century the Commissioner of Fisheries and Game of Indiana (1913) commented that in the Wabash River from the Terre Haute area ". . up to Huntington, and beyond, there are many splendid bass fishing places. But as a fishing stream, where one would choose to go to enjoy all the beauties of nature, most people prefer its neighbor and tributary the Tippecanoe, a few miles to the north."

Fish in the Upper and Lower Wabash River

Fish in the lower Wabash River have been studied in recent years mostly by Illinois Department of Conservation personnel. This data is currently being summarized (Day, personal communication).

Studies of the upper Wabash River by the Indiana Department of Natural Resources indicate that, while a variety of other species is present in small numbers, gizzard shad, carp, and carpsucker species constituted over 55% of numbers and 65% of biomass of fishes in the river above Huntington Reservoir (Pearson 1975). This section of the river is sluggish and ditch-like, with the surrounding landscape dominated by an extensive rowcrop agriculture. The mainstem and its tributaries are used primarily for the removal of excess water during the spring rainy period.

In 1989 Braun (1990) electrofished 18 stations from the Indiana-Ohio state line at rkm 746 (RM463) to Peru, Indiana at rkm 590 (RM366) and, upstream of Huntington Dam, found a "mediocre" fish population dominated by carp and white suckers. Catches of sauger were lower than in 1974, but walleye, which were first stocked into upper Wabash reservoirs in the early 1970's, had increased. During the intervening period the shovelnose sturgeon and spotted bass had extended their ranges northward.

Water clarity, water quality, and the overall quality of the fish community improve considerably below the Huntington dam. Assisting in the process are the Salamonie and Mississinewa rivers, both of which also pass through reservoirs before entering the Wabash mainstem.

In the upper mainstem "the river retains much of its native beauty" although algal densities sometimes color the river brownish and produce high mid-afternoon DO concentrations (Robertson 1975). The fish community between Huntington dam (rkm 658 = RM410) and Covington, Indiana (rkm 436 = RM 271) was found to be dominated by gizzard shad, carp, and carpsuckers (58% by numbers, 56.6% by weight), but redhorse and game fish (drum, sauger, and smallmouth bass) were also much more abundant below Huntington dam.

Collection Methods for Fish in the Middle Wabash River

The fish communities of the middle Wabash River have been studied annually since 1967. Studies from 1967 to 1973 focused mainly on effects of the Wabash Electric Generating Station (EGS) located north of Terre Haute, Indiana and the Cayuga EGS which began operating in the fall of 1970. During this period we experimented extensively to determine which collecting methodologies were most effective and efficient. The positive and negative attributes of hoop nets, D-nets, gill nets, A.C. electrofishing, and D.C. electrofishing were examined. The most versatile and effective sampling apparatus proved to be D.C. electrofishing using a Smith-Root Type VI electrofisher and this has been employed as the standard since 1972 (Gammon 1973).

Initially seines were not used because of a scarcity of suitable habitat in the vicinity of the electric generating stations. In most areas water depth increases rapidly from shore and the near-shore bottom substrate usually consists of a soft mixture of sand and mud. Log jams are also common throughout the river. During some years seining would have been impossible because of high summer discharge.

Seegert (personal communication) believes that a seining component should be included in the Wabash River sampling program because smaller fish species such as darters and minnows are otherwise omitted. Seining collections were part of the sampling protocol in 1977 and 1997 and, although new species of fish were found, environmental assessments were not clarified greatly beyond those shown by the electrofishing collections alone. Furthermore, during high-flow summers it is virtually impossible to find suitable seining sites anywhere. Although seining alone has been of limited value for providing a health index for the Wabash River itself, it is a worthwhile component in studies of somewhat smaller rivers.

In order to make valid comparisons of the fish community over time and space, we have excluded young-of-the-year (y-o-y) fish of all species of fish from the catch and community analyses. This was necessary because collections sometimes occurred during June, before the appearance of y-o-y, but more often were made during August when y-o-y were sufficiently large to be captured. Population studies which ignore this practical reality run the risk of making invalid comparisons.

During the early investigative years we found that diversity and abundance of fish differed from place to place (Gammon 1976), which suggested that environmental factors other than heated water were also influencing the fish community. In 1973 we began to systematically collect fish at stations scattered through 100 miles of river between Independence, Indiana and Terre Haute, Indiana. Over the next few years the range was extended to greater distances both up- and down-river. Based upon experience in a variety of mainstem habitats, most of the collecting stations were sited in relatively fast-water sections of river having good cover and an average depth of one meter or less. These locations were often, but not always, on the outside of river bends. These sites usually contained a greater variety of species and yielded larger catch rates than either sites located in deeper water or the slow, sandy-bottomed shallows on the inside of river bends.

For many years a total of 63 collecting stations have been regularly sampled between Delphi, Indiana and Merom, Indiana. Eight of these stations were concentrated in the area of the Cayuga Electric Generating Station and another five stations were concentrated above and below the Wabash River Electric Generating Station. The remaining stations were sprinkled throughout the river with greater densities at population centers of Lafayette, Indiana and Terre Haute, Indiana. Each collecting station initially consisted of a variably long section of shoreline, but after 1975 each station consisted of a 0.5-km section of shoreline as measured with a Leitz optical rangefinder.

Our annual monitoring objective was to execute three electrofishing passes at each of these regular collecting stations each summer. An ideal series of collections began at Delphi, Indiana and concluded four to five days later at Merom, Indiana or Hutsonville, Illinois. However, only one or two passes were possible because in some years unusually high discharge reduced collecting efficiency at some stations. In addition, weather and flow conditions often restricted the collections to shorter sections of river. Sometimes it was possible to launch a fourth sampling effort. During droughty summers such as 1988 and 1991, for example, the problem was often finding water which was sufficiently deep in which to collect.

For analytic purposes the river was divided into 12 Reaches as indicated in Table 7 and shown in Figure 40, with each Reach containing at least five collecting stations. Although eight stations were located at the

Table 7: Location of the study Reaches of the middle Wabash River.

Reach	River Mile (Rkm)	Description of Location
1	330 to 313 (531-502)	Delphi to north Lafayette, Ind.
2	312 to 302 (501-486)	Lafayette & W. Lafayette area
3	301 to 286 (485-460)	Lafayette to Attica, Ind.
4	285 to 269 (459-433)	Attica to Covington, Ind.
5	268 to 251 (432-404)	Covington to Coal Creek
6	249.6 (402)	Cayuga Elec. Generating Station
7	247 to 233 (396-375)	Cayuga EGS to E. Lilly, Clinton
8	232 to 218 (374-351)	E.Lilly, Clinton to Otter Creek
9	215.4 (347)	Wabash River Elec. Gen. Station
10	213 to 203 (343-327)	Terre Haute, Ind. area
11	202 to 186 (325-299)	Terre Haute STP to Darwin, Ill.
12	185 to 160 (298-257)	Darwin, IL to Merom, Ind.

77

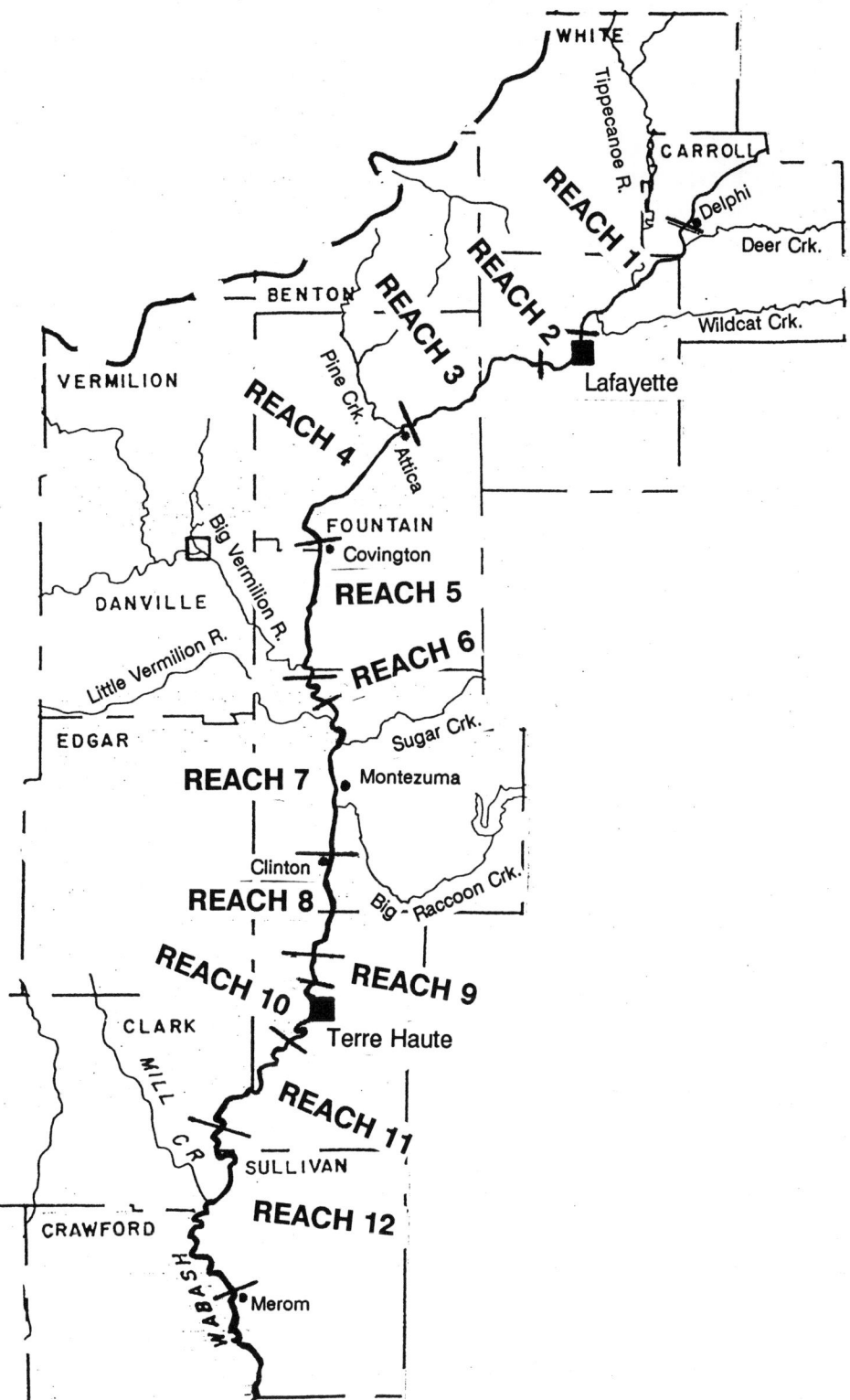

Figure 40: Locations of 12 collecting Reaches in relation to major streams and population centers.

Cayuga Electric Generating Station, only five stations having the desired physical characteristics were included in the overall analysis. At the Wabash River Electric Generating Station only two of the six stations were unheated. Community parameters and the individual species abundance was usually based upon 3 collections at each of 5 locations within each Reach for a total of 15 samples each summer.

Each collecting station was sampled near shore with a Smith-Root Type VI electrofisher producing 600 VDC at 6 amps and 60 pps. Pulsed DC current flowed into an electrode system consisting of two circlets of short stainless steel anodes suspended at the water surface by bow booms and two gangs of long woven copper cathodes off the port and starboard gunwales. A single netter worked from the bow of the boat using a long-handled dip net having one-quarter inch square mesh netting. All captured fish were placed in a holding tank and processed after the entire collecting station had been electrofished. Fish were identified to species, weighed, measured, and returned to the river unharmed as soon as possible.

Data Analysis

All data were entered into a computer data base and processed by a battery of analytic programs which were developed in-house and modified over time by a generation of students and DePauw Computer Center personnel. The data program permits the entry of some physical and chemical parameters as well as biological attributes. Output options include summarizations of the catch data for each species, calculations of community parameters, and length frequency and condition factor analysis.

Two measures of abundance of the fish community included the following: (1) numbers per unit collecting effort, which in this study is the number of fish captured per kilometer of shoreline electrofished, and (2) biomass per unit collecting effort or kilograms of fish captured per kilometer.

Three measures of diversity were derived including: (1) the mean number of species captured per collection, (2) the Shannon-Weaver or Shannon-Weiner index of diversity or heterogeneity based on numbers of each species captured per collection using natural logarithms (H'), and (3) the Shannon-Weiner index based upon weight of each species using natural logarithms.

Equitability of species abundance in each electrofishing collection was derived from the S-W index values as $J = H'/H'_{max}$.

Early in the study it was observed that all of these measures of community, except equitability, exhibited the same basic pattern over space. Therefore, a Composite Index of Well-Being (Iwb) was constructed, which appeared to best represent the catches in terms of both diversity and abundance (Gammon 1980). The composite "Index of Well-Being" (Iwb) was calculated as:

Iwb = 0.5 lnN + 0.5 lnW + Div(no) + Div(wt)

where

N = number of fish captured per km
W = weight in kg captured per km
Div(no) = Shannon diversity based on numbers
Div(wt) = Shannon diversity based on weight

In some studies a modification of the Iwb is calculated, in which an Iwb value is derived with carp and other tolerant fish deleted from the numbers and biomass captured, but included in calculating the diversity values. This modified Iwb has been of value in some state-wide studies where rivers are much more polluted than the Wabash River (Ohio EPA).

Part of the Wabash River data base has also contributed to the development of a Great River Index of Biotic Integrity (IBI). Some of the extensive analyses will be discussed in a later section.

Species of Fish in the middle Wabash River

The Wabash River possesses a diverse and abundant ichthyofauna. Table 8 summarizes all of the species which have been collected during the regular collecting series

since 1967 and, in addition, includes species of smaller fishes which have been taken in special studies utilizing the seine. Scientific terminology follows Robins et al. (1991). Only a few of these species were distributed throughout the 274 km (160 mile) study segment and only a very few species of fish were truly abundant. A few other species populations were moderately abundant, but most were relatively small and were concentrated in certain areas of the river.

In the next few sections, we'll describe the distribution and abundance of some of the more important species by grouping them into clusters which share certain characteristics. The most common and abundant fishes will be examined first and then a great variety of other species.

An electrofishing run by Graduate students Brandon Kulik, Ernest Roggelin, and Rick Wright.

Table 8: List of fish species collected from the middle
Wabash River from 1967 to 1997.

FAMILY, Common Name, and *Scientific Name*

Lamprey Family - PETROMYZONTIDAE

American brook lamprey - *Lampetra appendix* (DeKay)
Silver lamprey - *Ichthyomyzon unicuspis* Hubbs and Trautman
Chestnut lamprey - *Ichthyomyzon castaneus* Girard

Sturgeon Family - ACIPENSERIDAE

Shovelnose sturgeon - *Scaphirhynchus platorynchus* (Rafinesque)

Paddlefish Family - POLYODONTIDAE

Paddlefish - *Polyodon spathula* (Walbaum)

Gar Family - LEPISOSTEIDAE

Longnose gar - *Lepisosteus osseus* (Linnaeus)
Shortnose gar - *Lepisosteus platostomus* Rafinesque
Spotted gar - *Lepisosteus oculatus* (Winchell)

Bowfin Family - AMIIDAE

Bowfin - *Amia calva* Linnaeus

Mooneye Family - HIODONTIDAE

Goldeye - *Hiodon alosoides* (Rafinesque)
Mooneye - *Hiodon tergisus* Lesueur

Freshwater Eel Family - ANGUILLIDAE

American eel - *Anguilla rostrata* (Lesueur)

Herring Family - CLUPEIDAE

Skipjack Herring - *Alosa chrysochloris* (Rafinesque)
Gizzard shad - *Dorosoma cepedianum* (Lesueur)
Threadfin shad - *Dorosoma petenense*

Table 8: (continued)

Minnow Family - CYPRINIDAE

Stoneroller - *Campostoma anomalum* (Rafinesque)
Goldfish - *Carassius auratus* (Linnaeus)
Grass carp - *Ctenopharyngodon idella* (Valenciennes)
Spotfin shiner - *Cyprinella spiloptera* (Cope)
*Steelcolor shiner - *Cyprinella whipplei* Girard
Common carp - *Cyprinus carpio* Linnaeus
*Streamline chub - *Erimystax dissimilis* (Kirtland)
Mississippi silvery minnow - *Hybognathus nuchalis* Agassiz
Bighead carp - *Hypophthalmichthys nobilis* (Richardson)
Striped shiner - *Luxilus chrysocephalus* (Rafinesque)
*Redfin shiner - *Lythrurus umbratilis* (Girard)
Speckled chub - *Macrhybopsis aestivalis* (Girard)
Silver chub - *Macrhybopsis storeriana* (Kirtland)
River chub - *Nocomis micropogon* (Cope)
*Golden shiner - *Notemigonus crysoleucus* (Mitchill)
Bigeye chub - *Notropis amblops* (Rafinesque)
Emerald shiner - *Notropis atherinoides* Rafinesque
*River shiner - *Notropis blennius* (Girard)
*Ghost shiner - *Notropis buchanani* Meek
Silverjaw minnow - *Notropis buccata* Cope
Rosyface shiner - *Notropis rubellus* (Agassiz)
Sand shiner - *Notropis stramineus* (Cope)
*Channel shiner - *Notropis wickliffi* (Cope)
Suckermouth minnow - *Phenacobius mirabilis* (Girard)
Bluntnose minnow - *Pimephales notatus* (Rafinesque)
Bullhead minnow - *Pimephales vigilax* (Baird and Girard)
Creek chub - *Semotilus atromaculatus* (Mitchill)
*Blacknose dace - *Rhinichthys attratulus* (Hermann)

Sucker Family - CATOSTOMIDAE

River carpsucker - *Carpiodes carpio* (Rafinesque)
Quillback - *Carpiodes cyprinus* (Lesueur)
Highfin carpsucker - *Carpiodes velifer* (Rafinesque)
White sucker - *Catostomus commersoni* (Lacepede)
Blue sucker - *Cycleptus elongatus* (Lesueur)
Northern hog sucker - *Hypentelium nigricans* (Lesueur)
Smallmouth buffalo - *Ictiobus bubalus* (Rafinesque)
Bigmouth buffalo - *Ictiobus cyprinellus* (Valenciennes)
Black buffalo - *Ictiobus niger* (Rafinesque)

82

Table 8: (continued)

Sucker Family - CATOSTOMIDAE (continued)
Spotted sucker - *Minytrema melanops* (Rafinesque)
Black redhorse - *Moxostoma duquesnei* (Lesueur)
Golden redhorse - *Moxostoma erythrurum* (Rafinesque)
Silver redhorse - *Moxostoma anisurum* (Rafinesque)
Shorthead redhorse - *Moxostoma macrolepidotum* (Lesueur)
River redhorse - *Moxostoma carinatum* (Cope)

Catfish Family - ICTALURIDAE
Black bullhead - *Ameiurus melas* (Rafinesque)
Yellow bullhead - *Ameiurus natalis* (Lesueur)
Blue catfish - *Ictalurus furcatus* (Lesueur)
Channel catfish - *Ictalurus punctatus* (Rafinesque)
Stonecat - *Noturus flavus* Rafinesque
*Mountain madtom - *Noturus eleutherus* Jordan
*Brindled madtom - *Noturus miurus* Jordan
*Freckled madtom - *Noturus nocturnus* Jordan & Gilbert
Flathead catfish - *Pylodictis olivaris* (Rafinesque)

Pike Family - ESOCIDAE
Grass pickerel - *Esox americanus* Gmelin

Pirateperch Family - APHREDODERIDAE
Pirateperch - *Aphredoderus sayanus* (Gilliams)

Codfish Family - GADIDAE
Burbot - *Lota lota* (Linnaeus)

Killifish Family - CYPRINODONTIDAE
Blackstripe topminnow - *Fundulus notatus* (Rafinesque)

Livebearers - POECILIIDAE
Mosquitofish - *Gambusia affinis* (Baird & Girard)

Silversides Family - ATHERINIDAE
*Brook silversides - *Labidesthes sicculus* (Cope)

Temperate Bass Family - PERCICHTHYIDAE
White bass - *Morone chrysops* (Rafinesque)
Yellow bass - *Morone mississippiensis* Jordan and Eigenmann
White bass x Striped bass hybrid

Table 8: (continued)

Sunfish Family - CENTRARCHIDAE
Rock bass - *Ambloplites rupestris* (Rafinesque)
Green sunfish - *Lepomis cyanellus* Rafinesque
Pumpkinseed - *Lepomis gibbosus* (Linnaeus)
*Orangespotted sunfish - *Lepomis humilis* (Girard)
Longear sunfish - *Lepomis megalotis* (Rafinesque)
Warmouth - *Lepomis gulosus* (Cuvier)
Bluegill - *Lepomis macrochirus* Rafinesque
Redear sunfish - *Lepomis microlophus* (Gunther)
Smallmouth bass - *Micropterus dolomieu* Lacepede
Spotted bass - *Micropterus punctulatus* (Rafinesque)
Largemouth bass - *Micropterus salmoides* (Lacepede)
White crappie - *Pomoxis annularis* Rafinesque
Black crappie - *Pomoxis nigromaculatus* (Lesueur)

Perch Family - PERCIDAE
Eastern sand darter - *Ammocrypta pellucida* (Putnam)
Bluntnose darter - *Etheostoma asprigene* (Forbes)
Greenside darter - *Etheostoma blennioides* Rafinesque
Rainbow darter - *Etheostoma caeruleum* Storer
*Fantail darter - *Etheostoma flabellare* Rafinesque
Slough darter - *Etheostoma gracile* (Girard)
Johnny darter - *Etheostoma nigrum* Rafinesque
*Orangethroat darter - *Etheostoma spectabile* (Agassiz)
Logperch - *Percina caprodes* (Rafinesque)
*Blackside darter - *Percina maculata* (Girard)
Slenderhead darter - *Percina phoxocephala* (Nelson)
*Dusky darter - *Percina sciera* (Swain)
*River darter - *Percina shumardi* (Girard)
Sauger - *Stizostedion canadense* (Smith)
Walleye - *Stizostedion vitreum* (Mitchill)

Drum Family - SCIAENIDAE
Freshwater drum - *Aplodinotus grunniens* Rafinesque

* Species collected with a seine during special studies
by Rogellin (1979), EA Science and Technology (1988, 1989),
Gammon and Gammon (1998), or Cinergy Inc. monitoring
program in vicinity of EGSs (Lewis, personal communication).

DOMINANT FISHES OF THE WABASH RIVER

Some of the most ecologically important fishes in the Wabash River are among the least well-known species to most people. They are dominant for two primary reasons: 1) large numbers of individuals are supported by the Wabash River and/or 2) their population contributes significantly to the total fish biomass. With the exception of the carp, these species are not well known because they are rarely caught by anglers. These most common species are gizzard shad (*Dorosoma cepedianum*), northern river carpsucker (*Carpiodes carpio*), and carp (*Cyprinus carpio*).

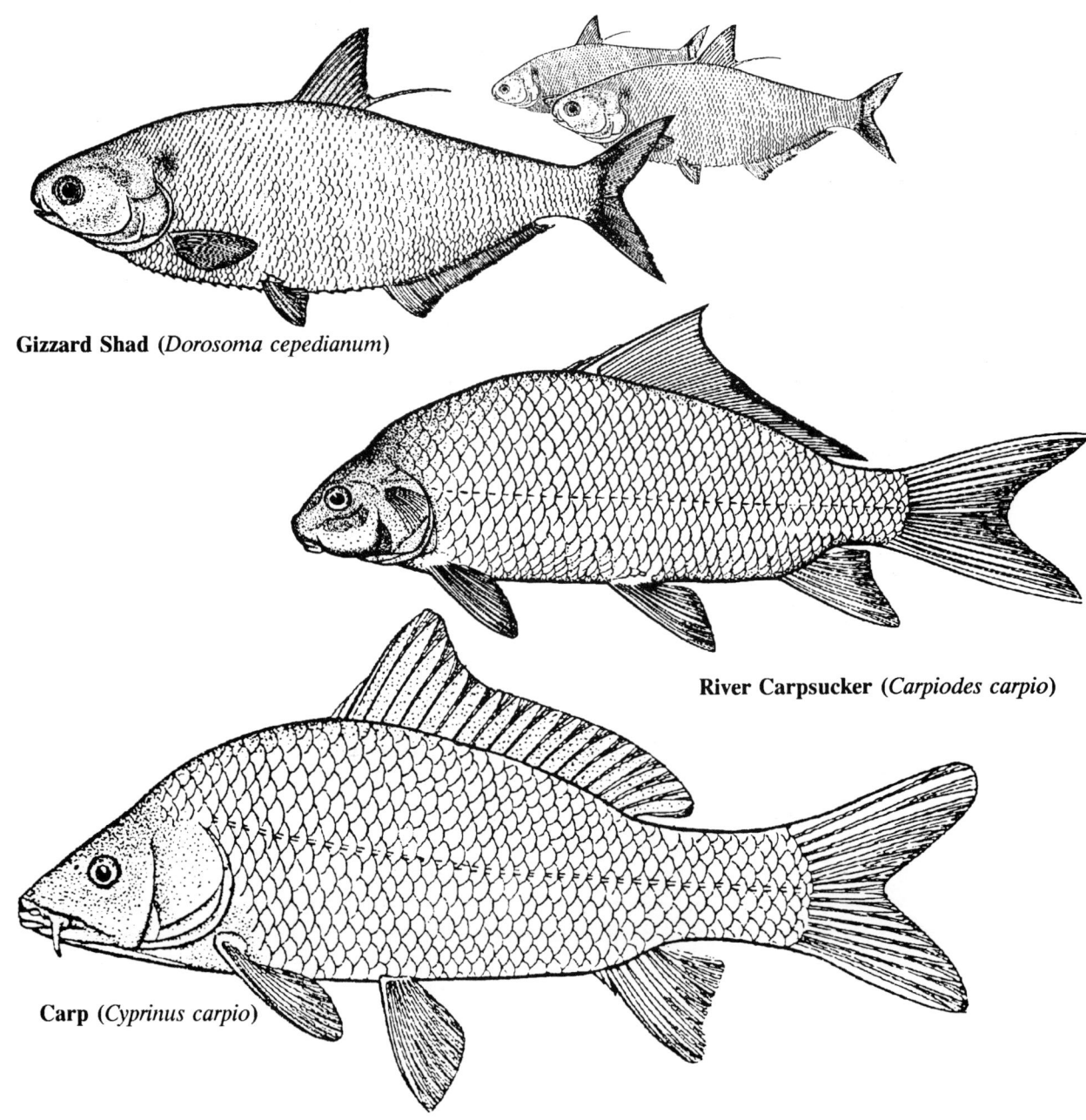

Gizzard Shad (*Dorosoma cepedianum*)

River Carpsucker (*Carpiodes carpio*)

Carp (*Cyprinus carpio*)

The gizzard shad is the most abundant member of large Wabash River fishes just as it is in most large midwestern U.S. rivers and reservoirs. In the Wabash River, dense schools of shad prefer the quiet waters of backwater areas and shallow, sandy insides of river bends. They are much less common in faster water where most of the collecting sites were located.

Prior to 1981 gizzard shad constituted about half of the total numerical electrofishing catch and about 20% of the biomass. They declined in abundance after 1981 for reasons to be discussed later and have comprised about 30% of the total numbers and 10% of the biomass since that time.

Small young-of-the-year (y-o-y) shad less than 35 mm long feed almost exclusively on microcrustaceans (Warner 1940; Kutkuhn 1958; Dalquest and Peters 1966; Cramer and Marzold 1970). Larger individuals are well equipped morphologically to feed on and digest detritus (Mundahl and Wissing 1988), but they also feed on zooplankton, phytoplankton, and other live foods when it is available and abundant (Jester and Jensen 1972; Jude 1973; Drenner et al. 1982, 1984). In tributaries of the Wabash River we have observed schools of shad feeding on tufts of aufwuchs/periphyton growing on submerged tree branches.

Gizzard shad usually spawn in early summer. Mancini (1974) found the highest density of shad dry in his ichthyoplankton tows during mid-June, with a smaller secondary peak in early July. Very few ichthyoplankton eggs or fry were tken after July 20, 1973. Ichthyoplankton drift studies near the Wabash River electric generating station indicated that gizzard shad fry constituted 40.1% and 84.1% of the total number of fry in 1973 and 1974 respectively (Gammon 1976).

Because of their abundance during early summer, small gizzard shad fry and y-o-y are probably an important food source for smaller piscivores and perhaps for larger ones as well. However, Rud (1982) found that larger shad were not an important food source for larger piscivores in the Wabash River.

Catch rates of gizzard shad one year old and older fluctuated considerably from year to year (Figure 41). Catch rates were quite stable at 10-30/km from 1973 to 1979. Relative abundance declined to 5-20 /km from 1980 through 1991. An enormously successful hatch in 1991 produced the largest shad population ever found one year later in 1992. As will be shown later, low shad densities for several years after 1984 were almost certainly the result of a large popu-

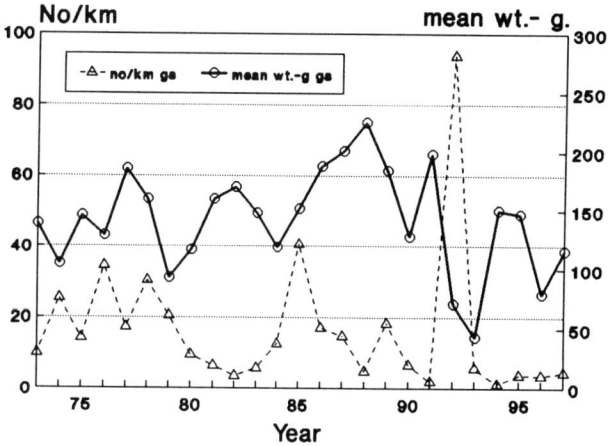

Figure 41: Annual catch rates and mean weight of gizzard shad from 1968 through 1997.

population of piscivorous fishes. The piscivore population had diminished by 1991 and no longer cropped off young shad. This reduction in biological control, together with a favorable discharge pattern during summer, combined to generate a hugh 1991 year-class of shad.

This 1991 year-class was virtually exterminated by the prolonged high water in 1993, as was the entire 1993 year-class of shad and most other species.

The temporal pattern of average weight is a mirror image of the catch rate. During years with low catch rates the shad were generally larger, older fish while high catch rates included more successful, younger year classes. The shift in size is best shown by examining the changes in length frequency over time. Figure 42 shows the large number of yearling shad in 1974, 1979, and 1992, the result of unusual spawning success each of the previous years. Figure 42 also indicates the preponderance of larger shad in the catches from 1986 through 1990.

Reaches 4 and 5 supported more shad than other Reaches (Figure 43) and Reach 8 consistently yielded fewer, but larger individuals than most other Reaches. The cause(s) for this scarcity in this particular segment of the river of an otherwise ubiquitous species is unknown.

The fish community of the middle Wabash River has been so stable during some years that predicting results for the upcoming year seems to be an easy matter. However, marked changes have occurred suddenly and almost without warning and it has been necessary to examine certain phenomona retrospectively for an explanation.

The year 1996 was a very unusual year for gizzard shad. The fish collecting effort was delayed until mid-summer because of wet weather and high discharge rates. When they finally were able to collect fish the field crew reported seeing schools of smallish shad which were viewed as young-of-the-year fish and, therefore, there was no attempt to capture them. During the second collecting series several specimens were preserved and examined in greater detail in the laboratory. These fish were much too large to be y-o-y fish. On the other hand, with total lengths ranging from 50 to 75 mm in early August, they were much too small for yearlings. All scales examined had a well-developed annulus testifying to their yearling status, albeit stunted yearlings.

One hypothesis for this unusual occurrence is that this cohort was spawned after a high-flow event in late June-early July 1995 (Figure 16) and had limited growth both because of a shortened growing period in 1995 and an inability to find sufficient food in early summer 1996 during a prolonged high discharge period. Whatever the actual cause there were small schools of small gizzard shad all along the Wabash River throughout the summer months of 1996.

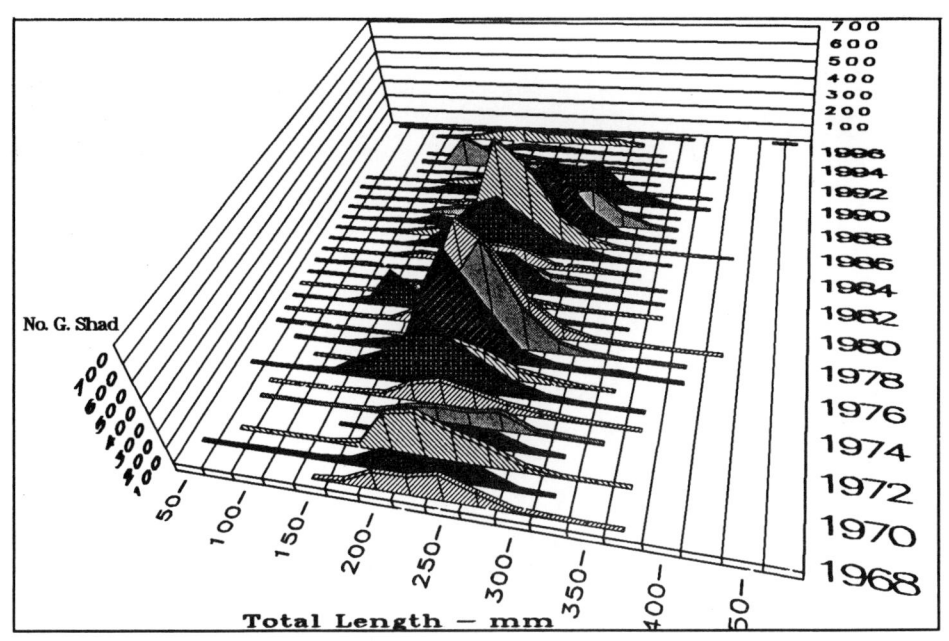

Figure 42: Length frequency of gizzard shad from 1968 through 1996.

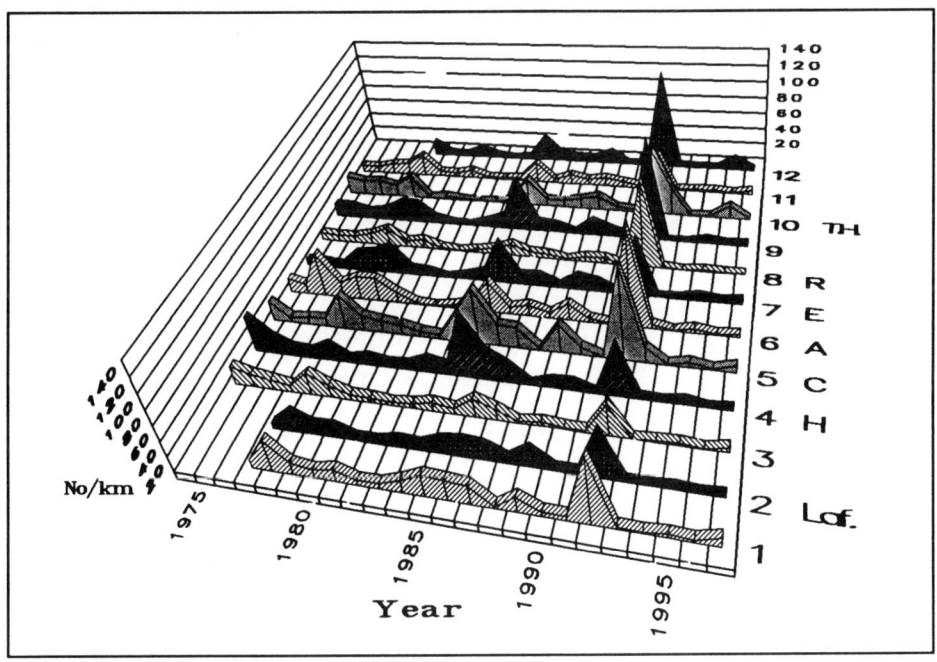

Figure 43: Catch rates (No/km) of gizzard shad from 1974 through 1997.

The carp (*Cyprinus carpio*) is another ubiquitous member of the Wabash River fish community. From the time trends summarized in Figure 44, it would appear that carp have steadily decreased in abundance and increased in mean size over the period of study. That, however, would not be a completely correct assumption.

The relatively high catch rates indicated for the period 1974 through 1978 and the correspondingly small mean size for much of that period were both the result of an unusual mainstem spawning event in 1973. An ichthyoplankton drift study was underway at that time. On June 11, 1973 tiny carp fry averaging 7.5 mm in total length appeared in the drift-net catch for the first time. These fry continued to be captured in until June 29, growing at a rate of 1.18 mm each day (Gammon 1976). Their appearance as ichthyoplankton coincided exactly with a period of late flooding from June 5 to June 10, 1973 during which time flood waters submerged the riparian vegetation bordering the river thus creating ideal conditions for spawning.

It appears that this spawning event was primarily responsible for the abundance of small carp in the Wabash River from 1974 through 1977. Since 1977 the mean size appears to have fluctuated around a horizontal plateau while the catch rate decreaseed overall. Glander (1984, 1987) also noted a general decrease in the magnitude of the commercial catch of carp from 1977 through 1986.

An abundance of carp is often regarded as an indicator of degraded environmental conditions in which only the most tolerant species can be supported. In

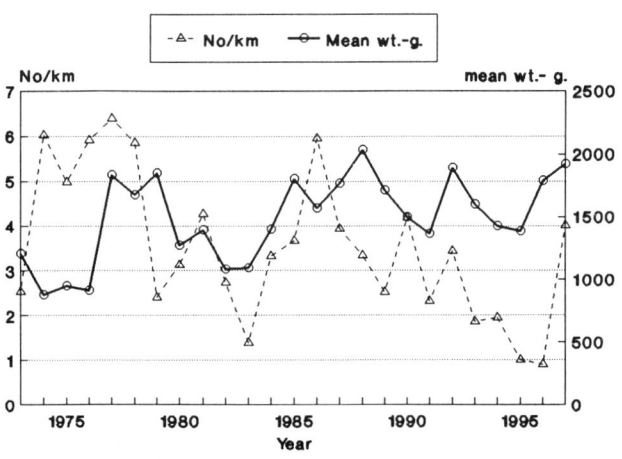

Figure 44: Annual mean catch rate (No/km) and mean weight of carp from 1968 through 1997.

the middle Wabash River, however, the largest catches of carp generally were found in the stretch of river from Delphi, Indiana through Covington, Indiana (Reaches 1 - 5) as well as in the lower reaches downriver from Terre Haute, Indiana (Reach 10) (Figure 45). Catch rates were lower from the Cayuga electric generating station to Terre Haute, Indiana (Reaches 6 - 9).

Three species of carpsuckers are found in the Wabash River. By far the most common species is the river carpsucker pictured earlier, *Carpiodes carpio*. Two other members of the same genus were also taken in smaller numbers; the quillback (*Carpiodes cyprinus*) and the highfin carpsucker (*Carpiodes velifer*).

Both of these species of carpsucker have deeper bodies and higher dorsal fins than the river carpsucker, but otherwise resemble it. The quillback is much more common in medium-sized tributaries such as

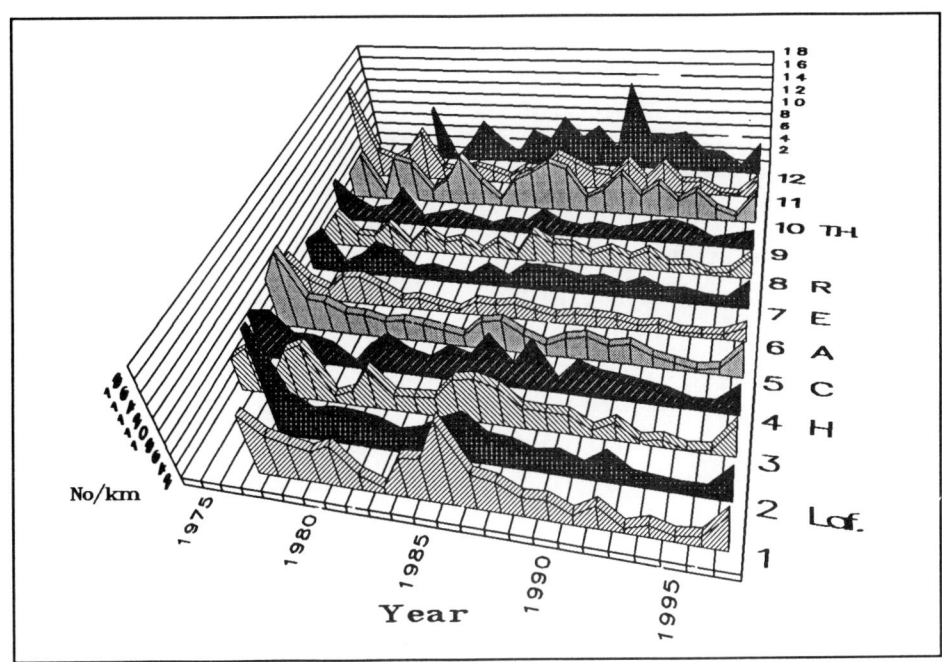

Figure 45: Catch rates (No/km) of carp in 12 Reaches of the Wabash River
from 1974 through 1997.

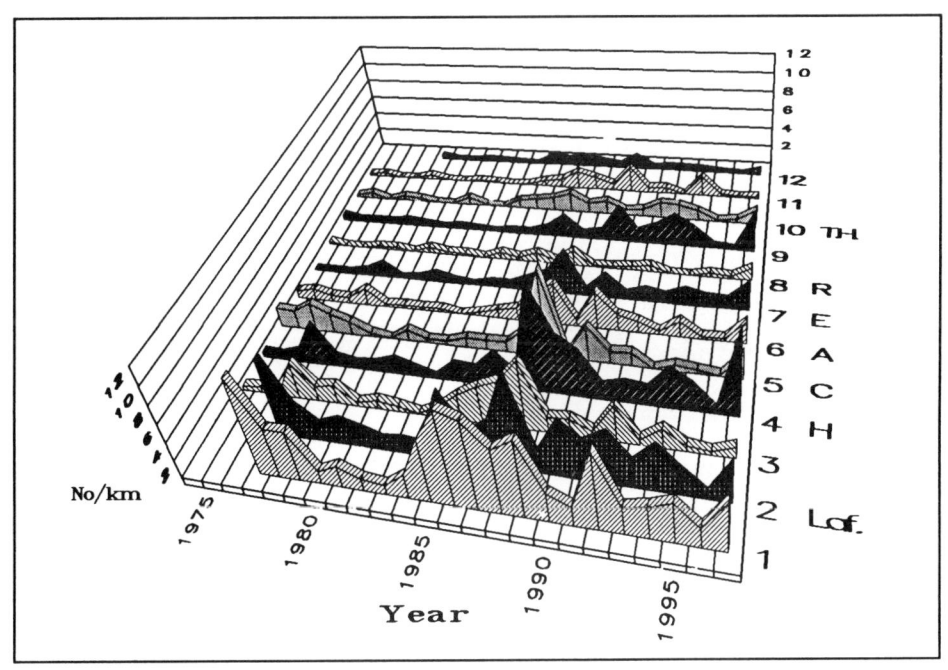

Figure 46: Catch rates (No/km) of northern river carpsucker
in 12 Reaches of the Wabash River from 1974 through 1997.

Sugar Creek and Big Raccoon Creek. All three species of carpsuckers are similar in habits as well as general appearance. They travel in sizeable schools feeding on the bottom detritus and small invertebrates.

By far the most common carpsucker in the Wabash River collected by electrofishing is the river carpsucker *Carpiodes carpio*. It is a big river species which rarely enters tributaries. As with many other species in the Wabash River, it increased in abundance after 1984 (Figure 46). It is also highly susceptible to capture by electrofishing and peak catches were made during 1977 and 1988. The average size appears to have decreased very gradually since about 1977 (Figure 47). Although this species was taken throughout the middle Wabash River it was consistently more abundant in the upper six

Reaches of the study segment and especially from Delphi, Indiana through Lafayette,

During the August 1988 drought, well after the regular electrofishing catches had been completed, we observed many large dead carpsuckers floating downriver in the Coal Creek area. River discharge was very low at the time and ambient temperatures exceeded 32° C (90° F).

Prior to 1984 the quillback was rarely found downriver from Montezuma, Indiana although after that year small numbers were captured nearly everywhere. The highfin carpsucker comprised a negligible part of the annual catch from the Wabash River with no obvious preference of one part of the river for any other section.

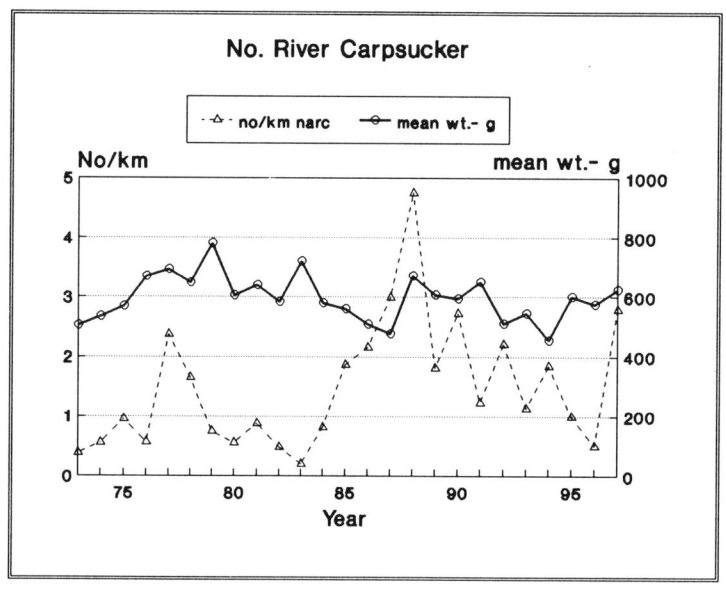

Figure 47: Annual mean catch rate (No/km) and mean weight of northern river carpsucker from 1968 through 1997.

REDHORSE OF THE WABASH RIVER

Five species of redhorse were regularly taken by electrofishing in the middle Wabash River. In order of numeric abundance, they were (1) golden redhorse *Moxostoma erythrurum*, (2) shorthead redhorse *M. macrolepidotum*, (3) silver redhorse *M. anisurum*, (4) black redhorse *M. duquesnei*, and (5) river redhorse *M. carinatum*.

One additional redhorse species, the greater redhorse *Moxostoma valenciennesi* Jordan, is probably also present in small numbers. This species has been collected from two Wabash River tributaries, Otter Creek by Whitaker (1976) and the Eel River by Gammon and Gammon (1990). It could well have occurred in our catches, but may have been misidentified as river redhorse, a species which it closely resembles.

The *Moxostoma* species are all similar in size and basic appearance. Their bodies are streamlined and either round in cross-section or slightly flattened laterally. In color they are green-gold dorsally, bronze-colored laterally with white bellies. Silver, golden, and black redhorse have caudal fins which are greyish in color while in all others they are bright red. All have ventrally located mouths with large lower lips.

The river redhorse is the largest species on the average, while the slender black redhorse is the smallest. Most redhorse range from 300 to 600 mm (1 to 2 feet) in total length and weigh 0.5 to 1.5 kg (1 to 3 pounds), but some live long enough to reach six to seven pounds in size.

All redhorse feed upon invertebrates living on and in the bottom substrate. Their large fleshy lips are abundantly provided with taste buds enabling them to taste and then eat succulent food items as they cruise over the bottom.

All species of redhorse are among the most thermally sensitive species of fish in the Wabash River (Gammon 1973, 1976). They are also intolerant of a wide variety of environmental disturbances including poor water quality and habitat degradation (Ohio Environmental Protection Agency 1987, Simon 1992).

By far the largest concentration of redhorse was found in the upper part of the study section, especially upriver from Lafayette, Indiana where their numeric abundance approached that of gizzard shad. They increased in overall abundance during the 1980s and then declined in numbers. Part of their demise in the 1990s is attributable to the flood of 1993 after which all redhorse species populations were seriously decimated. However, golden redhorse and river redhorse began decreasing in abundance well in advance of that major flood event.

Golden redhorse are widely distributed throughout Indiana's medium and large streams. They occur in sizeable schools in the pools of streams where they feed upon small invertebrates living on and in the bottom substrate. They undertake spring spawning migrations into tributaries and are often sought by anglers at this time.

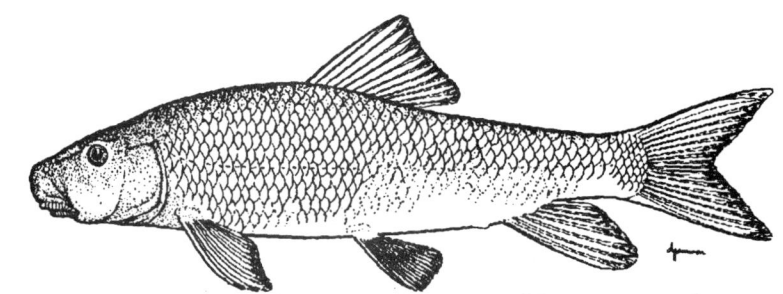

Golden Redhorse - *Moxostoma erythrurum*

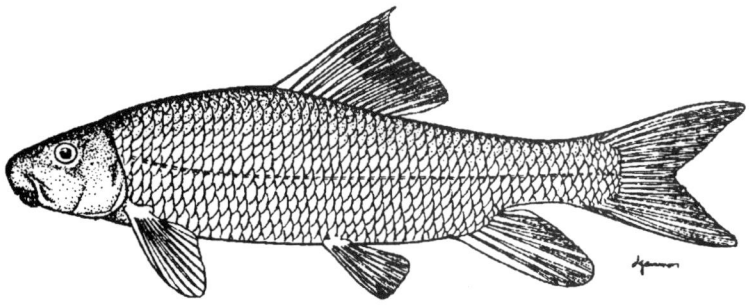

Shorthead Redhorse - *Moxostoma macrolepidotum*

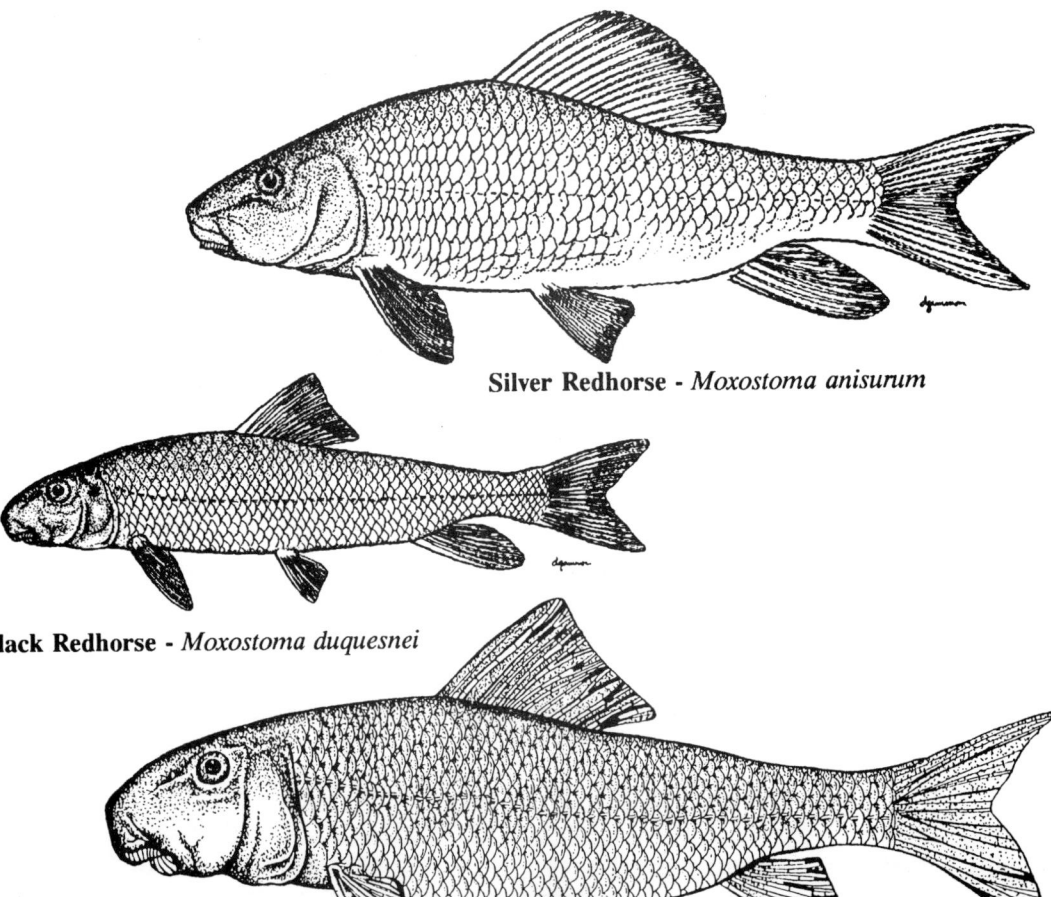

Silver Redhorse - *Moxostoma anisurum*

Black Redhorse - *Moxostoma duquesnei*

River Redhorse - *Moxostoma carinatum*

The electrofishing catches and the mean weight of golden redhorse increased from 1973 through 1986 and declined thereafter (Figure 48A). This species population was decimated by the 1993 flood.

Of the five *Moxostoma* species which live in the Wabash River, the golden redhorse is among the most abundant and common together with the shorthead redhorse. They are both much more abundant upriver from Covington, Indiana than elsewhere (Figure 49). They are thermally sensitive species and were formerly common in Reach 6 before the Cayuga Electric Generating Station began operating.

Golden redhorse participated in the population expansion of the middle 1980s and even expanded into Reach 11 downriver from Terre Haute, Indiana during that period.

Golden redhorse from the Wabash River mainstem probably move into tributaries in spring to spawn. The young normally reside in clean tributaries and then enter those sections of the Wabash River with better water quality and suitably cooler temperatures after maturing.

The shorthead redhorse population abundance is about the same as golden redhorse. However, it prefers to live and feed in faster-moving microhabitats of the river rather than in pooled sections. This "redtail" is prized by spring anglers for both its sporting attributes and its edibility.

The catch rate of shorthead redhorse approximately doubled from 1984 to 1992 (Figure 48B). Remarkably, the average size increased during this same period, except for 1990 and 1991. The sharp decline in catch rate after 1993 reflects the negative impact of the 1993 flood and not, as was previously postulated, an artifact of poor collecting conditions. There is, however, an indication that a population of smaller individuals has persisted since about 1990 and as of 1997 is again well established.

The shorthead redhorse is fairly common north of Montezuma, Indiana except in the vicinity of the Cayuga EGS (Figure 50). Some expansion of population density appears to have occurred since 1980 in Reaches 3, 4, and 5, between Lafayette and Cayuga, Indiana. They even appeared in small numbers at the Cayuga EGS during the high flow summers of 1989 and 1990. As with golden redhorse, this species is very scarce downriver from Clinton, Indiana.

The silver redhorse population of the Wabash River appears to be less than half as abundant as either golden redhorse or shorthead redhorse. The catch rates have been quite variable, but the mean size increased considerably over the period of study (Figure 48D). Silver redhorse seek out areas in the river where there is deeper water and less current than most other redhorse species.

Silver redhorse were most common in Reach 1 upriver from Lafayette (Figure 51), but they were consistently found in good numbers in Reaches 2 through 7. They were relatively scarce in Reaches 4 through 7 until 1979-1980, when their numbers increased in the upper Reaches. They began to appear in small numbers even in downriver Reaches 8 through 12 after that date.

Increasing numbers were captured at the Cayuga EGS (Reach 6) after 1985. The increase in numbers at this site was particularly large in 1988 when river flow was very low and temperatures often high. The increase here may be related to the altered operation of the Cayuga EGS, a subject which will be examined later. Catches were about half as great after the 1993 flood than before, but the flood's negative effects were proportionally less for this species than for other species of redhorse.

The river redhorse was originally described from specimens collected from the Wabash River by Cope (1871) at Lafayette, a section of river which still harbors a healthy population. This species superficially resembles a large golden redhorse, but has a red caudal or tail fin. It prefers swift, deep riffles where it feeds heavily on mollusks such as snails and small clams. As with other redhorse species, the pharyngeal arch is adapted to the food preferences of the species. For river redhorse the heavy pharyngeal arch possesses molarized teeth for grinding mollusks.

Although never common, the abundance of river redhorse increased somewhat from 1980 to 1989 (Figure 52) as did the average size (Figure 48). River redhorse averaged 800-1200 g in weight before 1982 and 1500-2000 after 1983. The 1993 flood devastated the population, but two individual fish were collected in 1997.

River redhorse were rarely seen downriver from Lafayette, Indiana until about 1980, after which time increasing numbers appeared in the catches between Lafayette, Indiana and Covington, Indiana (Reaches 2, 3, and 4). The extension in range of river redhorse, and the blue sucker to be considered later, may be correlated with improved water quality in the Wabash River during the 1980s.

The black redhorse is restricted almost exclusively to the upper four Reaches from Delphi to Covington, Indiana where they were taken in small numbers. Rarely were they found further downriver (Figure 53). This slender species almost certainly enters into the Wabash mainstem from its normal haunts in smaller tributary streams.

95

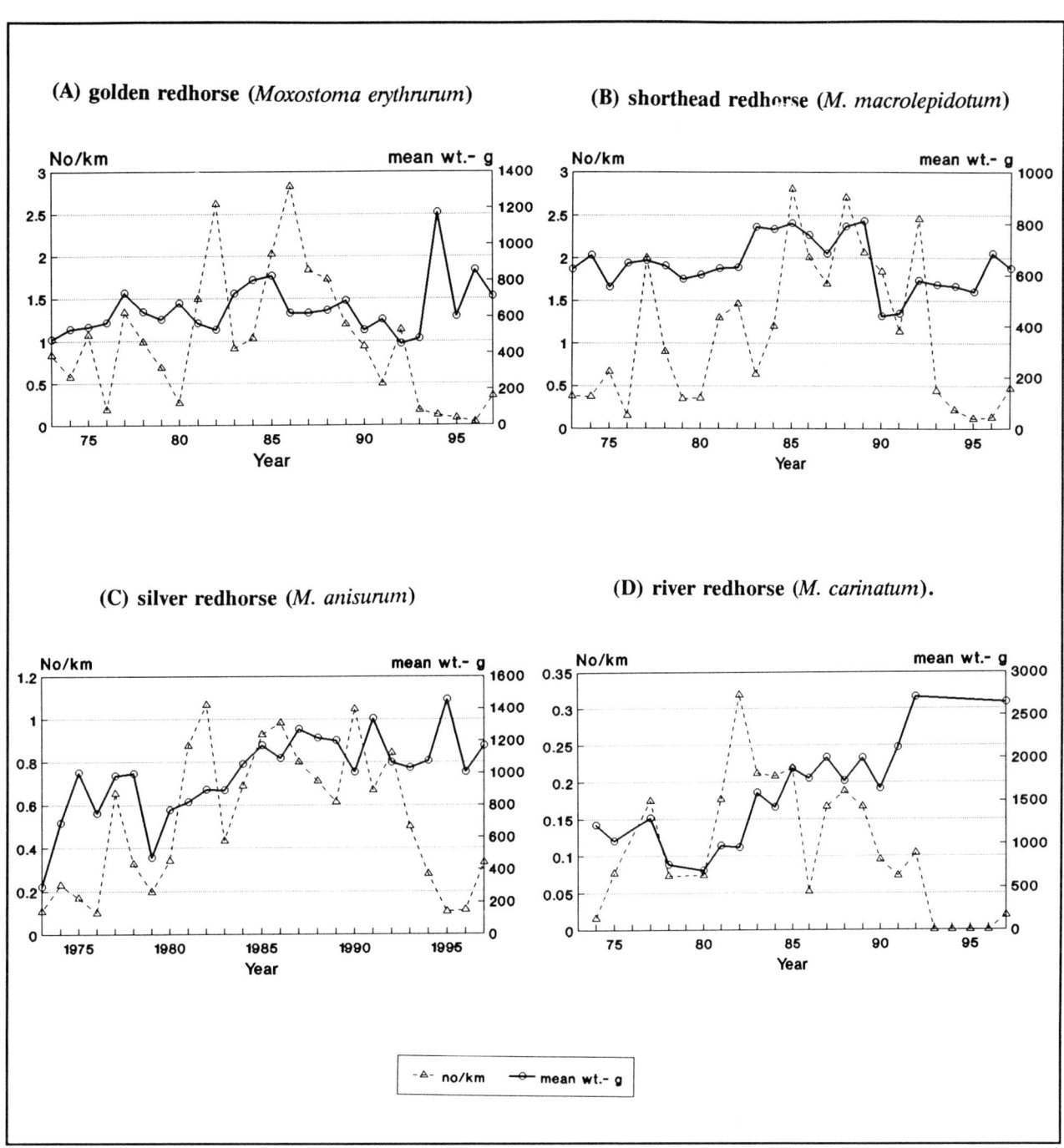

Figure 48: Annual mean catch rate (No/km) and mean weight from 1968 through 1997:
(A) golden redhorse (*Moxostoma erythrurum*), (B) shorthead redhorse (*M. macrolepidotum*),
(C) silver redhorse (*M. anisurum*), and (D) river redhorse (*M. carinatum*).

96

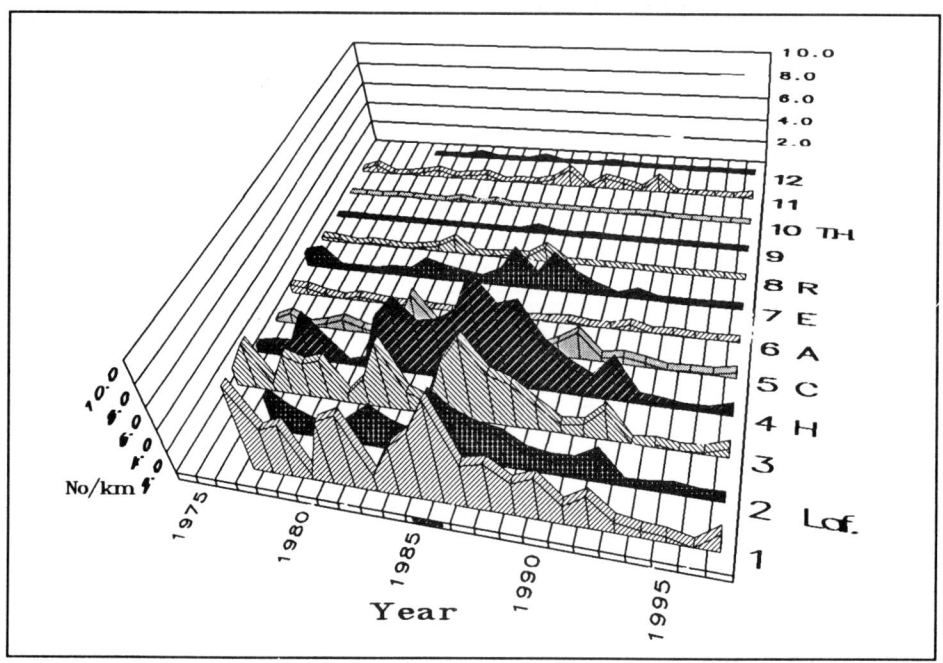

**Figure 49: Catch rates (No/km) of golden redhorse
in 12 Reaches of the Wabash River from 1974 through 1997.**

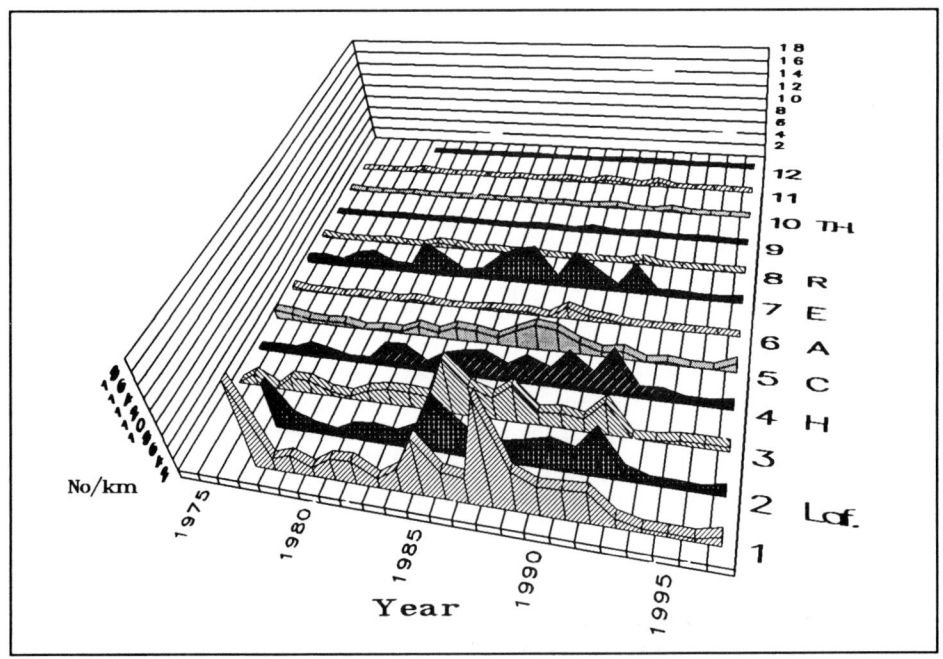

**Figure 50: Catch rates (No/km) of shorthead redhorse
in 12 Reaches of the Wabash River from 1974 through 1997.**

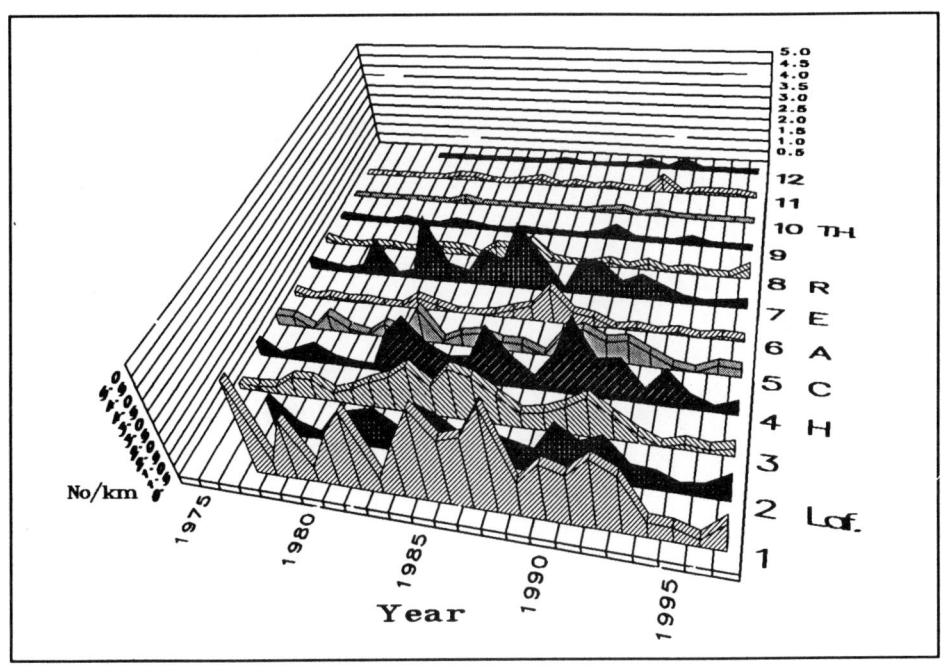

**Figure 51: Catch rates (No/km) of silver redhorse
in 12 Reaches of the Wabash River from 1974 through 1997.**

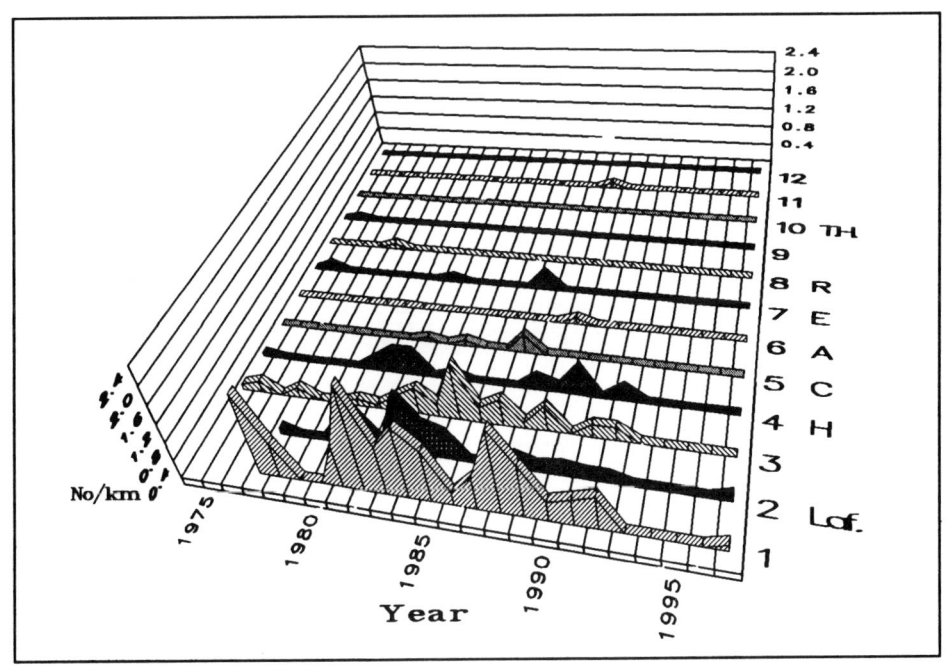

**Figure 52: Catch rates (No/km) of river redhorse
in 12 Reaches of the Wabash River from 1974 through 1997.**

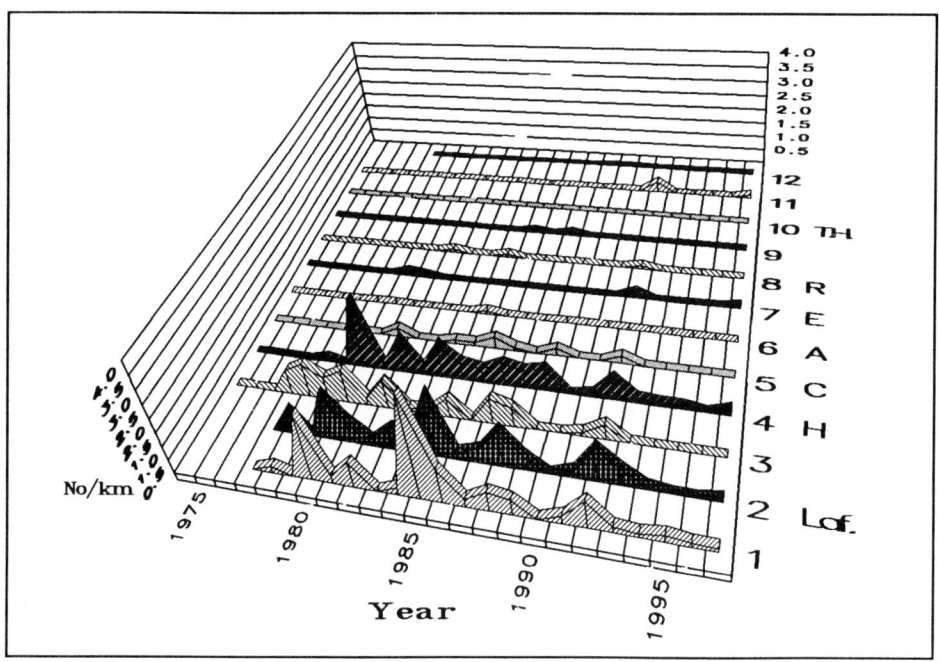

Figure 53: **Catch rates (No/km) of black redhorse**
in 12 Reaches of the Wabash River from 1974 through 1997.

THE BLUE SUCKER AND BUFFALOFISH

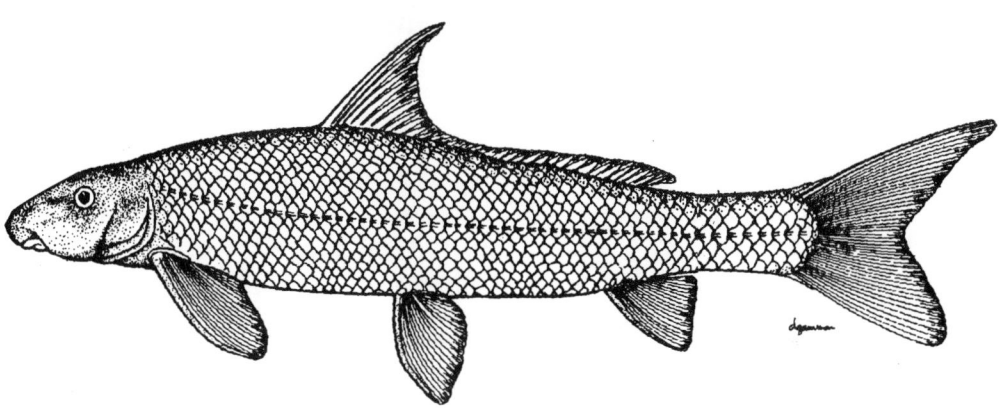

Blue Sucker - *Cycleptus elongatus*

The blue sucker is widely distributed in larger rivers from the Ohio and Missouri Rivers southward to the Rio Grande River where it inhabits deep, swift main channels of large streams and big rivers. Although it is locally abundant, Williams et al. (1989) indicate that it is declining throughout its range mainly because of habitat alteration. They recommended placing blue sucker on the species list of Special Concern, as did Whitaker and Gammon (1988) with regard to its status in Indiana.

Lesueur first described blue sucker in 1817 from specimens taken near Pittsburgh, Pennsylvania. It was common in the Ohio River before 1850, but declined dramatically from 1900 to 1950 (Pearson and Krumholz 1984). The Wabash River supports one of the few remaining sizeable populations.

There is little doubt that blue suckers spawn and live their entire lives in the Wabash River mainstem. Beyond that attribute, little has been learned about their spawning preferences or early life history

because few small, young individuals have been found. Concentrations of adults and, perhaps, subadults occur in deep, swift chutes and they are especially abundant near the mouths of several larger tributaries, especially Sugar Creek and Coal Creek.

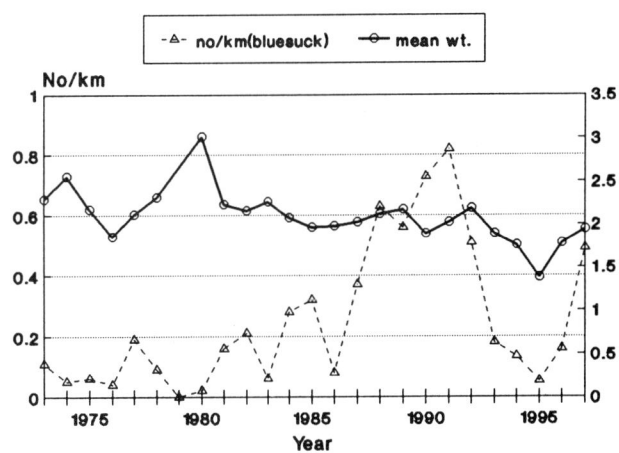

Figure 54: Mean annual catch rate and mean weight of blue suckers from 1968 through 1997.

Prior to 1984 electrofishing catch rates were less than 0.2/km (Figure 54). From 1985 to 1991 catch rates quadrupled. Catch rates fell to less than 0.2/km after the 1993 flood and then rebounded in 1997. The average size also appears to have declined gradually over time with 1995 marking the lowest observed mean weight ever (N = 13).

Before 1978 most blue suckers were captured within a 65 km (40 mile) section of river between Cayuga and Terre Haute, Indiana (Figure 55). Very few individuals were found prior to that time either downriver from Terre Haute, Indiana or upriver from Covington, Indiana. Since then blue suckers have extended their range throughout the middle Wabash River and have also grown in density since 1984.

With the expansion in distribution and the increase in population size has come a greater diversification of sizes (Figure 56). Total lengths ranged from about 500-650 mm during the period 1968-1978. By 1988 this expanded to about 425-750 mm. Two yearlings less than 100 mm in total length were taken for the first time in 1997.

A typical Wabash River blue sucker.

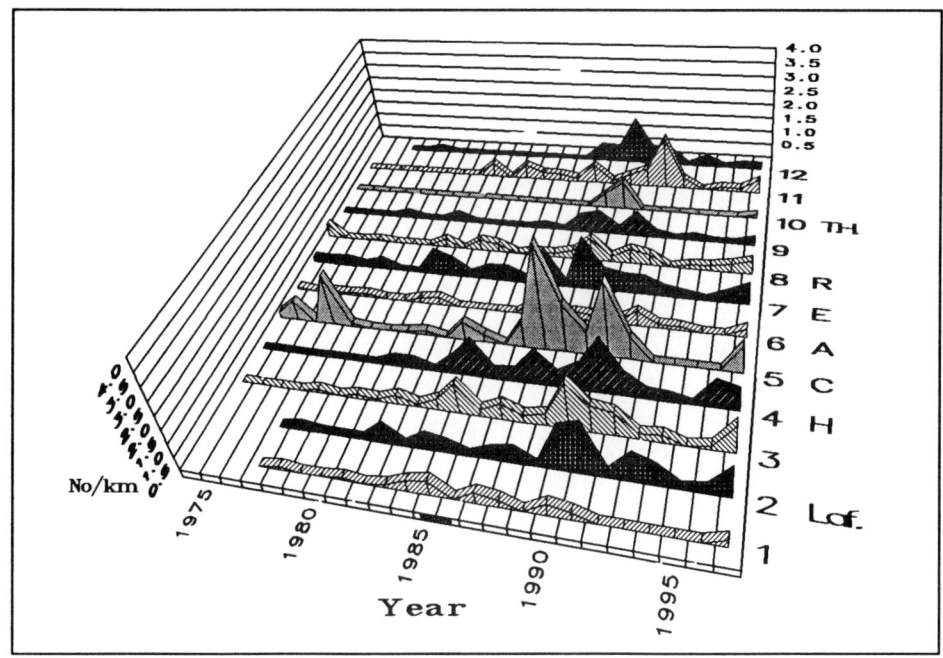

Figure 55: Catch rates (No\km) of blue sucker in 12 Reaches of the Wabash River from 1974 through 1997.

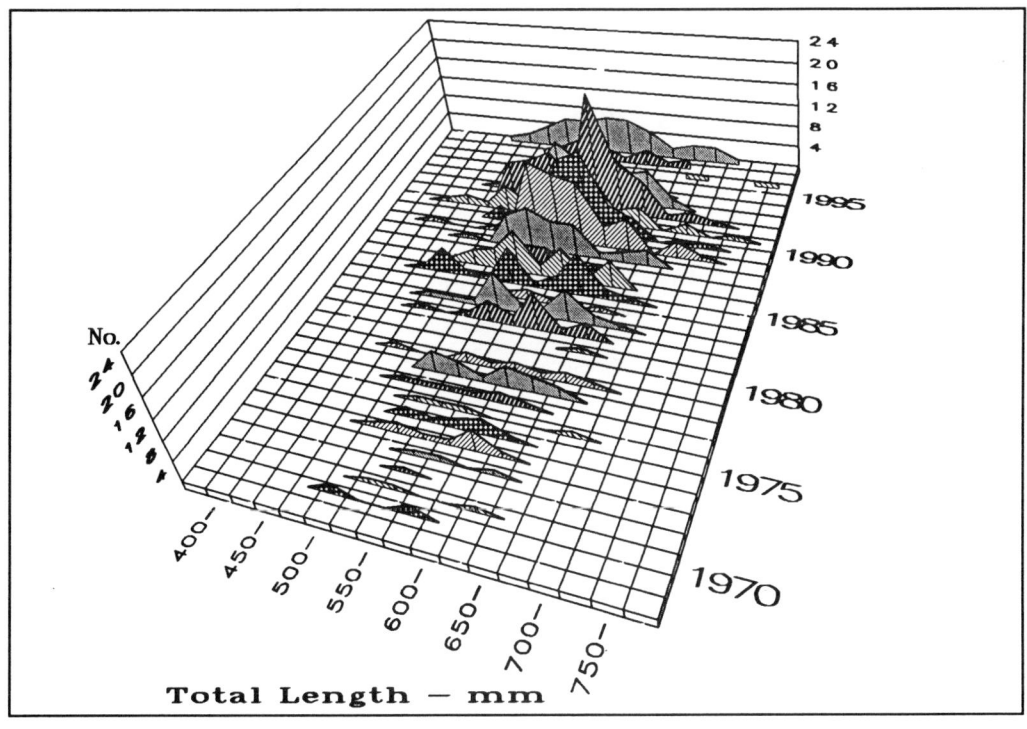

Figure 56: Length frequency of blue suckers 1968 - 1997.

Smallmouth Buffalo - *Ictiobus bubalus*
Bigmouth Buffalo - *Ictiobus cyprinellus*
Black Buffalo - *Ictiobus niger*

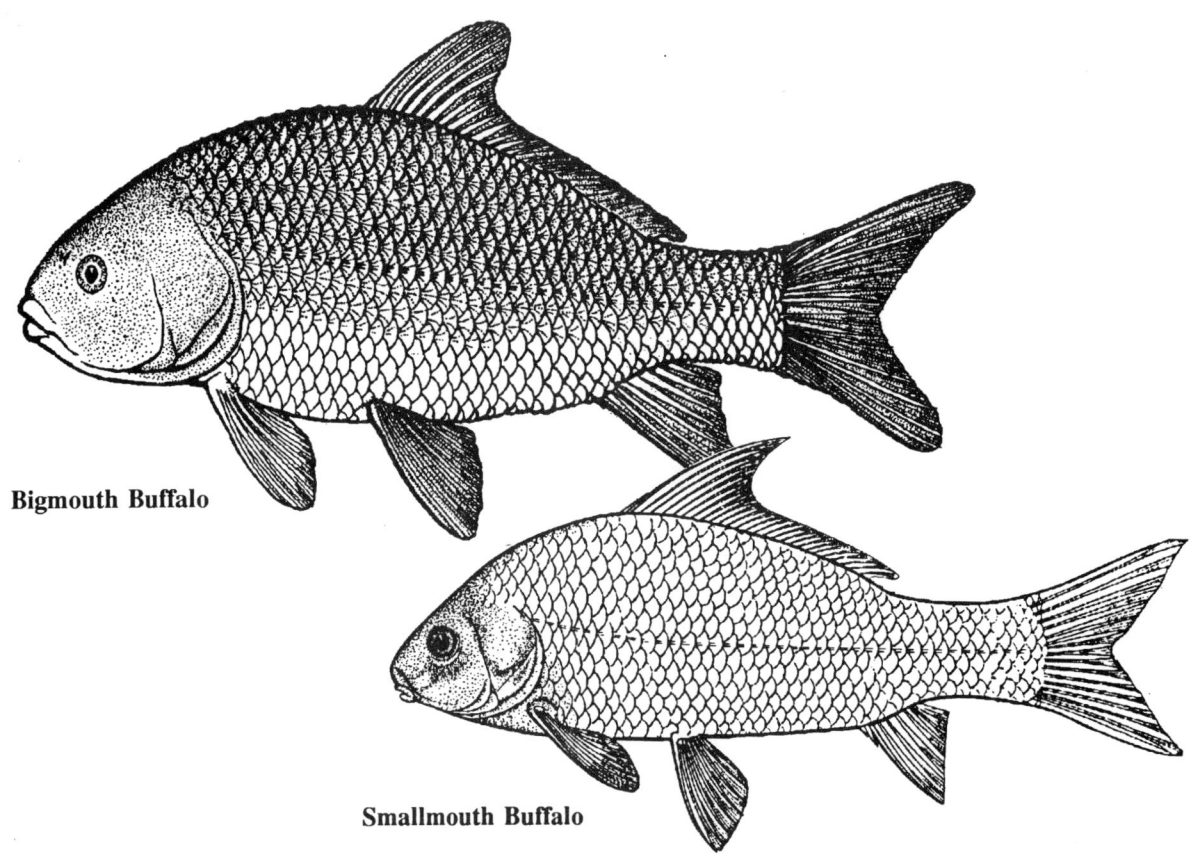

Bigmouth Buffalo

Smallmouth Buffalo

The buffalo or buffalofish are among the largest species of fish native to the Wabash River. Bigmouth buffalofish (*Ictiobus cyprinellus*) reaches a size of 25 kg (50 to 60 pounds); smallmouth buffalo (*Ictiobus bubalus*) is usually much smaller, but still may grow to 20 kg (40 to 45 pounds). The larger specimens are less easily collected by electrofishing than most other species simply because their large size and great strength often propel them beyond the electric field before they can be netted. Approximately twice as many smallmouth buffalofish have been taken by electrofishing than bigmouth buffalo over the period of study.

Buffalo are fairly good food fish, especially smallmouth buffalo. The larger individuals may be stuffed and baked. Large bigmouth buffalo are often smoked. Neither species is sought by anglers, but they do enter into the commercial catch and are regularly marketed in Indiana at the seafood section of grocers.

Although they belong to the same genus, smallmouth and bigmouth buffalo are adapted for quite different life-styles. Smallmouth buffalo (*Ictiobus bubalus*) superficially resemble large, muscular, dark-colored carpsuckers. They were most often found in deep, fast-water chutes in company with river redhorse, blue sucker, and shovelnose sturgeon. They are opportunistic bottom feeders and eat a variety of foods including zooplankton, algae, insect larvae, and detritus (McComish 1967).

The average Wabash River small-mouth buffalo weighed 2-3 kg (4-7 pounds) (Figure 57). The catch rate of this species increased substantially after 1983, but the size remained fairly constant until recently. Smallmouth buffalo were most common upriver from Montezuma in Reaches 1 through 6 (Figure 58), and especially in Reach 3 from Lafayette to Covington, Indiana.

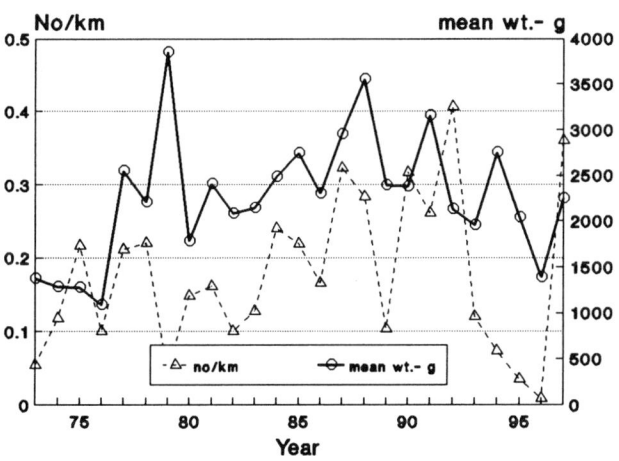

Figure 57: Mean annual catch (No/km) and mean weight of smallmouth buffalo from 1968 through 1997.

The bigmouth buffalo (*Ictiobus cyprinellus*) feeds primarily by filtering microcrustaceans from the water column. The mouth is located terminally rather than subterminal as with all other members of the Catostomidae. The gills are fitted with long, comb-like gill rakers for feeding on small planktonic and bottom invertebrates.

Bigmouth buffalo were most often encountered in the Wabash River near ashpond effluents and backwater areas in which zooplankton populations flourished. Elsewhere in Indiana, they are especially common and large in waters below the larger dams and reservoirs where they fatten on zooplankton suspended in release water. The catch rate of bigmouth buffalo in the Wabash River was low and evenly distributed nearly everywhere (Figure 59). Catches of this species have been extremely low throughout the middle Wabash River since about 1988.

Another species of buffalo, the black buffalo (*Ictiobus niger),* was considerably less common than either of the other species with only 22 specimens having been positively identified since studies began in 1967. It is most similar to the smallmouth buffalo, but it is darker in coloration and somewhat more slender. Trautman (1981) states that it prefers a type of habitat that is intermediate to the other two species.

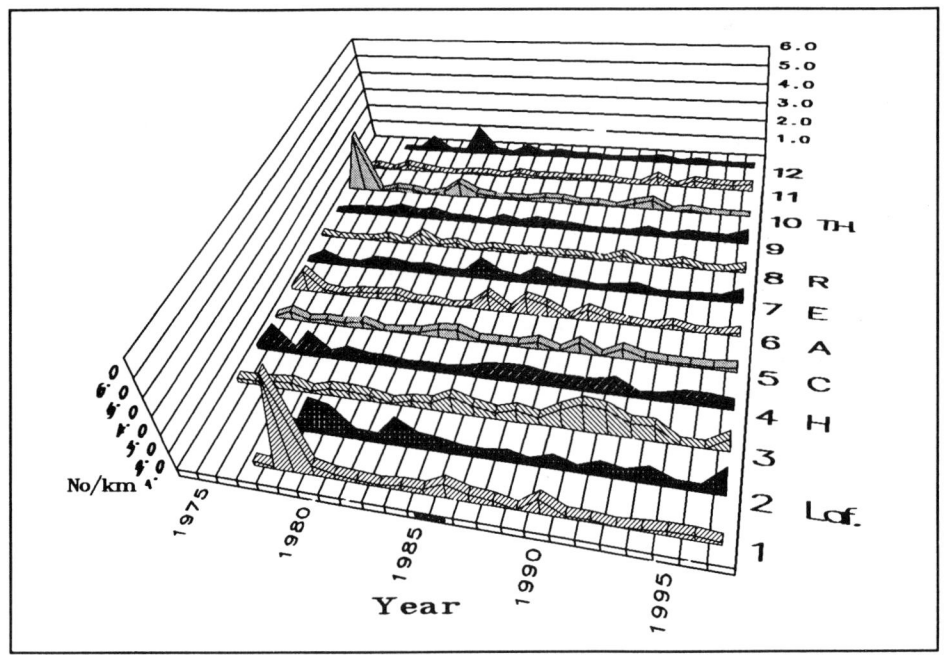

**Figure 58: Catch rates (No\km) of smallmouth buffalo
in 12 Reaches of the Wabash River from 1974 through 1997.**

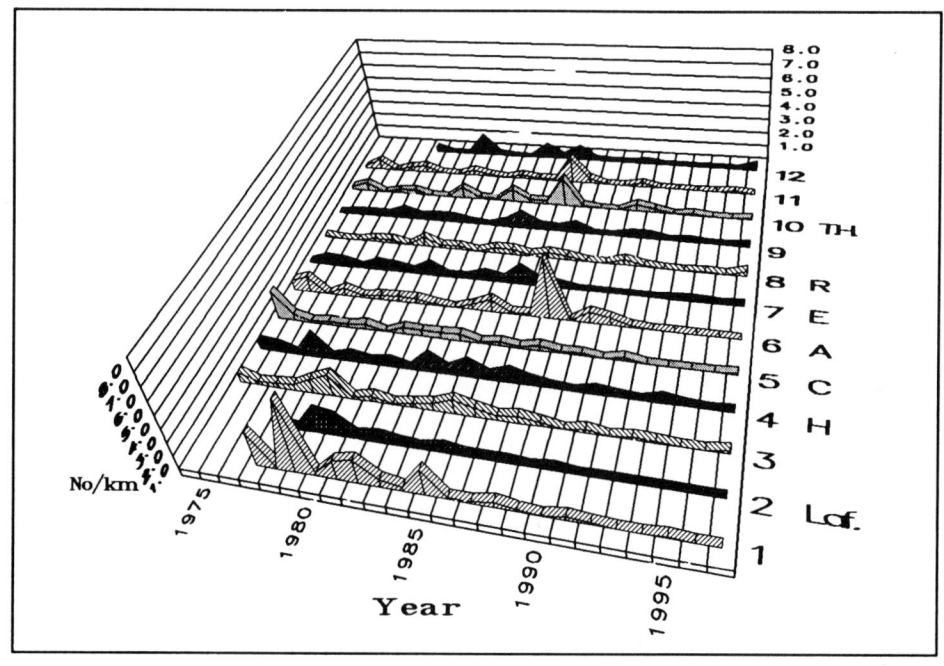

**Figure 59: Catch rates (No\km) of bigmouth buffalo
in 12 Reaches of the Wabash River from 1974 through 1997.**

THE GAR AND BOWFIN

Fossil gar specimens have been found in North America, Europe, Africa, and India, but the seven living species are now restricted to the Western Hemisphere from Costa Rica to southern Canada (Wiley 1976).

The Wabash River contains four species of these ancient fishes, but we have taken only three species in the middle Wabash River; longnose gar (*Lepisosteus osseus*), shortnose gar (*Lepisosteus platostomus*), and spotted gar (*Lepisosteus oculatus*). Alligator gar (*Lepisosteus spatula*) was recorded in the lower Wabash River at New Harmony by Jordan (1890). It had not been taken again until 1993 when specimens were reportedly found in the lower White River (Lydy, personal communication) and the lower Wabash River (T. Simon, personal communication). Alligator gar has declined in abundance in the Ohio River according to Pearson and Krumholz (1984), although it was never common there.

Gar inhabit larger rivers, lakes, and swamps within their range and even frequent brackish coastal waters. Their air bladder functions as a "lung", enabling them to take in air at the water surface and exist in waters with low dissolved oxygen such as swamps and weedy backwater bayous. Gar are carnivores which feed primarily on living and dead fish. Because of their supposed competition with more desirable sport species gar are often regarded with contempt by anglers, especially since they are usually regarded as inedible. The greenish-colored eggs of the female are toxic, but the flesh of adults is sometimes smoked and eaten.

The spotted gar is uncommon in the middle Wabash River and only a few specimens were collected each year. Those which were taken may have strayed from their normal habitat in the northern lakes or from the oxbow lakes and weedy ditches of the lower Wabash River basin.

Longnose gar and shortnose gar are both common and widely distributed. They have been collected in approximately equal numbers over the years. The numbers of both species have fluctuated considerably from year to year (Figures 60 and 61). Changes in catch rates are similar for both species with lower catch rates generally coinciding with summers having lower rates of discharge. The density of shortnose gar has increased slightly over the period of study and the average size of longnose gar has increased somewhat since 1984.

Longnose gar are equally common nearly everywhere (Figure 62). However, shortnose gar were particularly abundant south of Terre Haute, Indiana, but scarce north of Lafayette, Indiana (Figure 63).

These distributional patterns of the adults and subadults may be related to differences in spawning preferences and available spawning habitat. Longnose gar ascend tributaries to spawn in the spring. Sugar Creek, for example, sometimes has a large migration with some fish entering even some of its smaller tributaries. By late summer small young-of-the-year longnose gar may be seen in the sandy shallows of the inside bends of the river.

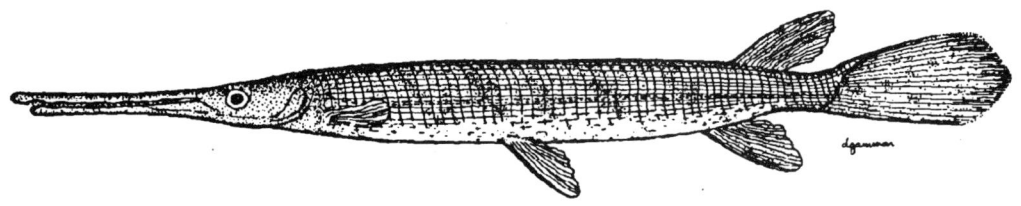

Longnose Gar - *Lepisosteus osseus*

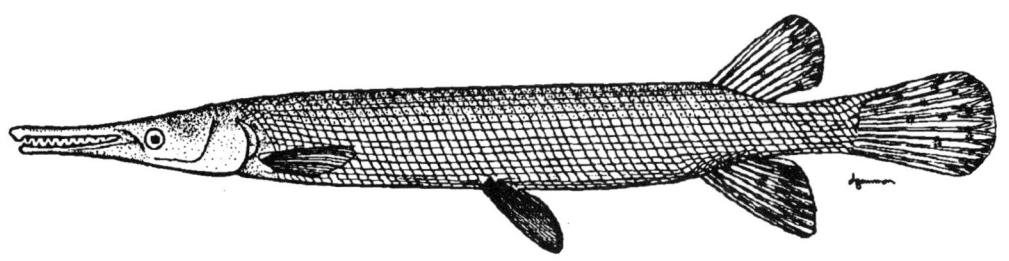

Shortnose Gar - *Lepisosteus platostomus*

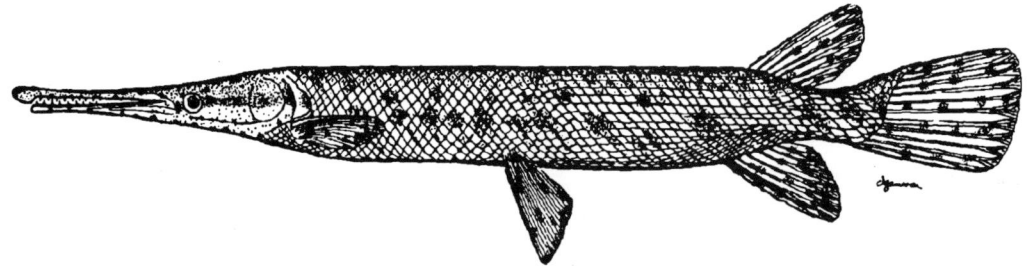

Spotted Gar - *Lepisosteus oculatus*

Shortnose gar, on the other hand, spawn in the sluggish, weedy tributaries and oxbows which make their appearance south of Terre Haute, Indiana. Large numbers of smaller shortnose gar can often be observed basking near the surface in weedy tributary backwaters south of Terre Haute, Indiana during the summer.

Since 1984 both species of gar became more common than before in the lower part of the study section (Figures 62 and 63). Increased numbers of both longnose and shortnose gar are now found at and downriver from the Cayuga Electric Generating Station (Reaches 6 through 12). However, there was little change in the densities of either species in the upper Reaches (Reaches 1 through 5) until the devastating flood of 1993, which reduced gar populations to very low levels.

A species closely related to the gar, the bowfin (*Amia calva*), also lives in the Wabash River. Its habitat preferences and ability to live in swampy bayous are similar to that of the shortnose gar, but they were not common anywhere in the Wabash River. The total electrofishing catch of bowfin since 1967 was only about 120 individuals.

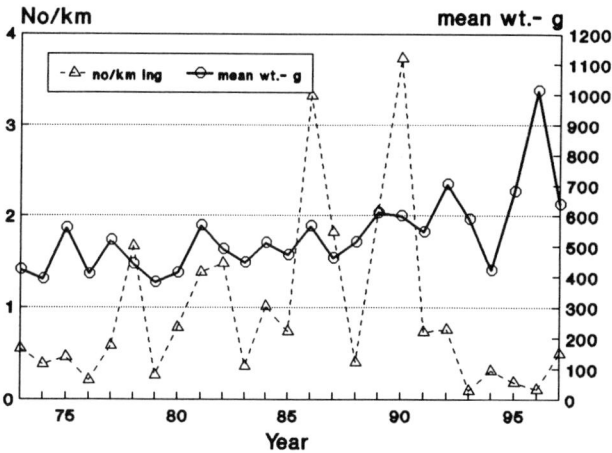

Figure 60: Mean annual catch (No\km) and mean weight of longnose gar from 1973 through 1997.

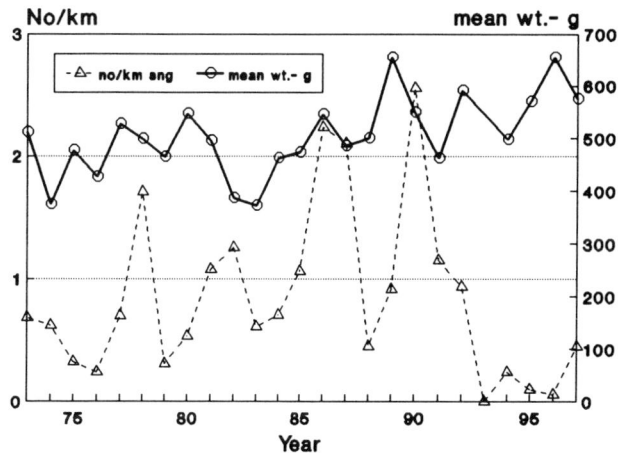

Figure 61: Mean annual catch (No\km) and mean weight of shortnose gar from 1973 through 1997.

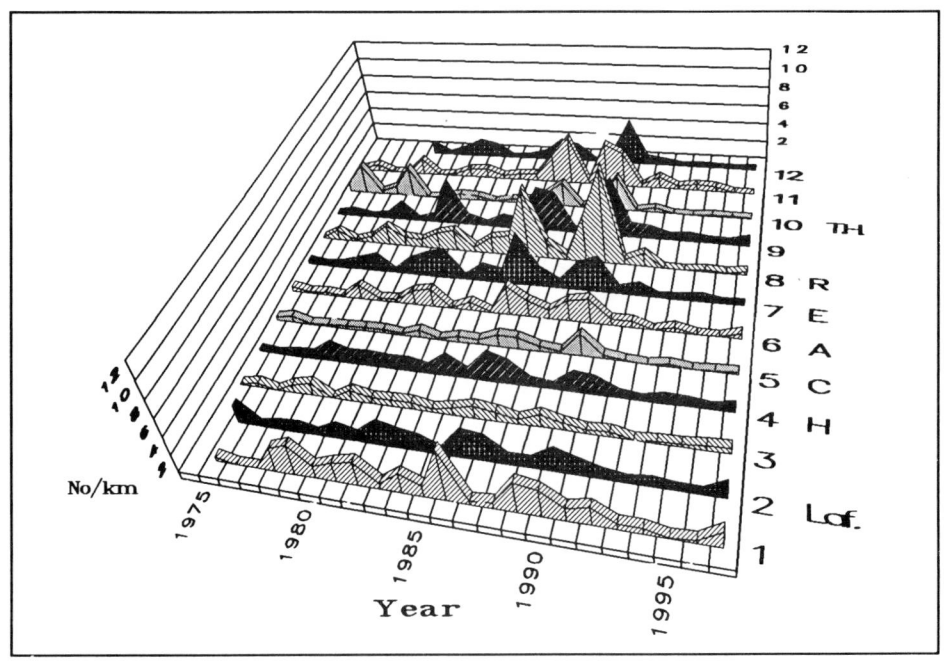

**Figure 62: Catch rates (No\km) of longnose gar
in 12 Reaches of the Wabash River from 1974 through 1997.**

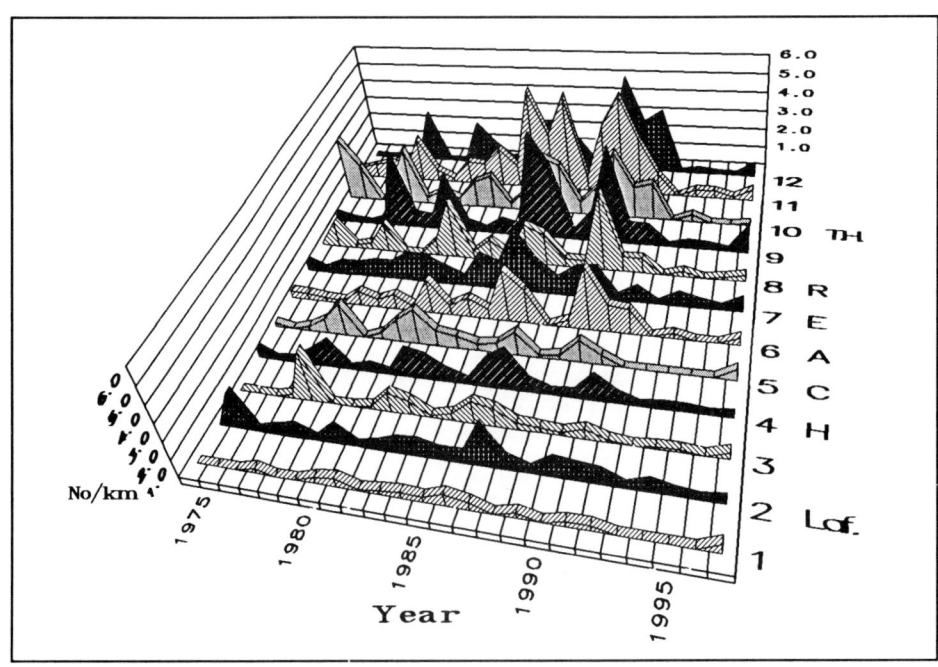

**Figure 63: Catch rates (No\km) of shortnose gar
in 12 Reaches of the Wabash River from 1974 through 1997.**

CATFISH AND FRESHWATER DRUM

Flathead catfish (*Pylodictis olivaris*) ranked third numerically in electrofishing catches after gizzard shad and carp primarily because small flathead catfish were unusually susceptible to capture by the D.C. electrofisher. A majority of flathead catfish captured were small yearlings, but some individuals, up to 60 pounds, were taken when the river discharge was low.

Strong year classes were produced in 1978, 1982, and 1983 and large catches of yearlings were collected the following years (Figure 64). Catch rates generally averaged one to three per kilometer, but the mean weight has increased since 1985. A new electrode system with four foot cathodes greatly increased their catch rates in 1997.

Higher water temperatures after startup of the Cayuga EGS (Reach 6) in 1972 stimulated reproduction and expanded populations of flathead catfish downriver. The flathead catfish population is mainly concentrated downriver from Reach 6 (Cayuga EGS) (Figure 66). They are one of the few species to be unusually abundant in Reach 8. There is also an indication of slight population increases upriver from the Cayuga EGS after its startup in 1971-72. This species was relatively unaffected by the 1993 flood (Figure 66).

Rud (1982) examined stomach contents of flathead catfish, most of which were fish three years old or less. Important food by weight included crayfish (42.4%), fish (23.7%), and aquatic insects (13.6%). Crayfish and fish become more important foods in older, larger specimens, while insect larvae were most important to younger fish.

Small individuals of other ichtalurids were frequently eaten by larger fish.

Meikle (1975) found fast initial growth with average total lengths of 164 mm at age 1 year, 261 mm at age 2, 337 mm at age 3, and 388 mm at age 4. He also estimated rates of parasitization by two intestinal helminths: 33.0% for *Corallobothrium* and 11.2% for *Dacnitoides cotylaphora*.

Flathead and channel catfish are the most sought after fish in the Wabash River. Trot lines are mostly employed by the casual angler, although there is some hook and line bait fishing. Commercial fishing is directed primarily at catfish with fishermen mostly using hoop nets. Flathead catfish harvest ranked second to channel catfish with catch rates increasing slightly between 1977 and 1986 (Glander 1984, 1987).

The channel catfish (*Ictalurus punctatus*) was not very susceptible to capture by electrofishing and, consequently, were generally taken in small numbers relative to their actual population abundance. It should be emphasized that they are much more abundant in the Wabash River than the catch rates shown here would indicate (Figure 65).

For many years they contributed little to the total catch. Catch rates were typically less than 0.5 fish per kilometer (Figure 67). In 1984, however, they began to rise spectacularly to a maximum of about 8 catfish per kilometer in 1986. The actual population density of channel catfish at this time must have been extremely large and extended into most tributaries as well as the

110

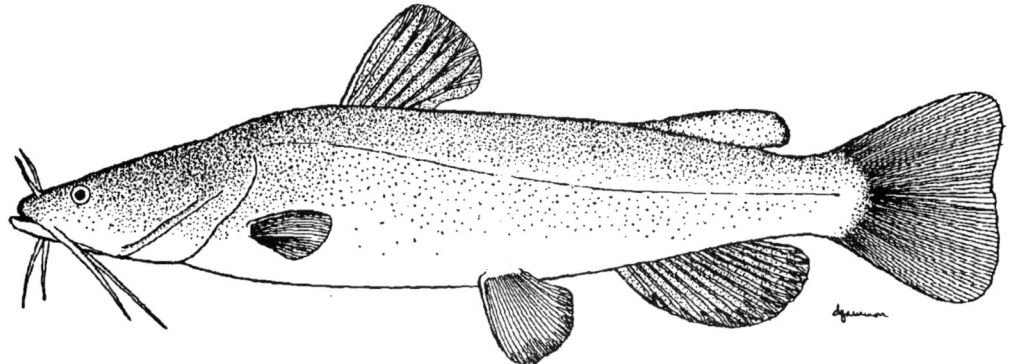

Flathead Catfish - *Pylodictis olivaris*

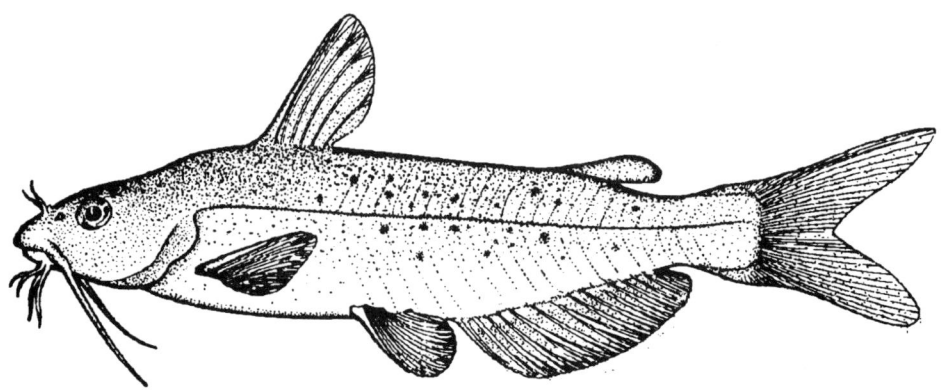

Channel Catfish - *Ictalurus punctatus*

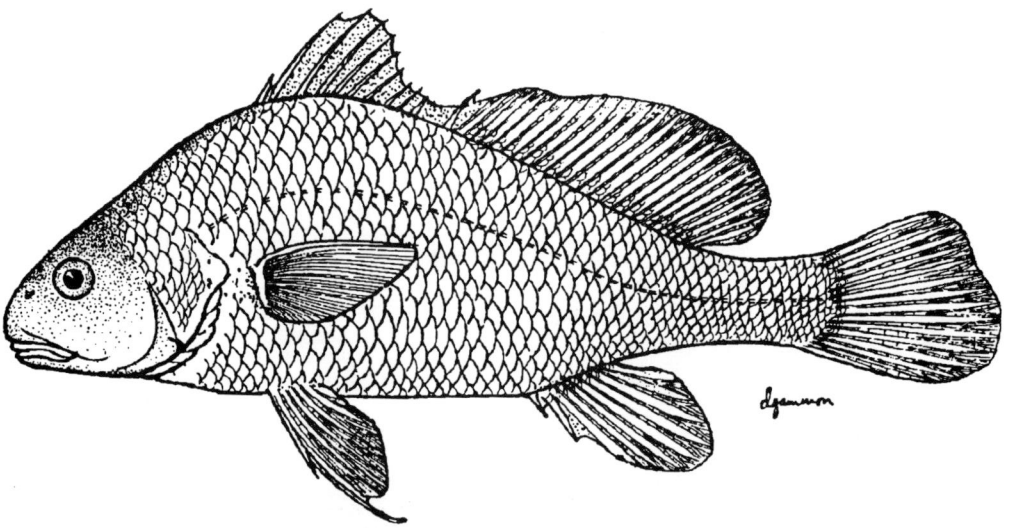

Freshwater Drum - *Aplodinotus grunniens*

111

mainstem of the Wabash River. This large increase also influenced the commercial harvest of channel catfish inasmuch as Glander (1984, 1987) reported elevated catches between 1977 and 1986.

The initial cause of this explosion of channel catfish, as with many other species, was the exceptional spawning success enjoyed during the summer of 1983. They grew to lengths of 100 to 150 mm by 1984 and 200 to 250 mm long by the summer of 1985. Spawning was successful again in 1984.

The large numbers of channel catfish generated by these two successive years of fruitfulness then led to an enormous increase in catches of 200 to 400 mm long fish in 1986. The year 1987 was also a reproductively prosperous year for this species, but the population diminished gradually during 1988 and 1989 to return to more normal levels by 1990.

Catch rates have been stable since 1990, but the mean size continues to increase. The 1993 flood did not apparently adversely influence catch rates in 1993-94. However, the mean size of channel catfish was the highest on record, perhaps indicating reductions in smaller individuals.

Unlike flathead catfish, channel catfish are known to travel widely (Hubley 1963) and probably range throughout the mainstem and into tributaries. Nevertheless, the catch curves indicate somewhat greater concentrations of fish in Reaches 3, 4, 10, and 11 than in other Reaches (Figure 67). During the drought of 1988, however, much larger catches were made in Reaches 1 through 4 than in the lower Reaches.

Channel catfish were found to eat many different kinds of foods in the study by Rud (1983). Most of the food biomass consisted of fish (54.6%), crayfish (34.6%), and aquatic insects (1.2%). However, the fish component included a large amount of fish eggs.

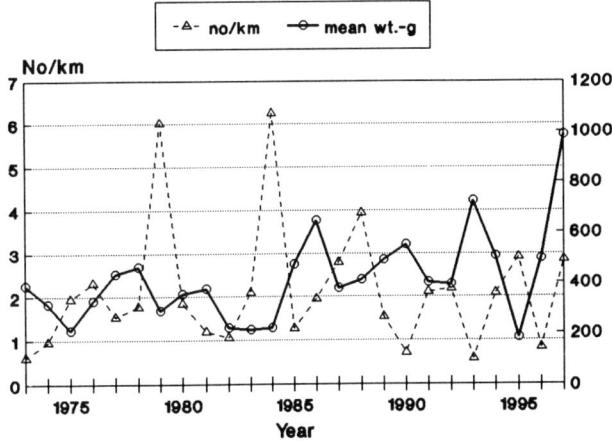

Figure 64: Mean annual catch (No/km) and mean weight of flathead catfish from 1968 through 1997.

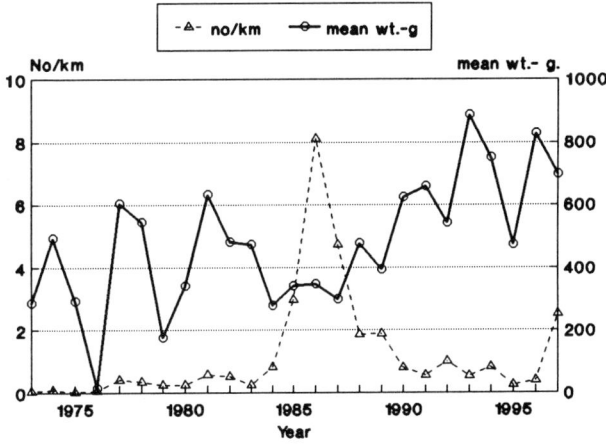

Figure 65: Mean annual catch (No\km) and mean weight of channel catfish from 1968 through 1997.

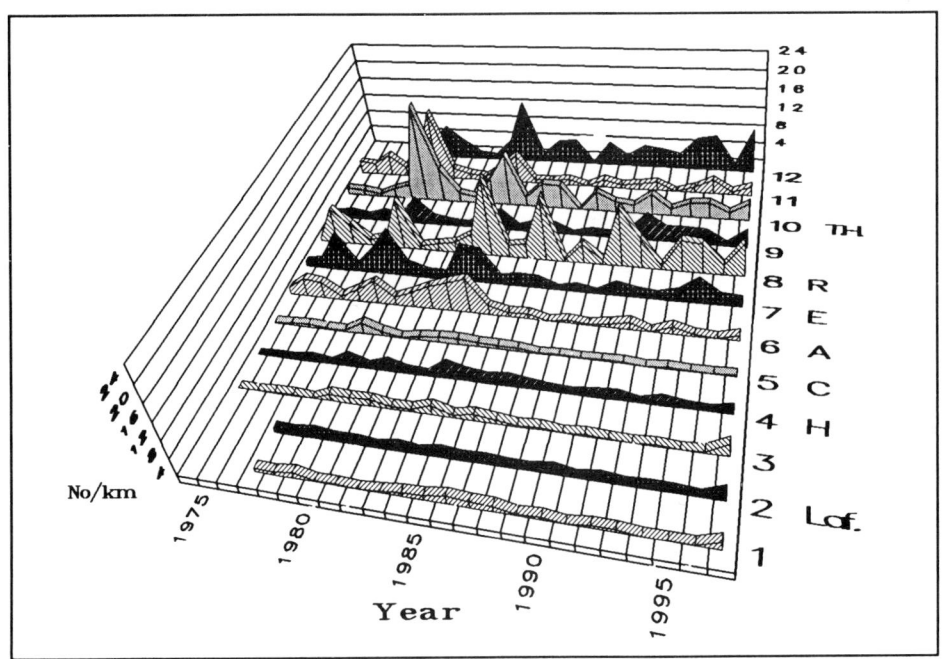

**Figure 66: Catch rates (No\km) of flathead catfish
in 12 Reaches of the Wabash River from 1974 through 1997.**

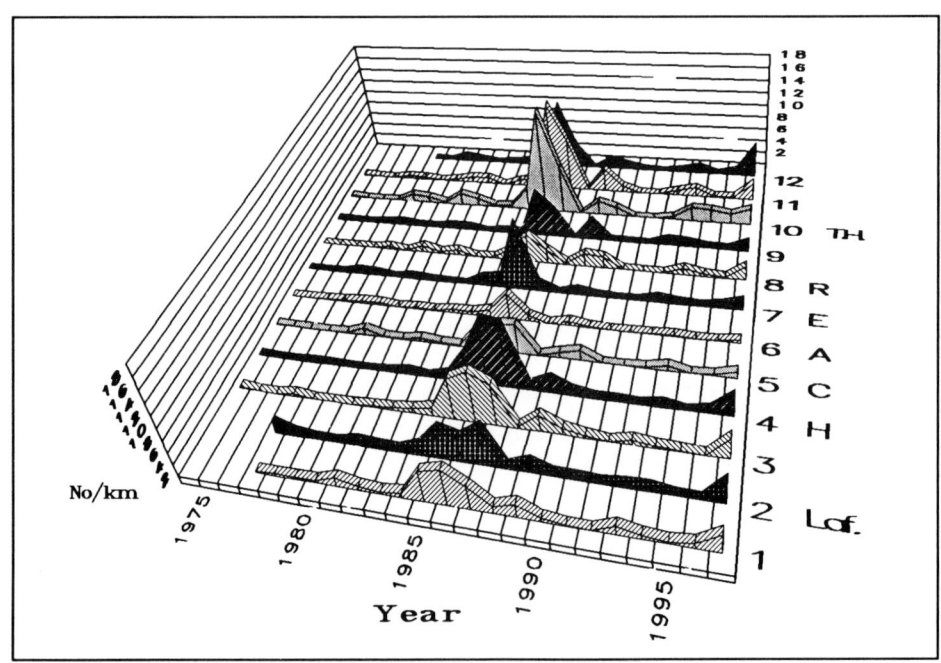

**Figure 67: Catch rates (No\km) of channel catfish
in 12 Reaches of the Wabash River from 1974 through 1997.**

Freshwater Drum (*Aplodinotus grunniens*) is an unusual species better known to Indiana fishermen as the "white perch" and to those a bit farther north as the "sheephead". It is the only freshwater member of a family of fish which is otherwise entirely marine. The males have the interesting ability to generate sound by means of a muscle stroking its air bladder during the spring reproduction season (Schneider and Hasier 1960).

The head contains large otoliths or earbones which have a L-shaped mark on one of the two flat surfaces. These "Lucky" stones are sometimes collected by fishermen when cleaning a large specimen and pocketed to foster good luck and, maybe, good fishing.

The drum is a large, handsome fish with a gray back, steely silver sides, and a white belly. The lateral line extends to the rounded rear edge of its caudal fin (tailfin). Drum are long lived and may exceed 15 kg (30 lbs) in weight. However, the mean weight of fish we caught in the Wabash River was less than 1 kg (2 lbs). It is a strong fighter for anglers using live bait, but it has soft flesh and is only a fair food fish.

Prior to 1985 the mean catch rate of drum rarely exceeded 0.5 fish per kilometer (Figure 68). In 1985 it increased to 1.0 fish/km and then stabilized at about 1.5 fish/km since 1986. However, in 1997 large catches were made with a new electrode arrangement. The average weight has oscillated between 400 and 800 grams (1 - 2 lbs) since 1973.

Large numbers of drum have been found in all Reaches of the Wabash River since 1985 (Figure 69). Only a modest expansion resulted in the immediate vicinity of Reach 6 (Cayuga Electric Generating Station), but catches in Reach 9 (Wabash Electric Generating Station) were nearly as great as in flanking Reaches.

Catches of drum in the lower Reaches were sharply lower during the 1988 and 1991 droughts, but since then they have increased even more in the lower Reaches while holding fairly steady in the upper five Reaches.

Catches of drum taken immediately after the 1993 flood and also in 1994 indicate that drum were impacted little or none by the 1993 flood.

Most of the drum whose stomachs were examined by Rud (1982) were 4- and 5-year-old fish which principally fed upon insect larvae, notably Ephemeroptera. Freshwater drum are generally believed to eat molluscs which they grind with heavy pharyngeal teeth. However, clams and snails made up only 10% of the weight of food found in stomachs, while fish comprised another 10%.

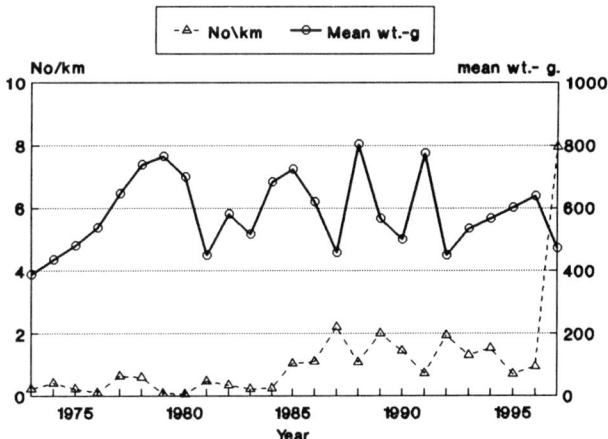

Figure 68: Mean annual catch (No\km) and mean weight of freshwater drum from 1968 through 1997.

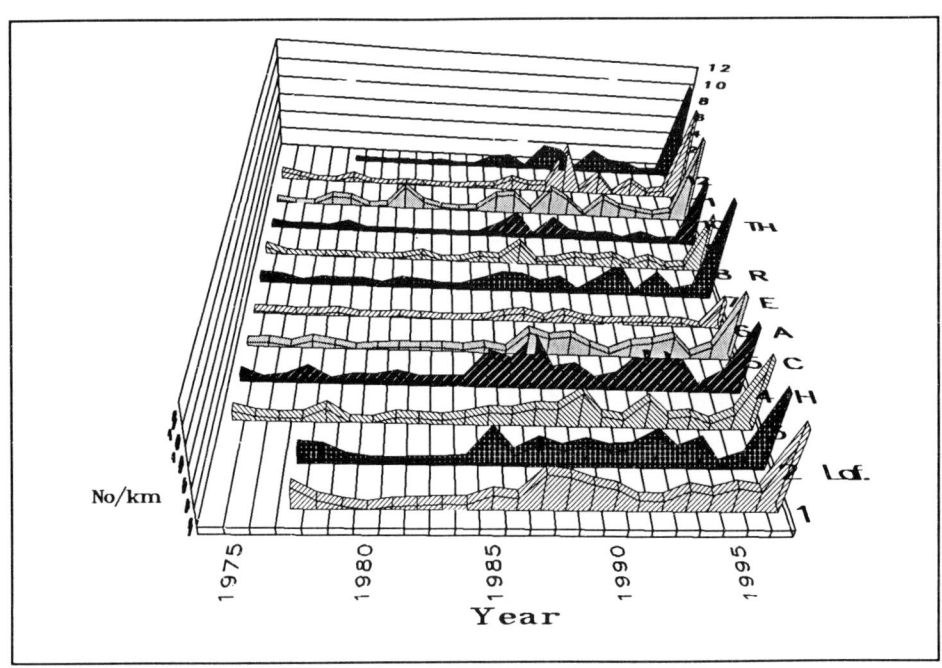

**Figure 69: Catch rates (No\km) of freshwater drum
in 12 Reaches of the Wabash River from 1974 through 1997.**

GOLDEYE, MOONEYE, AND SKIPJACK HERRING

Goldeye and mooneye are the only members of the family Hiodontidae, an ancient group restricted to North America. Their geographic range extends from the gulf of Mexico far north into Canada. Both species are schooling fish which prefer larger streams, rivers, and lakes. The eyes of both species are adapted to low light intensities. The goldeye is mainly nocturnal and appears to be more tolerant of turbid waters than mooneye. Both species feed on a variety of invertebrates and small fish. In the Wabash River they seem to prefer the near-shore areas over fairly deep pools.

The average Wabash River goldeye is an impressive and beautiful fish weighing about 475 g (1 lb.), more than twice the weight of the average mooneye. They bite occasionally on live bait or strike a noisy artificial worked along the surface. In Minnesota and Canada smoked goldeye is commercially available.

Catches of goldeye from the Wabash River increased considerably during the early and mid-1980's, but have declined to pre-1980 levels in recent years (Figure 70). However, the average size appears to have increased slightly since 1983. Until 1984 catches of goldeye outnumbered mooncye by a margin of nearly two-to-one. However, the catch rate of mooneye increased ten-fold after 1983 (Figure 71). Both the catch rate and average size of mooneye have declined since 1988. Populations of both species deteriorated after the 1993 flood and have not yet recovered.

Catches of goldeye were generally greater in the lower Reaches (Figure 72), while catches of mooneye have been higher in the upper Reaches (Figure 73), especially during the mid-1980's population boom.

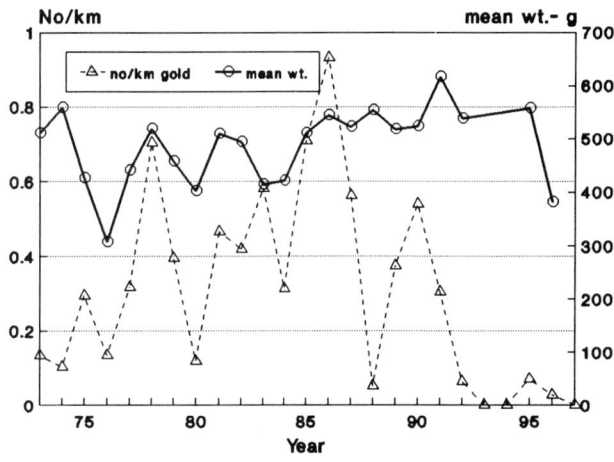

Figure 70: Mean annual catch (No\km) and mean weight of goldeye from 1968 through 1997.

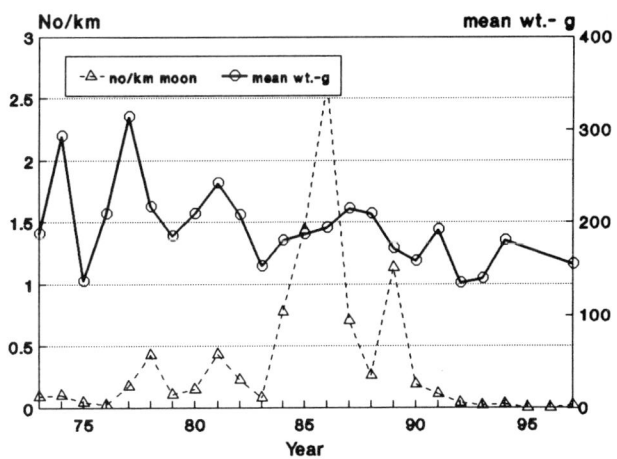

Figure 71: Mean annual catch (No\km) and mean weight of mooneye from 1968 through 1997.

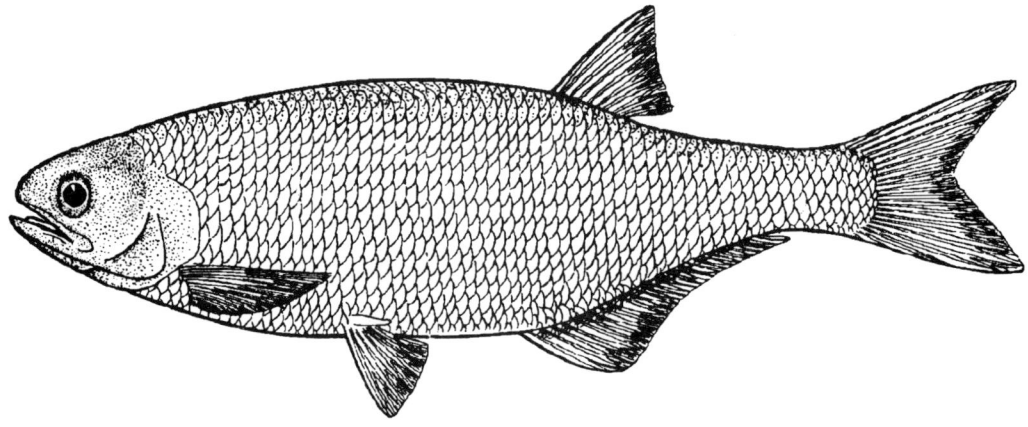

Goldeye - *Hiodon alosoides*

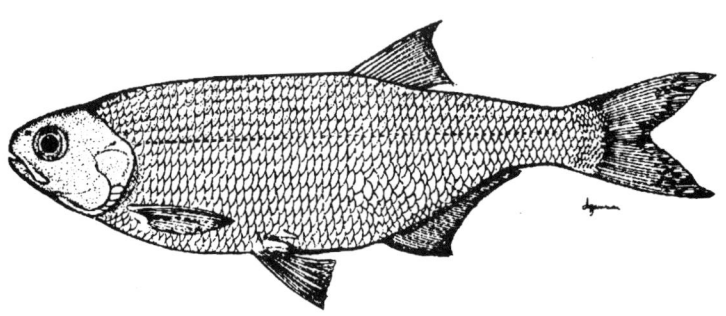

Mooneye - *Hiodon tergisus*

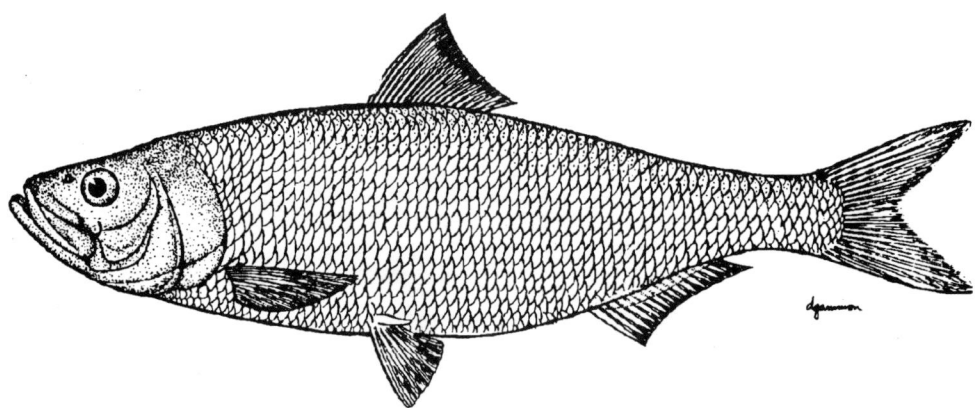

Skipjack Herring - *Alosa chrysochloris*

117

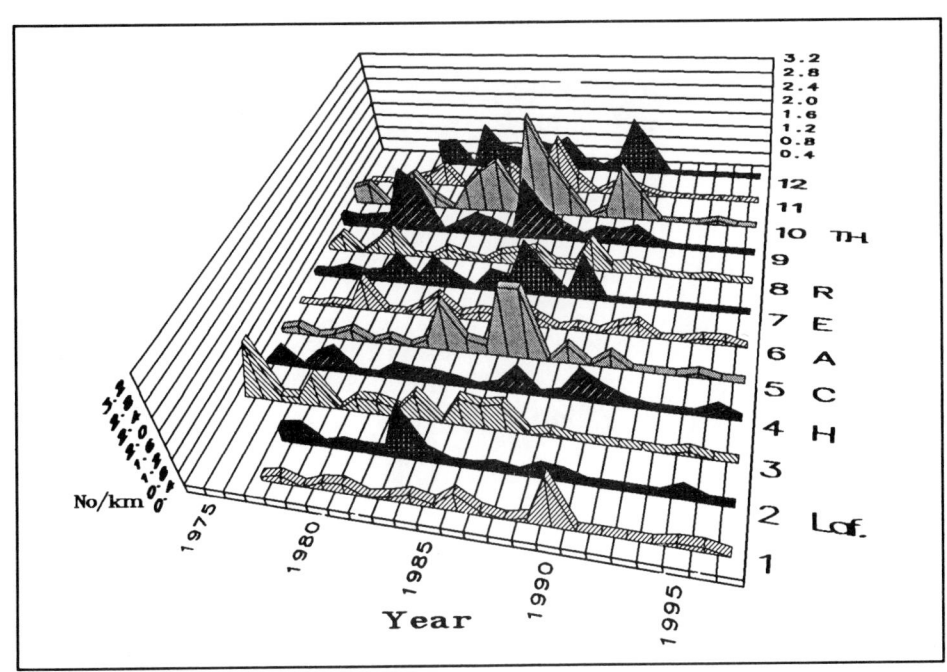

**Figure 72: Catch rates (No\km) of goldeye
in 12 Reaches of the Wabash River from 1974 through 1997.**

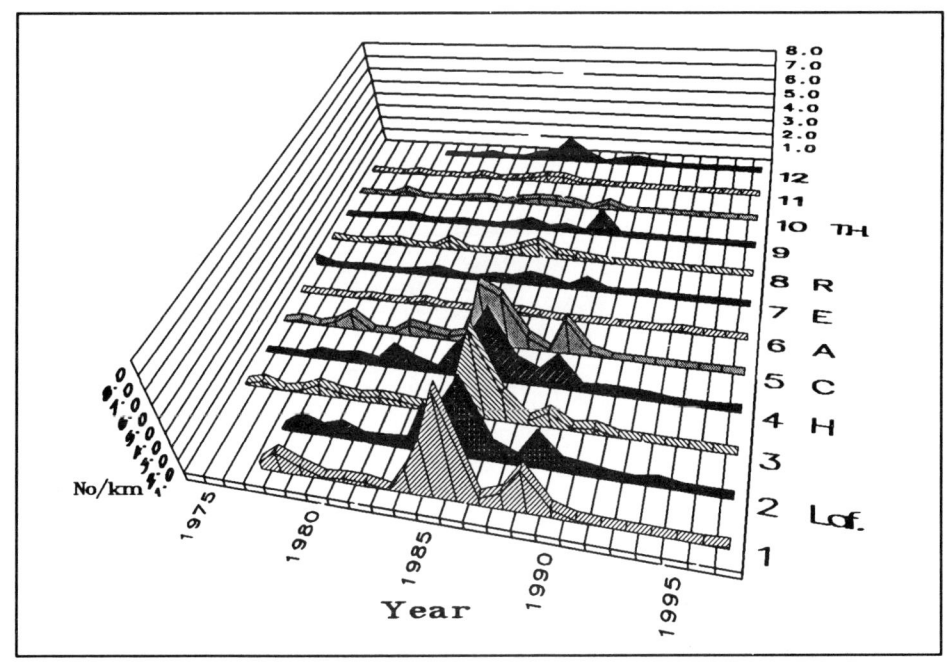

**Figure 73: Catch rates (No\km) of mooneye
in 12 Reaches of the Wabash River from 1974 through 1997.**

118

The skipjack herring (*Alosa chryso-chloris*) is a close relative of the gizzard shad, although it superficially resembles the goldeye and mooneye. It is much less common in the Wabash River than it is in the Ohio River and other more southerly rivers. It is a highly mobile species and may undertake extensive spring migrations from the Ohio River into and up the Wabash.

Skipjack feed primarily by sight on small minnows and aquatic insects. Larger individuals offer anglers excellent sport on light tackle, but the Wabash River population consists mainly of small fish. They are not a good food fish.

Our catch rates over the years were generally low with sporadic increases (Figure 74). The average size of skipjack herring has increased from about 150 grams in 1975 to about 250 grams in 1997.

Catches over the years have been quite evenly distributed throughout the 12 Reaches (Figure 75). Positive and negative

changes in population density generally occurred simultaneously throughout the study area. The population of skipjack herring may be expanding slightly in recent years in Reach 11 downriver from Terre Haute, Indiana, but in other Reaches the population appears to be quite stable.

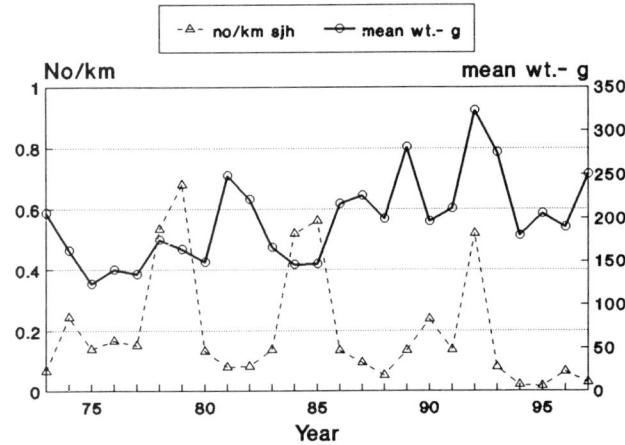

Figure 74: Mean annual catch (No\km) and mean weight of skipjack herring from 1968 through 1997.

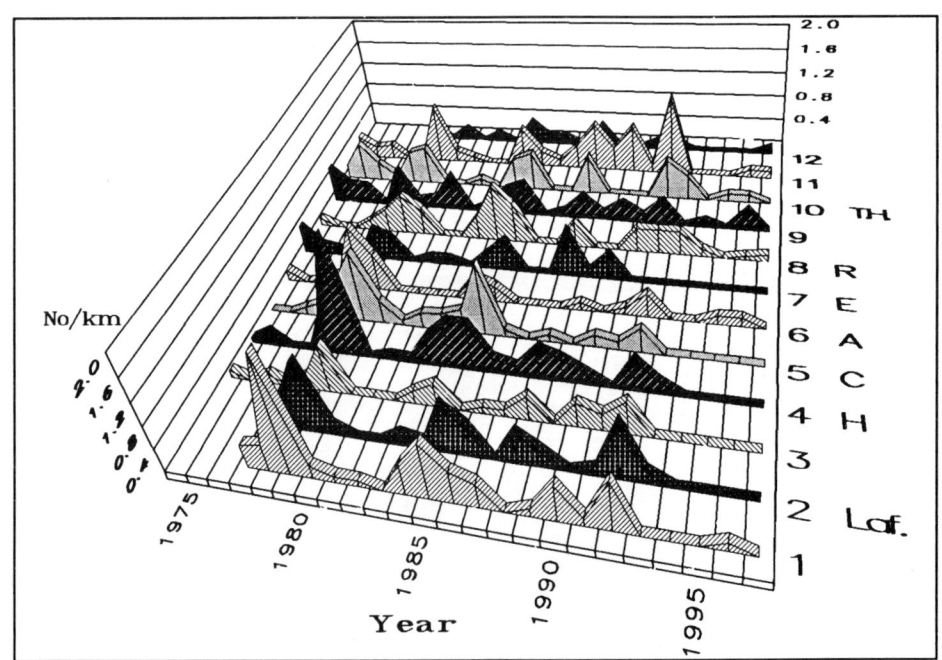

Figure 75: Catch rates (No\km) of skipjack herring in 12 Reaches of the Wabash River from 1974 through 1997.

THE SPORT FISHES

Smallmouth bass - *Micropterus dolomieui*
Spotted bass - *Micropterus punctulatus*
Largemouth bass - *Micropterus salmoides*

The smallmouth bass is one of the finest sport fish to be found anywhere. In the Wabash River system it is common today only in smaller clear, clean tributaries to the Wabash River, although it once occurred in abundance throughout most of Indiana's rivers including the Wabash. Water quality improvements may be benefitting smallmouth populations of both the White River and the Wabash River.

The life cycle of smallmouth bass begins with spawning in the spring when the water temperature reaches about 16°C (60° F). Males build a small nest in gravel and fertilize the eggs deposited by several females. He then guards the nest and drives away intruders. Normally, the small fry hatch within a week and remain in the nest for a few days before scattering throughout the stream. The jet black fry feed on small invertebrates, growing rapidly and gradually turning brown in coloration.

Smallmouth bass may spawn several times each summer. Juveniles grow to a length of 50 to 75 mm (2 to 3 inches) by fall, feeding primarily on small invertebrates. They grow to 150 to 200 mm (6 to 8 inches) by the second year and include more and more crayfish and fish in their diets. By the end of the third year of life they typically reach total lengths of 225 to 275 mm (6 to 11 inches) and will be ready to spawn the next spring.

Catch rates of smallmouth bass were usually low, about 0.1 to 0.2 fish per kilometer, except for the years immediately after the low-flow summers of 1988 and 1991 and for three years after 1983 (Figure 76). Young smallmouth bass hatched in 1983 entered the Wabash River by 1984 and remained there for at least three more years. During that period the size of smallmouth steadily increased.

Other than during the peak years, smallmouth bass were usually found close to the mouths of tributaries. The catch pattern of smallmouth bass over the period of study was very similar to that of mooneye with the greatest expansion of the population occurring in the upper four Reaches (Figure 77).

Rud (1982) examined only 16 smallmouth bass stomachs, mostly one- and four-year-old fish. Crayfish were found in about two-thirds of stomachs containing food, while insect larvae were found in half the stomachs. Fish made up a negligible part of their diet.

Many rivers and streams in central and southern Indiana contain spotted bass, although most fishermen might not recognize them. Here on the northern fringe of their geographic range, they grow slowly and seldom reach 300 mm (12 inches) in length. Physical characteristics are somewhat intermediate between smallmouth and largemouth bass. They possess the head and smaller mouth of the former, but the general coloration of the latter although they usually have linear rows of spots below the broad dark band along the lower flanks.

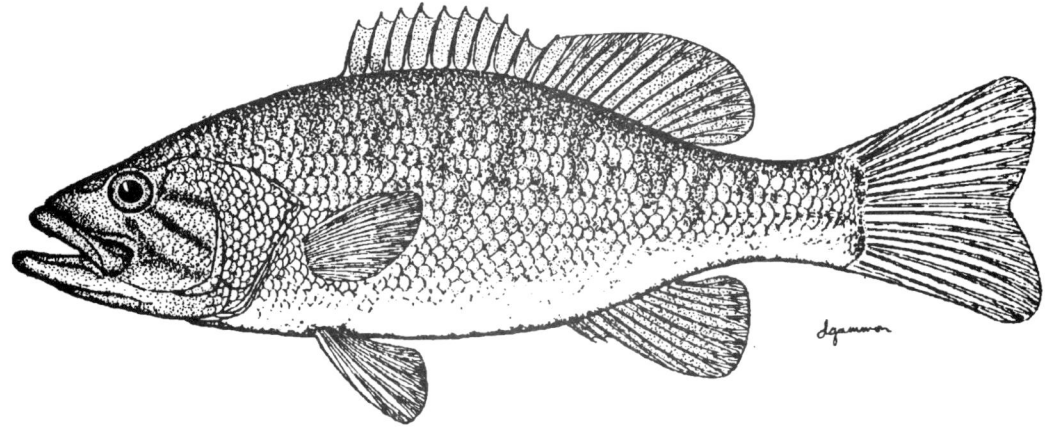

Smallmouth Bass - *Micropterus dolomieui*

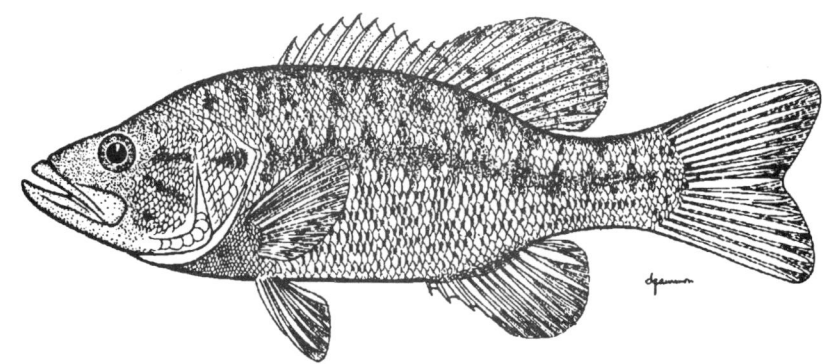

Spotted Bass - *Micropterus punctulatus*

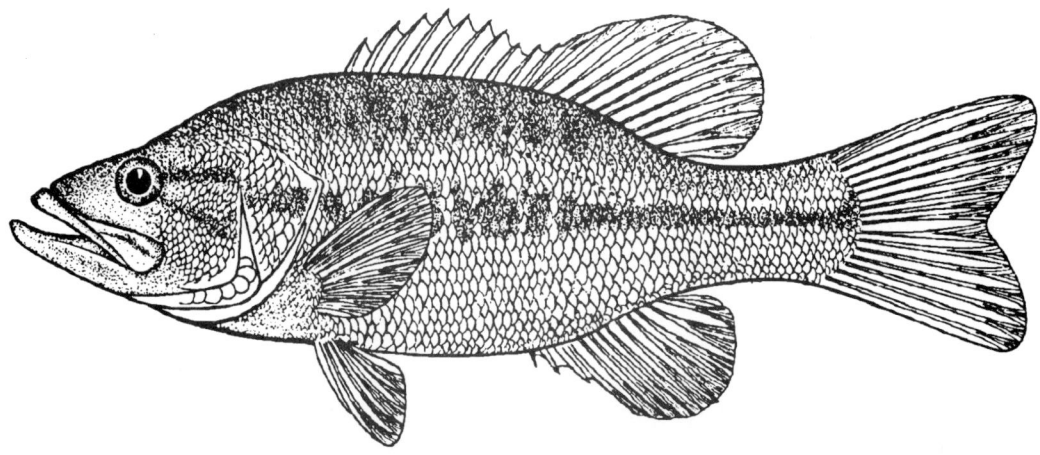

Largemouth Bass - *Micropterus salmoides*

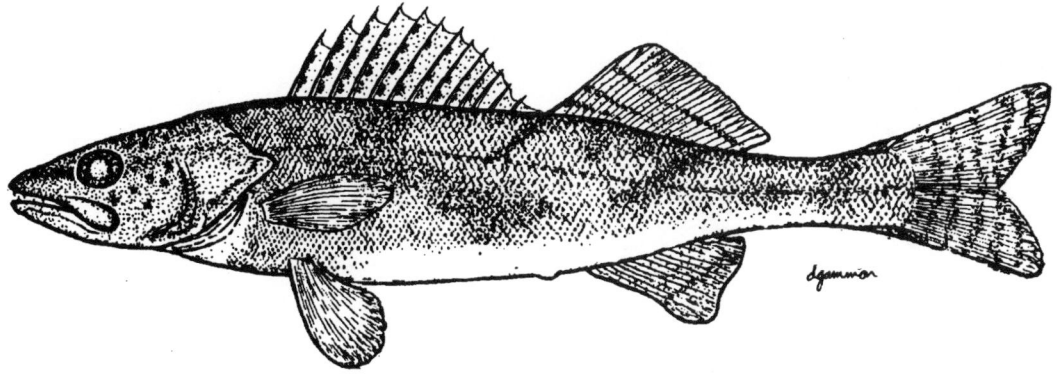

Sauger - *Stizostedion canadense*

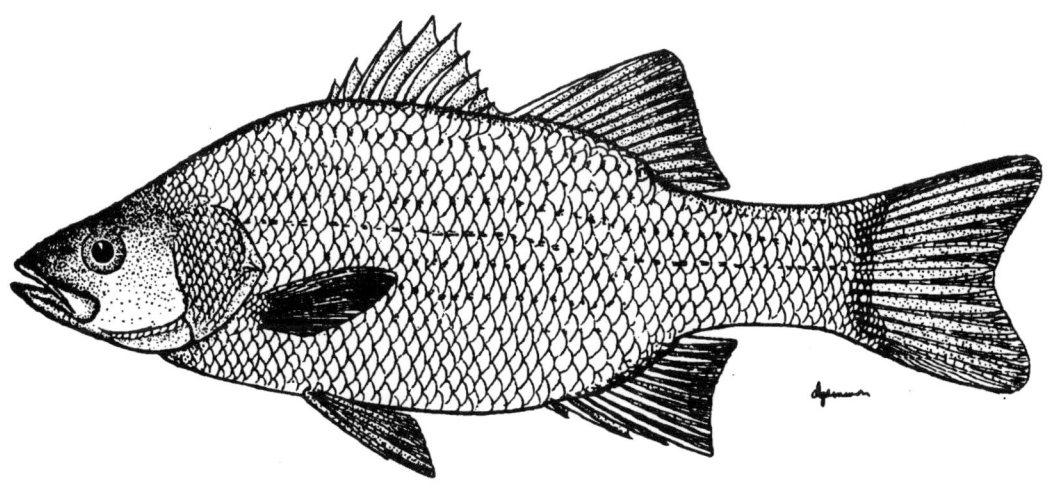

White Bass - *Morone chrysops*

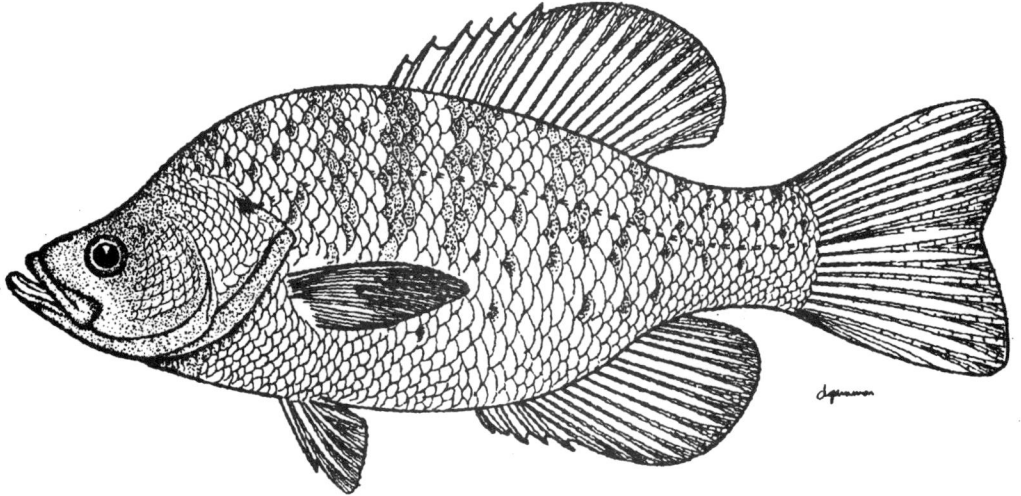

White Crappie - *Pomoxis annularis*

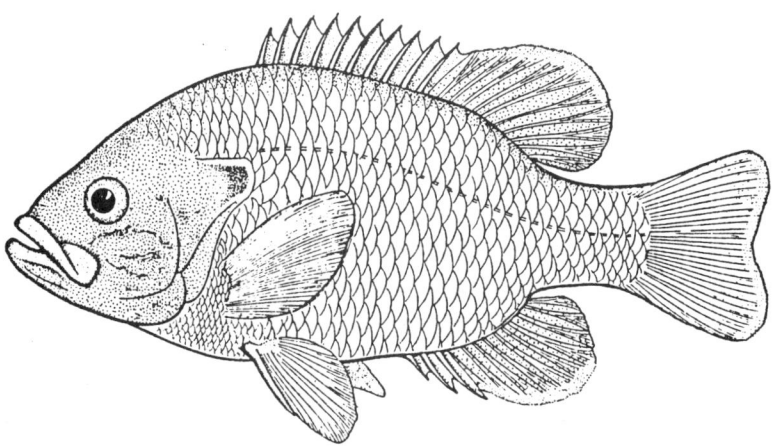

Green Sunfish - *Lepomis cyanellus*

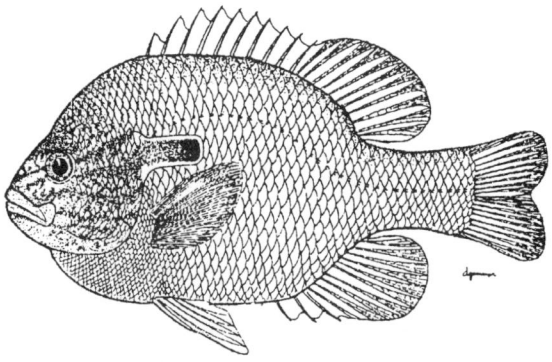

Longear Sunfish - *Lepomis megalotis*

Catch rates of spotted bass exhibited the same basic pattern as for smallmouth bass (Figure 76). The catch rate never exceeded 0.2 fish per kilometer until after 1985 and peaked in 1986, 1989, and 1992, again following summers of low-flow. The average weight was usually less than about 300 grams (0.66 lbs).

The distribution of spotted bass within the middle Wabash River changed considerably during the period of investigation. Prior to about 1980 nearly all specimens were found in the lower Reaches south of Covington, Indiana (Figure 78). Over the next decade they moved northward to Lafayette, to Delphi, and then continued on upriver. In 1990 we found that this species was widely distributed in the Eel River system (Gammon and Gammon 1990, 1993).

As with some other species, the catch rate of spotted bass increased in the lower as well as upper Reaches, with a particularly large increase in Reach 12 south of Darwin, Illinois. This dramatic increase, in particular, may be indicative of a genuine improvement in water quality of the lower Wabash River. The avoidance of thermally unacceptable areas (Reaches 6 and 9) is approximately the same order of magnitude as for smallmouth bass.

It appears that both smallmouth and spotted bass populations were again expanding in 1992, but the flood of 1993 severely reduced the numbers of both species.

About half of the 36 spotted bass stomachs examined by Rud (1982) contained crayfish. Aquatic insect larvae were found in 39% of the stomachs, while fish were found in 30%.

Only a few individuals of largemouth bass were collected each year. These probably entered the Wabash River via tributaries and originated from populations both in reservoirs and farm ponds. The middle Wabash River lacks weedy backwater areas which would be the preferred habitat of this species.

White Bass - *Morone chrysops*

The white bass is the most common member of the "temperate basses" found in the Wabash River. Its close relative is the smaller yellow bass *(Morone mississippiensis)*, which also occurs in small numbers. Both species have dark, longitudinal stripes along their sides which are continuous in the white bass and discontinuous above the anal fin in the yellow bass.

A third member of this family has appeared in the Wabash River in recent years, a hybrid striped bass x white bass (*M. saxatilis* x *M. chrysops*) sometimes called a "wiper", sunshine bass, or palmetto bass. The Indiana Department of Natural Resources stocks this hybrid regularly in reservoirs as a trophy fish. Hybrid striped bass feed voraciously, grow fast, and have excellent fighting qualities.

The population sources of most of the white bass in the Wabash River are probably nearby reservoirs: Huntington, Salamonie, Mississinewa, Freeman, Shaffer, and Mansfield. This species prefers clear water in larger rivers, reservoirs, and lakes. Schools of white bass roam near the surface feeding on small fishes, zooplankton, and insects and are most active in early morning and late evening. In the Wabash River white bass prefer the quiet, sandy bottom areas on the insides of river bends.

During the spring white bass migrate from reservoirs into tributaries feeding into them to spawn. After these spring "runs" most of the fish return downstream to the reservoirs. It is likely that many white bass in the Wabash River originate from reservoirs and are carried into the Wabash River during the periodic drawdowns which stabilize reservoir water levels. White bass are a popular fish to catch and eat.

Three-quarters of white bass stomachs examined by Rud (1982) contained aquatic insect larvae, 75% of which were Ephemeroptera and Trichoptera. Fifty-nine percent contained fish, which made up nearly half the weight of food consumed. Seventeen percent of the fish had eaten a wide variety of crustaceans.

Electrofishing catches of white bass exhibited the characteristic increase following 1983, peaking in 1985 and then declining thereafter (Figure 76). As many individuals were captured from 1984 through 1986 as had been caught during the previous decade. After 1986 the catches everywhere gradually declined until they had returned to "normal" levels of abundance by 1990.

The upper six Reaches generally yielded more white bass than the lower six Reaches (Figure 79). Very few white bass were taken after the 1993 flood.

If the reservoir populations of white bass are the primary source of the fish found in the Wabash River, what factors are responsible for the dramatic increase in the white bass population of the river after 1983? White bass abundance might merely be a function of how frequently the reservoirs are drained. Another possibility is that environmental conditions in the Wabash River itself have been conducive to a thriving existence only in recent years. Whatever the

reasons the population boom appears to be over and reestablishing the population may require some years of more stable flows.

Sauger - *Stizostedion canadense*
Walleye - *Stizostedion vitreum*

The sauger maintains a low, but persistent population in the Wabash River. It is similar in basic appearance to its close relative the walleye but, unlike walleye, it has spots between the rays of the spinous dorsal fin.

Sauger are currently caught only occasionally by anglers usually while fishing for other species such as channel catfish. In the Ohio and Mississippi Rivers sauger are important sport fish. It could become an equally important asset of the Wabash River in the future if the river can be further improved in environmental quality.

The sauger is more tolerant of turbid water than the walleye, perhaps because they tend to feed during the daylight hours rather than at night as with walleye. They prefer swifter, deeper water than walleye and feed in deeper water than walleye.

The sauger population has exhibited great variability in abundance and average weight (Figure 76). Until 1977 catch rates were variable and averaged about 0.4/km. The average weight increased steadily from about 400 grams to about 700 grams. Between 1977 and 1984 catch rates were very low and consisted mostly of large individuals weighing an average of 800 to 900 grams. As with other species the catch rate increased markedly to 0.7 fish per kilometer from 1985 through 1987 and then declined to about 0.3 fish per kilometer since 1988. The weight has averaged about 600 grams (1.3 lbs) since 1983. The 1993 flood may have had an negative impact on the sauger population.

Catch rates have been much higher in Reaches 1 - 4 than in Reaches 6 - 9. Reach 5 appears to be a stretch of river with catches of an intermediate size. Sauger are about as thermally sensitive as redhorse and few are found downriver from the Cayuga or Wabash River electric generating stations. Sauger were scarce in Reaches 7 and 8 and were nearly absent from the Terre Haute, Indiana area and downriver until the mid-1980s.

The population increases of 1985-87 were most prominent in the upper four Reaches, but there was also an encouraging increase in catch rates in the lower three Reaches as well (Figure 80). The sauger population is currently at a low level of abundance, as low as it was in the late 1970s and early 1980s. This species has shown itself capable of rapid population growth, however, and it is hoped that it will soon rebound again.

Rud (1982) examined stomachs from 86 sauger, mostly four-year-olds, and found fish remains in all of them. Fish constituted more than 95% of the diet. Nearly 30% of food eaten were centrarchids, including centrarchid bass, followed by cyprinids (11% by weight) and white bass (11%). Neither gizzard shad nor catfish species were important food items.

The walleye, also known as walleyed pike, pikeperch, or jack salmon, was once abundant in the Wabash River. Descriptions of spring spawning runs into some of its larger tributaries indicate a large population of large fish was present at least into the early 1800s. However, the walleye was absent from the Wabash River and its tributaries until recently. It has survived in small numbers in the Ohio River, but Pearson and Krumholz (1984) indicate that it appears to be diminishing in numbers in recent decades.

We captured very few walleye until about 1985, after which they became a small, but regular component of the electrofishing catch. Their appearance can be traced to Indiana Department of Natural Resources program of raising walleye in fish hatcheries and stocking most of Indiana's reservoirs. Some of the stocked walleye escape and enter tributaries and eventually the Wabash River itself.

Walleye are rare during summer near the Cayuga and Wabash River electric generating stations because they are among the more thermally sensitive species. They are also rare in Reaches 8, 10, 11, and 12. However, they do appear occasionally in other segments of the river. As indicated previously, walleye are nocturnal fish which feed mainly in the evening, night, and early morning. The sauger, on the other hand, feeds primarily during the day. The turbidity of the Wabash River would make food location difficult for both species, but especially the walleye. Should the turbidity of the Wabash River be reduced in the future the walleye could once again become an important species.

White Crappie - *Pomoxis annularis*
Black Crappie - *Pomoxis nigromaculatus*

Both species of crappie are widely distributed in Indiana streams, rivers, reservoirs, and lakes. Both species prefer quiet waters near some kind of subsurface cover. White crappie are more tolerant to turbid waters than black crappie and perhaps this is why they are nearly ten times more abundant in the rivers and streams of westcentral Indiana. Neither species is very abundant in the Wabash River, but they are a regular component of the electrofishing catch. More than 85% of crappie taken from the Wabash River consist of white crappie.

Annual catches were low and changed little over the period of study (Figure 81). Reaches 1 and 5 were most consistent producers of crappie, but crappie were nearly as common downriver from Terre Haute, Indiana. Catches from Reaches 6, 7, 8, and 9 were consistently lower than elsewhere. The population increases experienced by many species from 1985 through 1987 were barely evident for crappie.

Rud (1982) examined 39 white crappie stomachs and found that 53% contained fish, primarily gizzard shad, and 93% yielded insect larvae, primarily Trichoptera (caddis flies). The contents of 10 black crappie stomachs consisted principally of aquatic insect larvae, mostly Trichoptera. However, nearly 65% of the weight of recovered food consisted of fish, mostly centrarchids.

Longear Sunfish - *Lepomis megalotis*

The longear sunfish is much more abundant in small and medium-sized streams than it is in the Wabash River. However, it was a regular component of the annual catches, but only in the upper five Reaches (Figure 82). The overall pattern of catch rates indicates a fairly strong negative effect in and downstream from Lafayette (Reaches 2 and 3), then a recovery followed by reduction to near zero at the Cayuga EGS (Reach 6). This species is virtually absent in the main river downriver from the Cayuga EGS except perhaps near tributary mouths. Longear sunfish exhibited only a mild increase in abundance during 1985-86.

Rud (1982) found that longear sunfish, mostly two- and three-year olds, fed primarily upon aquatic insects, especially Diptera and Trichoptera larvae followed by Ephemeroptera larvae. Crayfish were also present in 24% of the fish.

Fishermen using a small hook baited with a worm or cricket might also find a number of other species of sunfish in the Wabash River, although none is common. The bluegill (*Lepomis macrochirus*) was the most common sunfish captured, most of which probably strayed into the Wabash River from the several connecting reservoirs.

The rockbass (*Ambloplites rupestris*) and green sunfish (*Lepomis cyanellus*) were also present, but even less common than the bluegill. On rare occasions we also collected an individual or two of the warmouth (*Lepomis gulosus*) and pumpkinseed (*Lepomis gibbosus*).

127

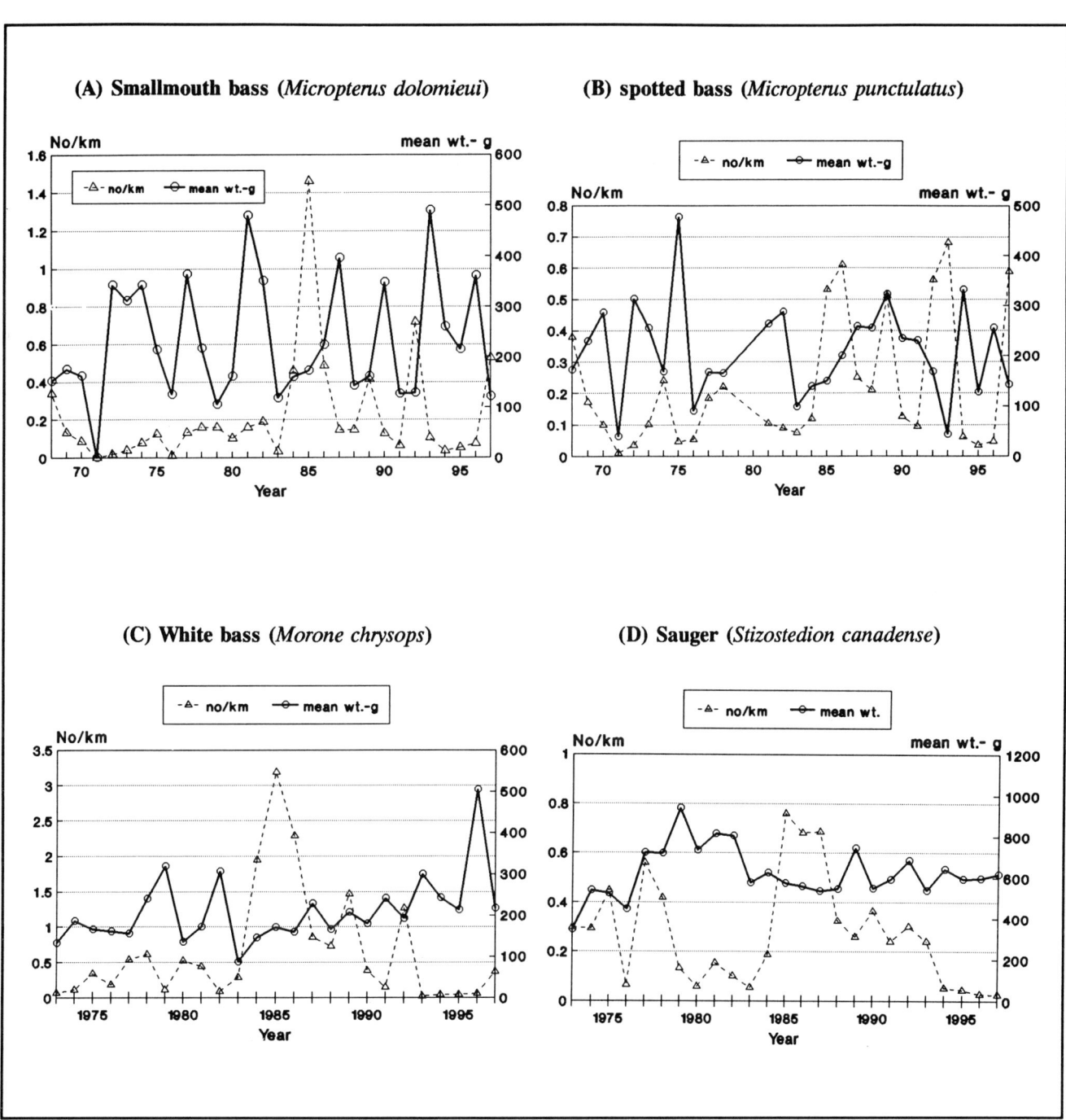

Figure 76: Annual mean catch rate (No/km) and mean weight from 1968 through 1997: (A) smallmouth bass (*Micropterus dolomieui*), (B) spotted bass (*M. puntulatus*), (C) white bass (*Morone chrysops*), and (D) sauger (*Stizostedion canadense*).

128

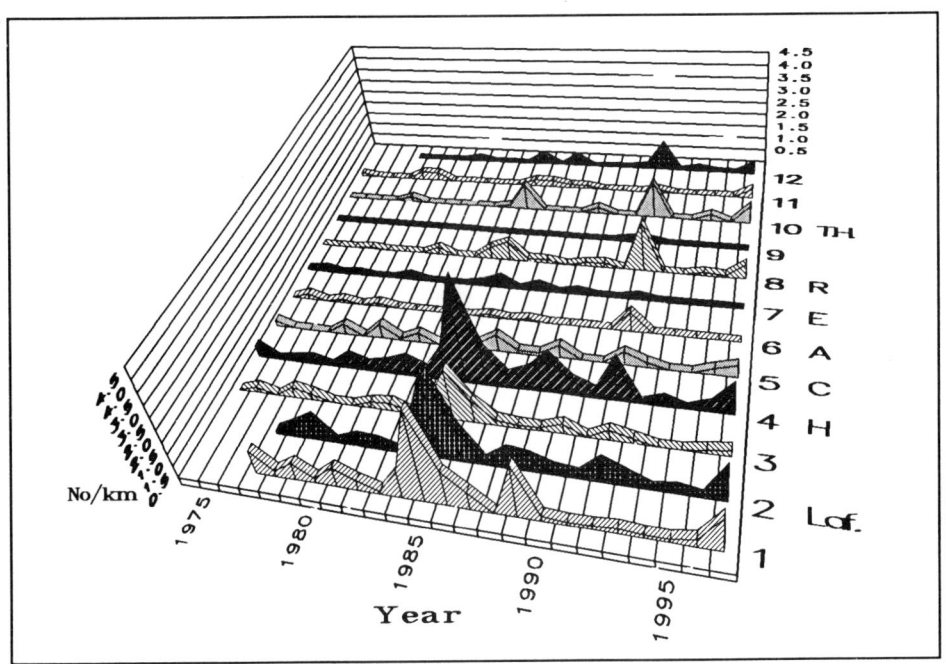

Figure 77: **Catch rates No\kg) of smallmouth bass in 12 Reaches of the Wabash River from 1974 through 1997.**

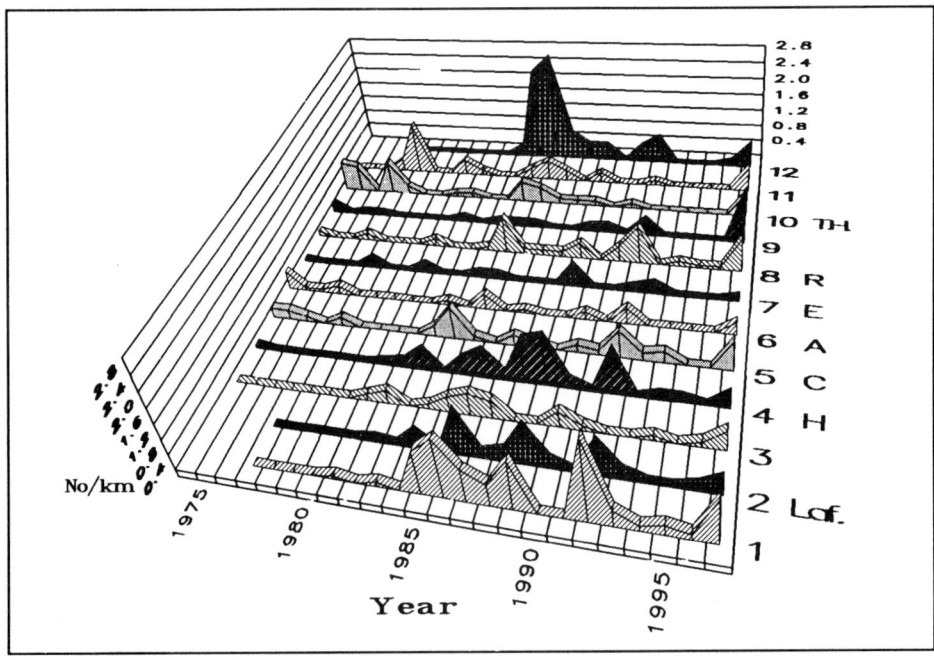

Figure 78: **Catch rates (No\km) of spotted bass in 12 Reaches of the Wabash River from 1974 through 1997.**

129

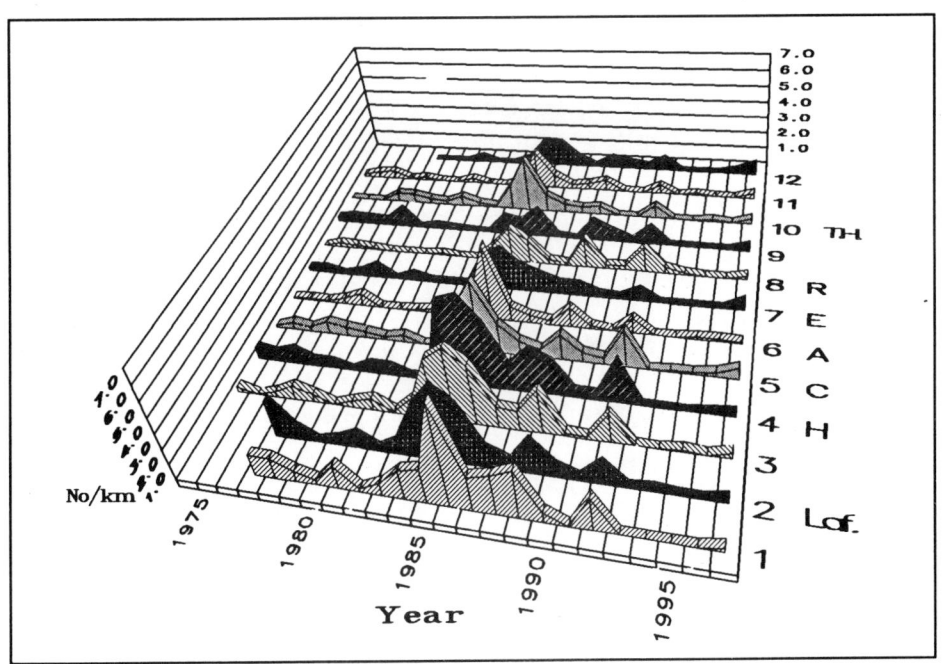

Figure 79: Catch rates (No\km) of white bass in 12 Reaches of the Wabash River from 1974 through 1997.

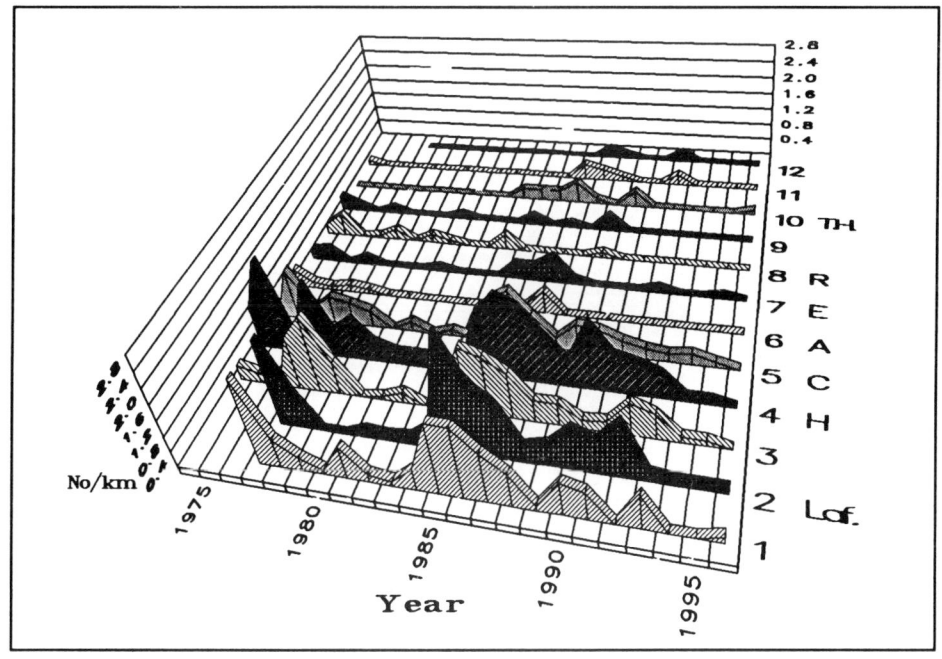

Figure 80: Catch rates (No\km) of sauger in 12 Reaches of the Wabash River from 1974 through 1997.

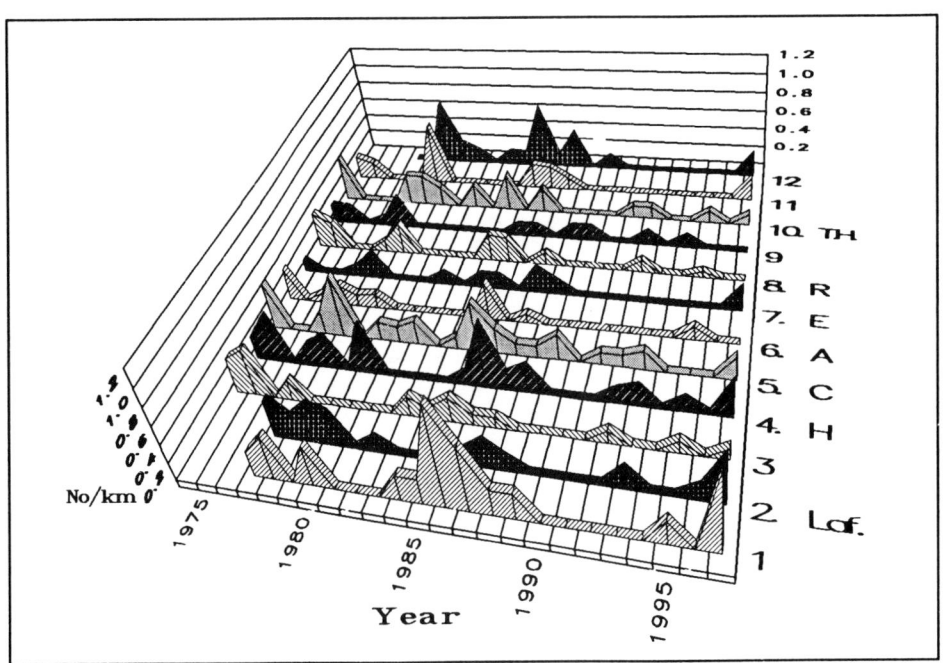

Figure 81: Catch rates (No\km) of white and black crappie in 12 Reaches
of the Wabash River from 1974 through 1997.

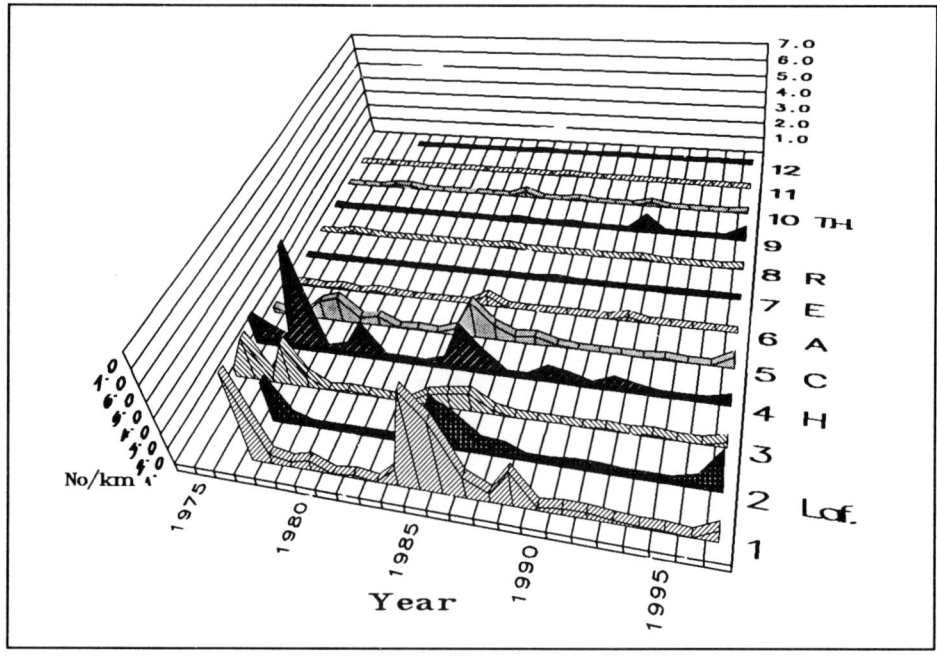

Figure 82: Catch rates (No\km) of longear sunfish in 12 Reaches of the Wabash River
from 1974 through 1997.

THE ANCIENT FISHES

All species of Wabash River fishes are ancient by human standards of time. Some are truly primitive. The lampreys, for example, are technically not really fish at all since they lack paired fins and a lower jaw. Some lampreys live most of their lives in the bottom mud of our rivers and streams and emerge only after they have matured for purposes of spawning, after which they die.

The silver lamprey (*Ichthyomyzon unicuspis*) and other closely related native species become parasitic on other larger fish when they mature. As with their larger relative the sea lamprey, they attach themselves to any large fish which happens by, use their toothy tongue to rasp a hole through scales and skin into the muscle, and feed on blood and juices.

We most often found them attached to large fish which were electrofished. They usually disengaged themselves from their host in the process of netting. They were often attached to carp, but we also found them on catfish, buffalo, and even gar. None of these native lamprey species is abundant and all are less than 10 inches long so that their host is usually not in any danger of dying.

The shovelnose sturgeon (*Scaphirhynchus platorynchus*) is one ancient species of fish which is fairly common and widely distributed in the Wabash River. However, they are not very susceptible to collection by electrofishing because they mostly cruise the river bottom in fast, deep chutes. Moving close to the bottom with barbels brushing the surface of the gravel, sturgeon "tastes" food items which are then sucked up with the extended vacuum-cleaner mouth.

Our electrofishing catches of sturgeon were strongly influenced by water level. When river discharge was strong the water was generally too deep for effective electrofishing. For example, no sturgeon were collected during the period of high summer discharge from 1979 through 1981. Catch rates improved considerably during summers with relatively low discharge rates.

Our catches of this small sturgeon generally consisted of single individuals from 700 to 900 mm in length and weighing 1 to 2 kg, although occasionally several would be netted. They contribute a minor part of the commercial catch on the Wabash River (Glander 1984, 1987). Overall, the abundance of sturgeon has increased over time and the average size also seems to have increased since 1982 (Figure 83).

The distribution of sturgeon appears to be almost discontinuous (Figure 84). No sturgeon were collected from Reaches 4 and 10 until 1992 and 1993. They were very rare in Reach 6 at the Cayuga EGS, although recent catches have been very good in Reach 9 at the Wabash River Electric Generating Station. Populations seem to have increased, however, in all the other Reaches since about 1982. As with other species, low annual catches were made for several years following the 1993 flood. Good numbers were again captured in 1997.

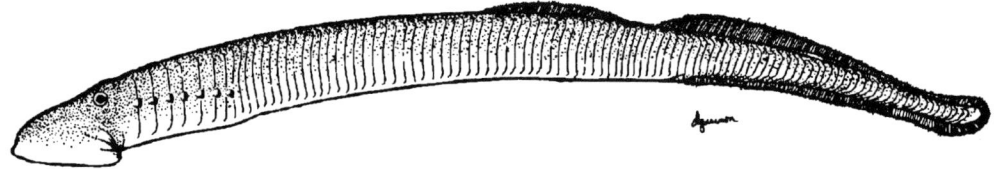

Silver Lamprey - *Ichthyomyzon unicuspis*

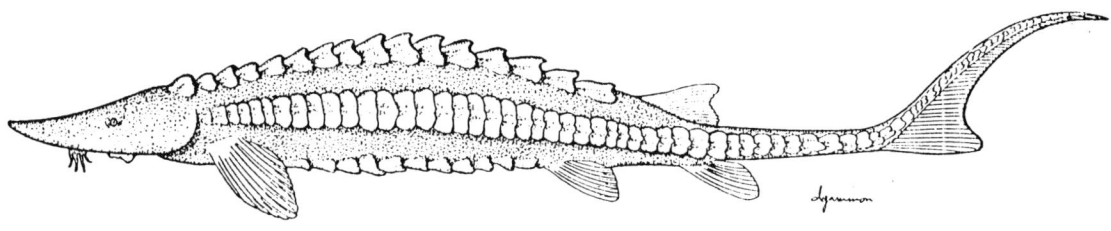

Shovelnose sturgeon - *Scaphirhynchus platorynchus*

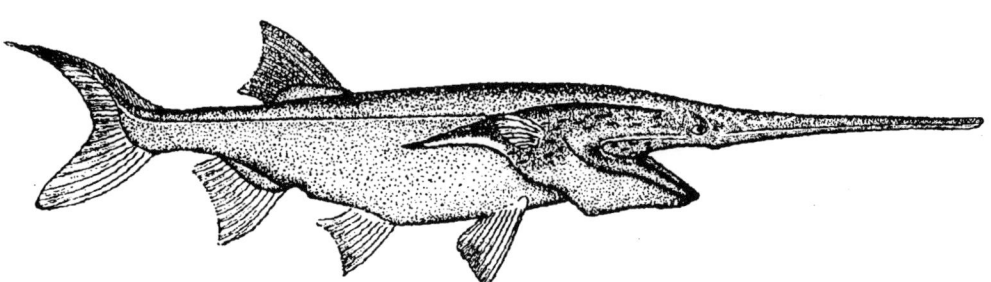

Paddlefish - *Polyodon spathula*

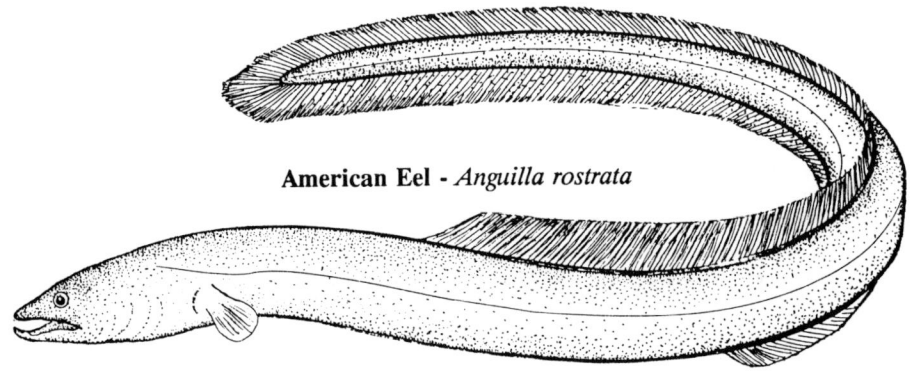

American Eel - *Anguilla rostrata*

The paddlefish (*Polyodon spathula*) is widely distributed throughout Indiana's waters, including the Wabash River where it is relatively scarce.

We usually found paddlefish and bigmouth buffalo wherever drainage from a side lagoon, lake, or pond entered the river. Both of these species essentially position themselves in the incoming stream of water with their mouths wide open filtering out zooplankton. This species was undoubtedly more common in pre-canal days when more numerous backwaters and bayous connected to the Wabash River mainstem. On occasion we have "shocked" five foot-long paddlefish which look very much like sharks as they heave their bodies away from the boat.

None of our resident species has a more fascinating life history than the American eel (*Anguilla rostrata*), although it is only a part-time resident. All of the American eels captured were large, immature fish. They are mostly nocturnal and feed upon dead and live fish and crayfish. They are incredibly hardy and can survive even in waters too polluted for carp.

After living several years in the Wabash River they migrate into the Ohio River, then down to the Mississippi River and ultimately out into the Gulf of Mexico. They must then move to their reproductive grounds, once thought to be the Sargasso Sea, but more recently believed to be south of the Bahama Islands (Vladykov 1964). Whatever their destination, they spawn from late winter to early spring and then the adults presumably die after spawning. The floating eggs hatch into planktonic larvae which are carried northward by the gulf stream toward North America and Europe. Eventually the growing eels reenter the Mississippi River and migrate northward to its tributaries, including the Wabash River, negotiating several dams in the process. Their continued existence far inland from their spawning grounds attests to their hardiness.

Many other smaller species are life-time residents of the Wabash River, all of which possess special attributes which have enabled them to successfully maintain their presence for thousands of years, despite the great changes in environmental conditions which have occurred over time.

Many species of minnows were collected by electrofishing, but only in small numbers relative to their actual population abundance for two reasons: (1) the relatively fast water habitats designated as collecting sites are not favored locations and (2) the bias of the electrofishing apparatus against collecting smaller fish.

During occasional special studies when a seine was employed for collecting a great many species of minnows and other fishes were collected together with numerous young-of-the-year of larger species.

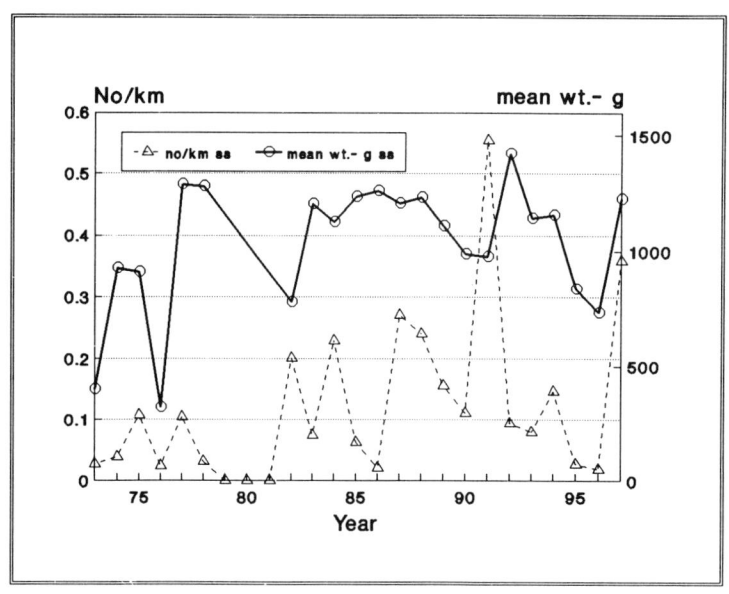

Figure 83: Mean annual catch (No/km) and mean weight of shovelnose sturgeon from 1968 through 1997.

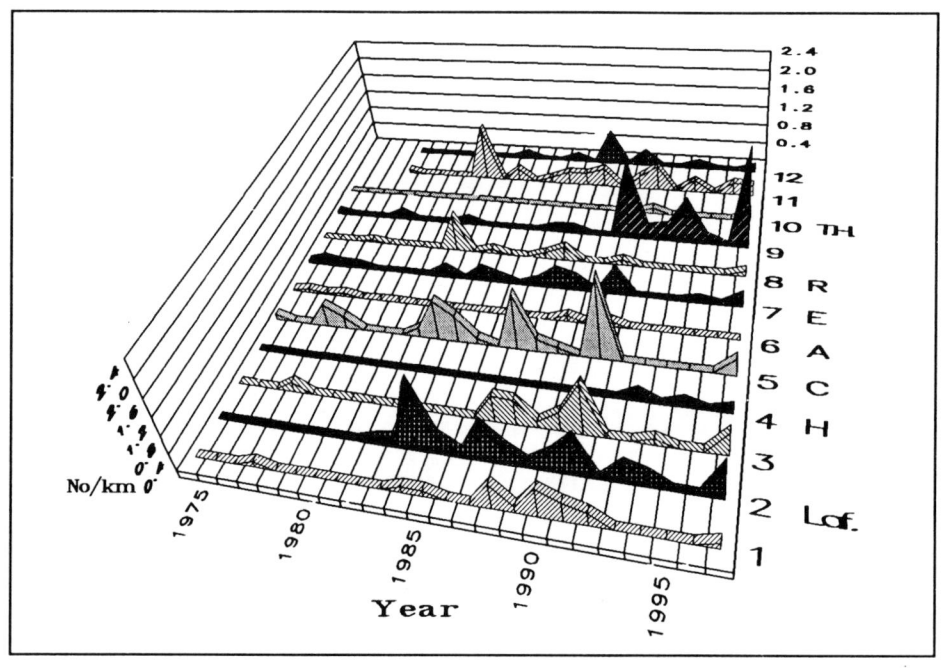

Figure 84: Catch rates (No/km) of shovenose sturgeon in 12 Reaches of the Wabash River from 1974 through 1997.

Cliff Gammon and Chris Tucker collecting fish with a seine.

SMALL DENIZENS OF THE WABASH

Not every fish in the Wabash River is large and long-lived. The river is also home to great numbers of nearly 60 other small species which typically have short life spans. Some of these dwell exclusively in the big river throughout their lives. Others stray or are washed into the Wabash from tributaries. Their great numbers compensate for their lack of size; each species playing a special ecological role in the ecosystem.

The presence and distribution of small fishes over the past 50 years has been charted using one of the most ancient of collecting gears - the seine. The seine is a simple, low-tech method of sampling small and mid-sized freshwater streams and rivers. The earliest comprehensive study of fish distribution in Indiana was that of Gerking (1945), who carefully and consistently used a very small seine. The seine was also used earlier by Jordan (1875, 1890), Jenkins (1886), and Evermann and Jenkins (1888). Technological advances eventually replaced the seine with more effective sampling methods such as electrofishing, but under the right conditions it is still useful for collecting very small or young fish, and it is possibly the most effective means of catching darters.

Gerking's distributional study in the early years of World War II included eight sites within our study segment of the Wabash River. Roggelin (1978) also sampled the middle Wabash mainstem and numerous tributaries with a seine in 1977 and 1978. EA Science and Technology (1988, 1989) extensively seined a decade later. The most recent seining series was made in August 1997 using a 30 X 6 foot, 1/4 inch mesh nylon seine with a 6 foot bag in the middle to sample 33 sites between Delphi, Indiana and Hutsonville, Illinois. More than 8,500 fish belonging to 46 species were collected from these sites. More than 80% of the total catch consisted of spotfin shiners (*Cyprinella spiloptera*) and young gizzard shad (*Dorosoma cepedianum*), both of which were widely distributed throughout the study segment.

Also widely distributed in smaller numbers were emerald shiner (*Notropis atherinoides*), river shiner (*N. blennius*), Mississippi silvery minnow (*Hybognathus nuchalis*), bluntnose minnow (*Pimephales notatus*), bullhead minnow (*P. vigilax*), log perch (*Percina caprodes*), and small freshwater drum (*Aplodinotus grunniens*).

A previously unrecorded species was collected from this section of the Wabash River; the streamline chub (*Erimystax dissimilis*).

Many other species lived within a more restricted section of river. For example, most young redhorse (*Moxostoma sp.*), bass (*Micropterus sp.*), sunfish (*Lepomis sp.*) and various species of chubs were mostly confined to the upper half of the study segment (Reaches 1-5). Young carpsuckers were, oddly, totally lacking from the 1997 seine catches although electrofishing indicated that they occurred in their usual abundance. The 1977 collections, in contrast, included many young carpsuckers.

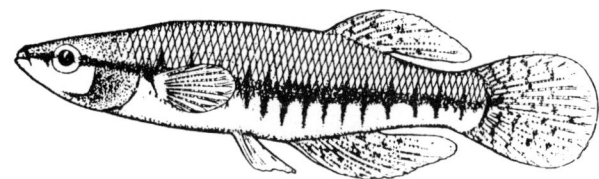

Blackstripe topminnow (*Fundulus notatus*)

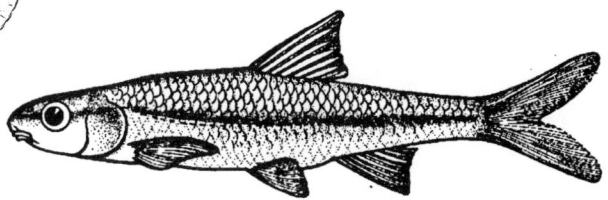

Bigeye chub (*Notropis amblops*)

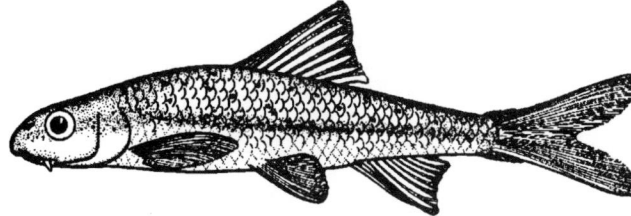

Speckled chub (*Macrhybopsis aestivalis*)

Silverjaw minnow (*Notropis buccata*)

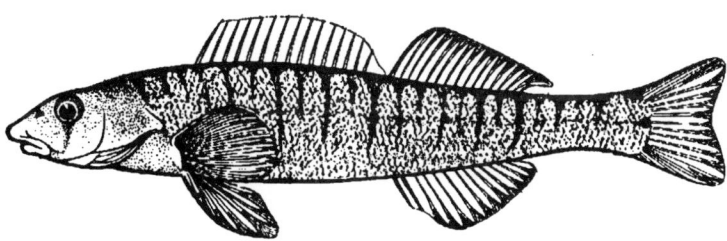

Logperch (*Percina caprodes*)

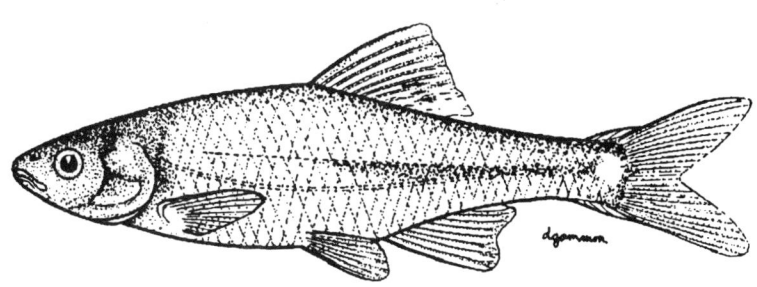

Spotfin shiner (*Notropis spilopterus*)

THE COMPOSITE INDEX OF WELL-BEING - Iwb

The foregoing discussion of patterns of relative abundance of important individual fish species indicates that most species populations of the Wabash River are highly variable over time and space. Considering all of these species in evaluating environmental conditions is a difficult, if not impossible, task. Nor can one focus only on "important" species, ie. the few dominant species or the sport fishes. Both the biotic and environmental components of nature are so complex that most ecologists have attempted to find some magical "Golden Fleece" to simplify the interpretative process.

Some ecological studies have attempted to reduce and simplify systems by designating some species as representative "indicator" species which somehow embody the tolerance and attributes of the entire community. Many other studies have sought ways to represent quantitatively the entire or some important part of the community (Pielou 1977, 1984; Ludwig and Reynolds 1988). The development of the composite index of well-being or Iwb described previously was partly motivated by the need to simplify data. Twenty-five years ago we endeavored to develop both a collecting protocol and methods for analyzing the resulting data for purposes of evaluating the environmental impact of electric generating stations (Gammon 1971, 1973, 1976). The environmental criteria which were being established at that time were to institute physical and chemical levels which would provide protection for resident aquatic biota.

The electric power industry sought and received from regulatory authorities an exemption from thermal limits if it could be shown that a healthy population of aquatic organisms lived in heated waters produced by generating plants. To achieve this exemption it had to be shown through the so-called 316(a) demonstration that the aquatic communities immediately downriver from electric generating plants were both diverse and numerous. In other words, it was necessary to demonstrate that the biotic integrity of the community was maintained despite excursions of temperature beyond the established criteria.

We examined several quantitative expressions of biotic diversity, including the total number of species and the Shannon-Weiner or Shannon-Weaver index of diversity for their utility in evaluating the impact of electric power plants. Quantitative expressions of abundance were also examined for their usefulness, including numeric abundance and aggregate weight or biomass. As will be illustrated shortly, all of these community parameters exhibited the same basic pattern through space (Gammon 1976). Therefore, it was decided to combine two indices of diversity and two suitably weighted indices of abundance into a single composite index of well-being or Iwb which would reflect both the diversity and abundance of the fish community:

$$Iwb = 0.5 \ln No/km + 0.5 \ln Kg/km + Div(no.) + Div(wt.)$$

When the individual electrofishing catches are converted to Iwb values and plotted over the length of river under study a scattering of points is generated, as

139

illustrated in Figure 85. The data set presented here is from the summer of 1988 during drought conditions and reduced river discharge. Equipment failure was minimal that year and the linear Iwb profiles were based on three optimal collections taken during June and July.

The scatter of points is considerable, but the spatial trend of separate collection runs is consistent. Indeed, the overall Iwb profile based on mean values is clear (Figure 86). The Iwb values are relatively high and steady in upriver areas. The shifts which do occur are for the most part gradual increases or decreases regardless of the direction of change. The Iwb values steadily decline and become much more erratic downriver from about RM 270 (Covington, Indiana) and the magnitude of differences between adjacent stations sometimes increase.

In order to further simplify interpretation the entire study section was subdivided into 12 smaller segments called Reaches. The delineation of each Reach was not completely arbitrary. Rather, it was based on a growing familiarity with the river and assessments that water quality was quite uniform throughout each Reach. Ultimately, each Reach was characterized annually by an average of electrofishing samples from 5 collecting stations, totalling 15 per Reach for most years. Figure 87 reduces the 1988 data to mean Iwb values for each Reach, thereby characterizing the diversity and abundance of the fish communities Reach by Reach throughout the middle Wabash River.

All individual components of the Iwb exhibit the same basic pattern of change over the 12 Reaches (Figure 88), as do species numbers. The average number of species taken per sample declined more steeply, but was of the same numerical magnitude and

exhibited the same spatial pattern as the Iwb. Relative numbers (No/km) and biomass (Kg/km) were somewhat more variable, but they also displayed the same fundamental pattern. The values for the two Shannon-Weiner diversity indices, while much lower and less variable, also followed suit. Diversity values based on biomass were always lower than diversity values based on numbers, even for most individual collections. Some other community attributes also appear to vary in similar fashion, the mean number of species per collection and the total number of species collected per Reach, for example.

All of the community values were somewhat depressed at the Cayuga electric generating station (Reach 6) and even more strongly depressed at the Wabash River electric generating station. The extent of this depression is somewhat exaggerated at both Reaches because all EGS collecting stations are included, not just those which were comparable in habitat to other collecting sites.

The fish community in Reach 1 was the most stable in terms of constancy of calculated community parameters. For most years the community values from the individual collecting runs were similar. However, departures from this relative stability sometimes occurred even in Reach 1. In 1983, for example, catches were unusually poor in June and early July and then improved slightly in late July (Figure 89). The cause of this depression was never discovered. Bait seiners also reported very poor catches during the early summer of 1988 (Spacie, 1989).

The community values from Reaches 6 and 7, on the other hand, sometimes fluctuated considerably from one electrofishing pass to another (Figure 8). A distinct decline in catches during early summer of 1977 preceeded a fish-kill in this section of river later on (Gammon and Reidy 1981). An even more severe depression extended through Reaches 7, 8, 9, and 12 in July 1988, indicating that much of the lower Wabash River was environmentally stressed. A similar phenomenon may have occurred in 1991, but equipment problems made it necessary to engage in piecemeal collections and the pattern generated was not clear.

Figures 90 through 95 summarize the changes in community parameters over time and space. The pattern for the composite index of well-being (Iwb) (Figure 90) is generally reflected by all of the other community indices. Considered as a whole, the fish community of the middle Wabash River improved measurably from 1973 to 1993.

The year 1983 was especially critical. Before 1983 the fish community was unremarkable and it was altogether poor in 1983. However, the widespread reproductive success that year and the survival of young through their first year of life resulted in a much improved fish community a few years later. Most species populations expanded in size after 1983 and some even changed in distribution abundance. This collective change, in turn, led to an increase in magnitude of the community parameters.

Low diversity and abundance characterized the fish community during the droughty summers of 1976-77, 1988, and 1991, as indicated by deep notches in the Iwb shown in Figure 90. Not all Reaches were affected, however. The 1976-77 and 1988 droughts did not affect either electrofishing catches or the derived community parameters

in the upper Reaches (Reaches 1-4), but clearly depressed them in all Reaches downriver from Covington (Reaches 5-12). The impact of the 1991 drought was apparently more severe and resulted in lowered catch rates and community parameters in all 12 Reaches. The summer of 1983 began with steady and moderate discharge rates which diminished to drought levels by late July and August.

The flood of 1993 was devastating to most species populations and strongly depressed the catches and derived community parameter values in subsequent years. The Wabash River was high, if not flooded, for most of that summer. Not until August did the water levels fall to acceptable levels for electrofishing. Post-flood catch rates were reduced to less than 50% of the 1992 catch rate for gizzard shad, golden redhorse, shorthead redhorse, river redhorse, blue sucker, smallmouth buffalo, longnose gar, shortnose gar, goldeye, skipjack herring, white bass, smallmouth bass, and spotted bass. Almost all other species were also reduced to a lesser degree.

It is likely that no reproduction occurred in 1993 whatever and, furthermore, many younger year classes were devastated. An enormous population of yearling gizzard shad were destroyed. Younger, immature members of other species populations were apparently also affected negatively because only the larger individuals remained for many species, including golden redhorse, flathead catfish, channel catfish, and smallmouth bass. Moderate discharge rates occurred during the summer of 1994, but spring and early summer 1996 were very wet. Perhaps the sharp upturn in 1997 catches is an indication that another cycle of population expansion is in progress. Time will tell.

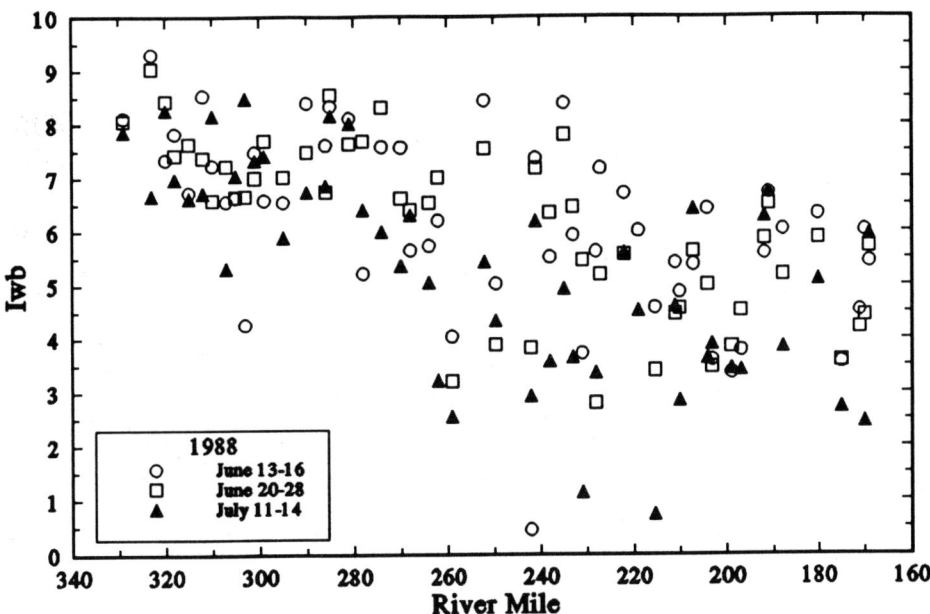

Figure 85: Iwb values for summer, 1988 based upon three electrofishing collections at each station. (All zones combined at Cayuga EGS (RM 249.6) and Wabash River EGS (RM 215.4).

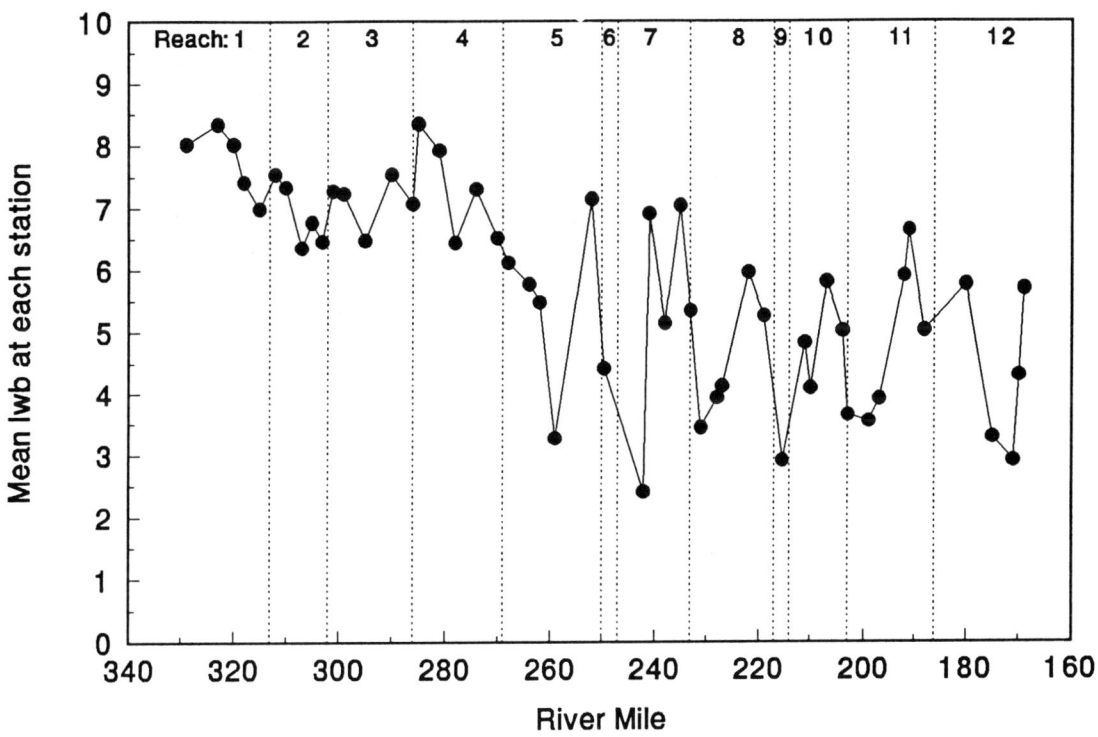

Figure 86: Mean Iwb values for 1988 at each collecting station. (All zones combined at Cayuga EGS (RM 249.6) and Wabash River EGS (RM 215.4).

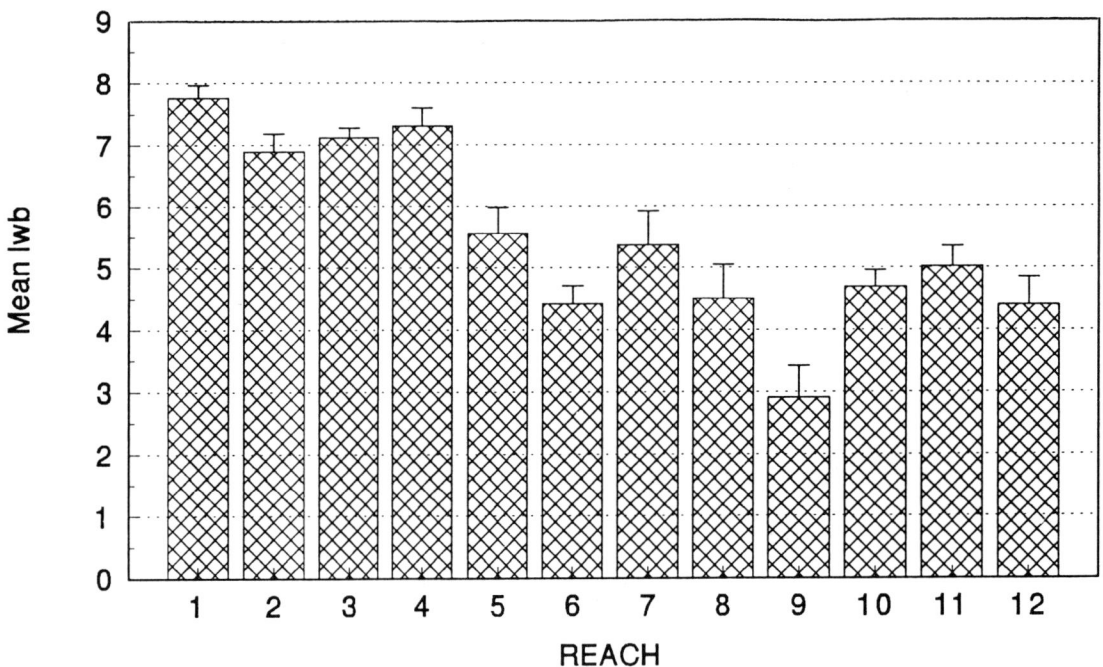

Figure 87: Mean Iwb values (+1S.E.) for each Reach in 1988.

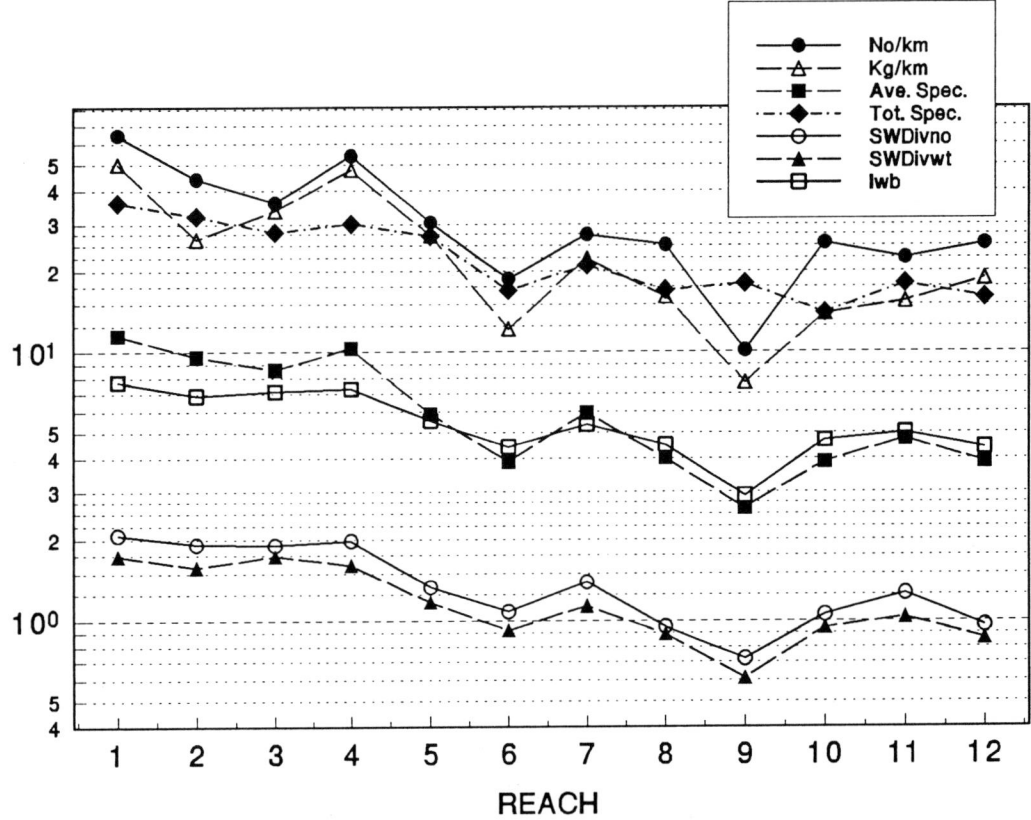

Figure 88: Spatial trends for community parameters derived from 1988 electrofishing catches.

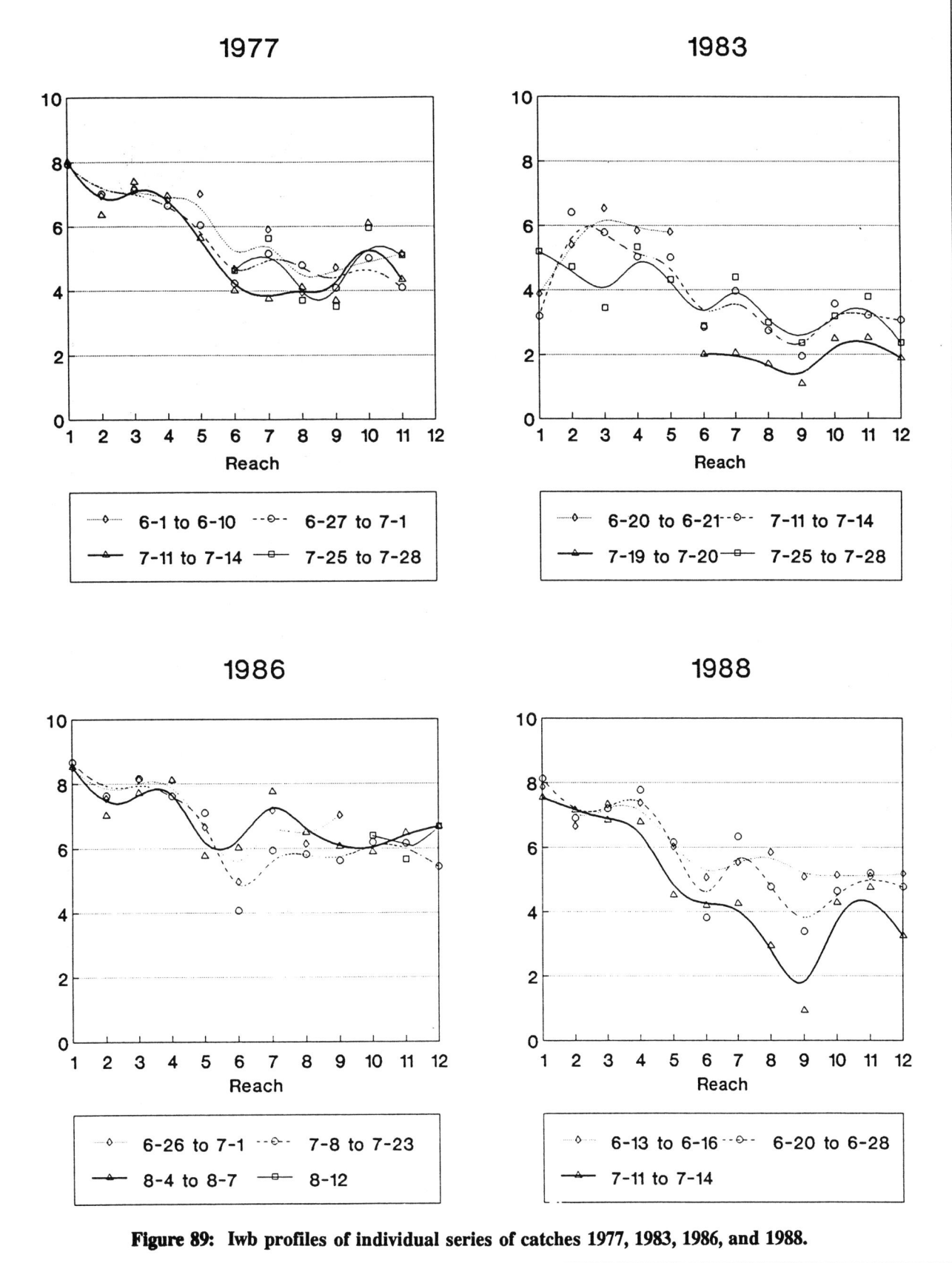

Figure 89: Iwb profiles of individual series of catches 1977, 1983, 1986, and 1988.

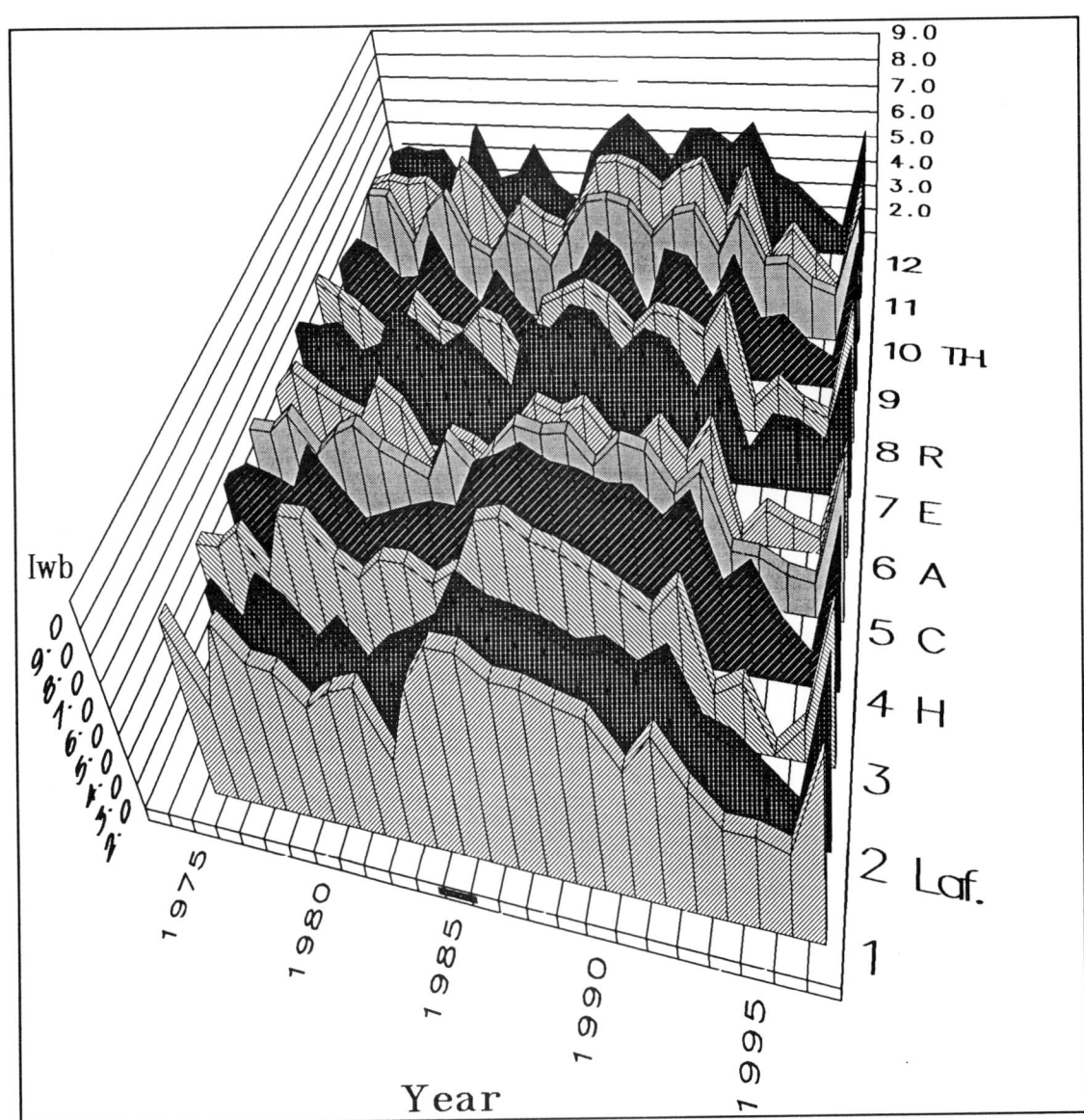

**Figure 90: Mean Composite Index of Well-being (Iwb) values
for 12 Reaches of the middle Wabash River 1973-1997.**

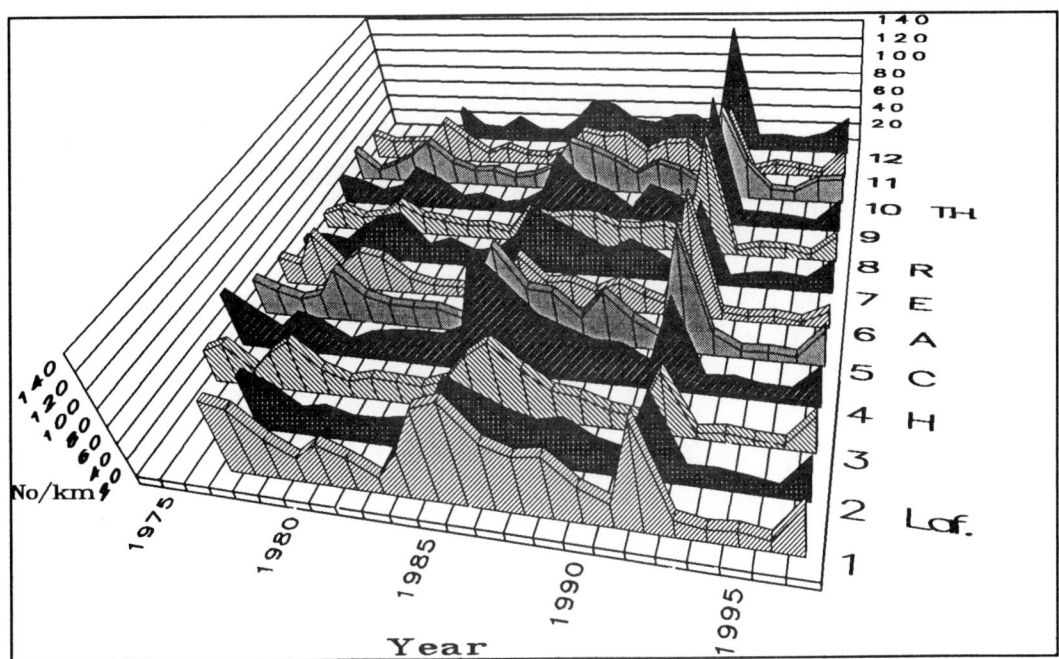

**Figure 91: Mean relative abundance values (No/km)
for 12 Reaches of the middle Wabash River 1973-1997.**

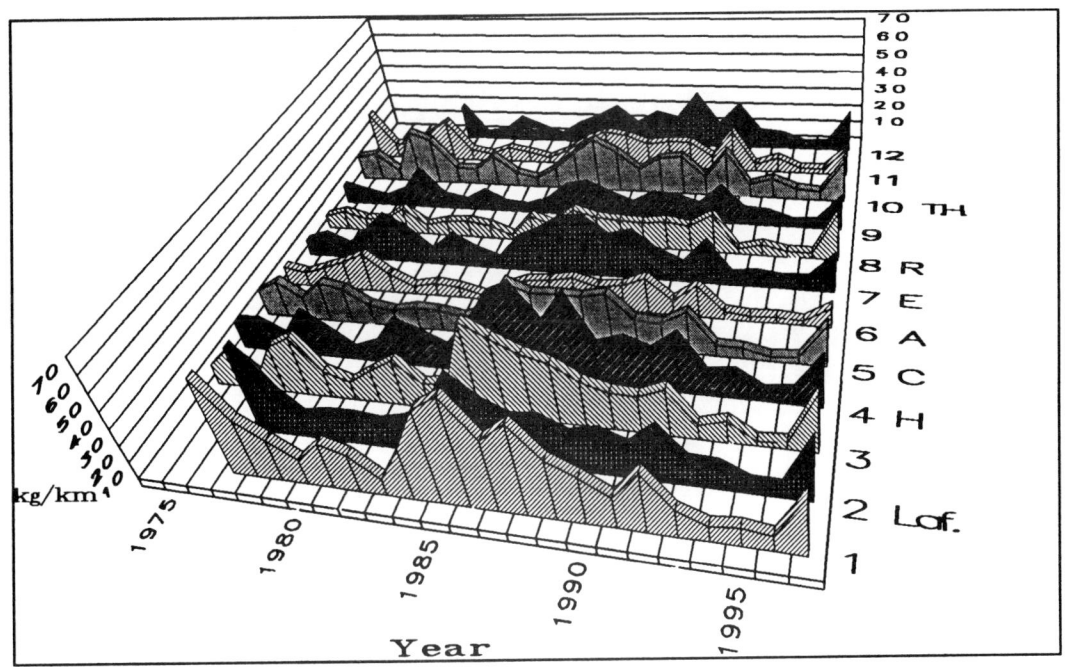

**Figure 92: Mean relative biomass values (Kg/km)
for 12 Reaches of the middle Wabash River 1973-1997.**

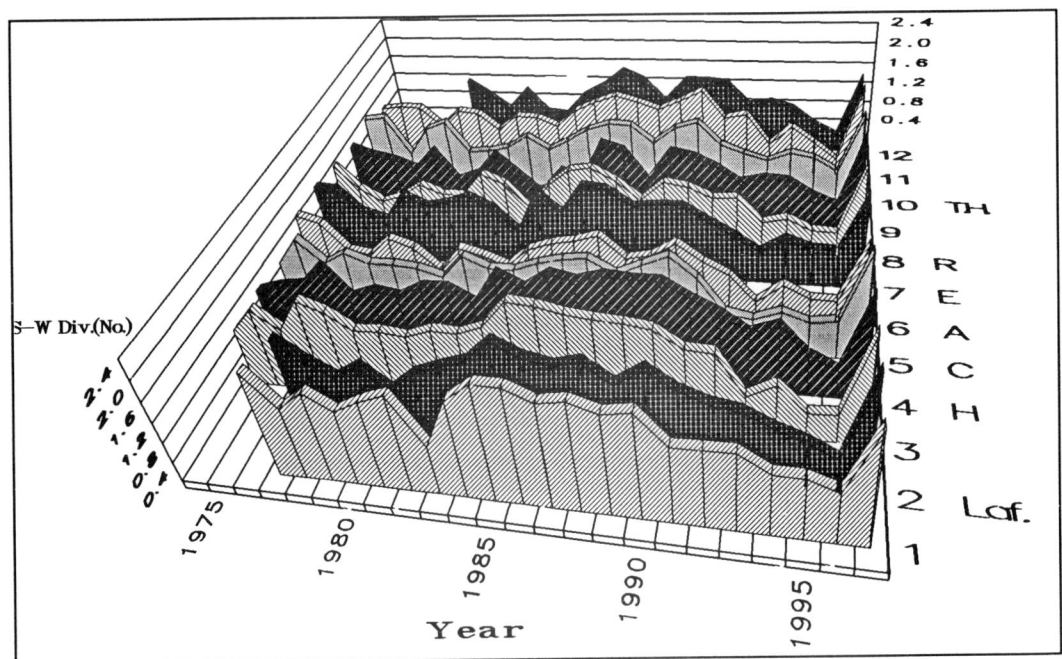

**Figure 93: Mean Shannon-Weiner Diversity values based on numbers
for 12 Reaches of the middle Wabash River 1973-1997.**

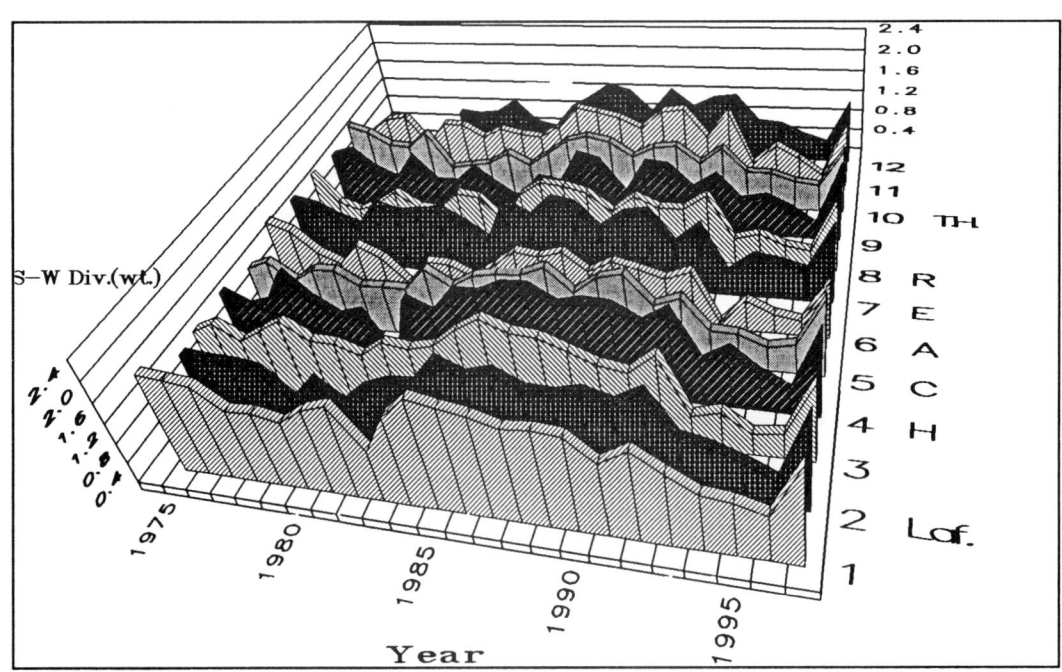

**Figure 94: Mean Shannon-Weiner Diversity values based on biomass
for 12 Reaches of the middle Wabash River 1973-1997.**

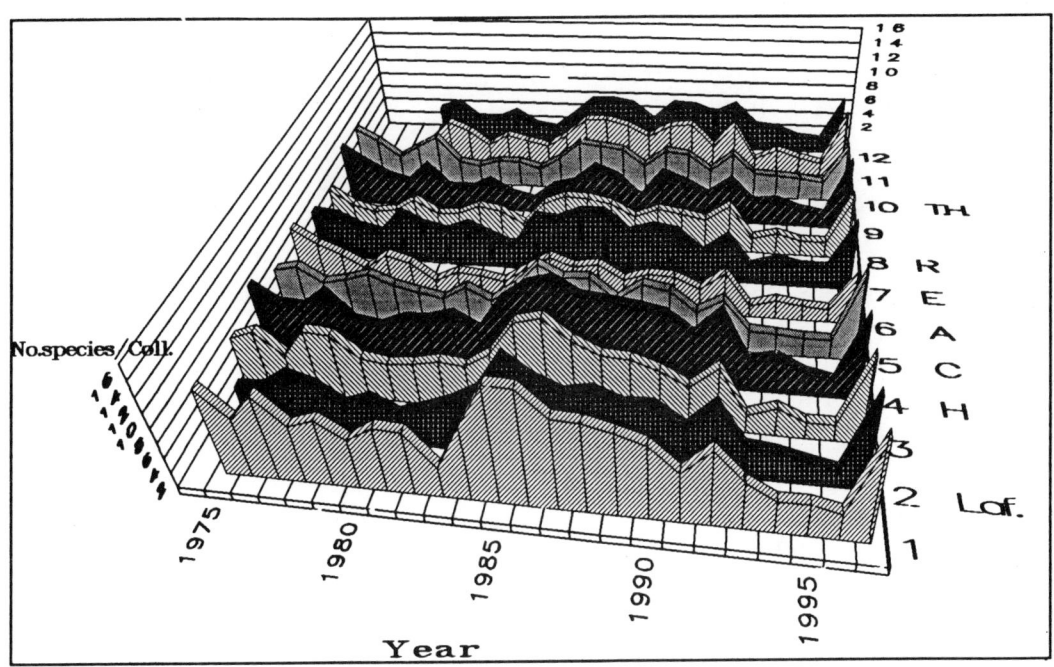

Figure 95: **Mean number of species of fish per electrofishing collection**
for 12 Reaches of the middle Wabash River 1973-1997.

Coefficients of variation (V) were calculated for all of the community indices over the years 1975 through 1990 in order to examine (1) the relative variability among the different community parameters and (2) community stability of the Reaches (Table 9).

Catch density (No/km) and catch biomass (Kg/km) were the most variable of all parameters (V range = 62 to 112). Iwb values and evenness values were the least variable community parameters (V range = 19 to 55). The variability of evenness by numbers and evenness by weights were similar. The two Shannon-Weiner diversity indices were equally variable and ranged between 27 and 65. The average number of species per collection varied considerably (V range = 42 to 55).

The community parameters were most stable in Reach 1 and least stable in Reaches 6 and 9 (Cayuga EGS and Wabash River EGS). Reaches 2 and 10 (Lafayette and Terre Haute, Indiana) were also slightly more variable than immediately flanking Reaches. There was, in general, a gradual increase in variability downriver. This analysis reinforces the assumption that Reach 1 was the best "reference" or ambient Reach and that the Iwb was the best community parameter to use for statistical analyses and evaluation.

Table 9: Coefficients of variation (V) for community parameters from 1975 through 1990.

Reach	No. per Km	Kg per Km	Ave. No. Spec	Div. No.	Div. Wt.	Even No.	Even Wt.	Iwb	Mod. Iwb
1	62.6	62.5	42.1	29.4	27.5	19.6	19.4	22.4	23.8
2	63.8	77.7	47.8	34.6	37.8	19.4	25.2	27.3	37.7
3	73.6	73.2	45.6	37.5	33.6	19.1	21.6	26.7	29.5
4	88.2	85.6	43.2	32.6	32.0	23.2	23.1	26.1	27.6
5	91.2	72.6	42.7	40.1	38.4	31.8	28.4	26.5	29.6
6	112	94.8	54.8	65.3	58.4	54.9	48.2	38.0	39.9
7	77.3	81.1	48.5	44.8	41.8	32.4	30.1	28.5	29.9
8	68.5	77.2	50.6	48.3	49.8	32.8	34.3	35.1	37.6
9	107	95.5	57.0	60.1	58.8	49.6	48.2	45.9	49.0
10	80.5	107	51.4	48.0	49.7	33.9	38.6	35.8	39.4
11	73.0	79.9	53.8	45.6	50.6	31.7	37.6	37.1	41.5
12	72.1	104	52.7	47.0	52.0	33.0	39.3	35.2	41.5

CHANGES IN ENVIRONMENTAL QUALITY
OF THE WABASH RIVER

The Wabash River has shown itself capable of supporting a very good, if not excellent, community of warm-water fishes. Before 1984 the annual electrofishing catches produced relatively few sport or game fish, ie. species of fish sought by recreational anglers either for their sporting abilities or their edibility. Prior to 1973 the entire collecting effort was concentrated at two restricted locations, the Wabash River electric generating station and the future Cayuga electric generating station which was then under construction. The electrofishing catch rates averaged only one to two game fish per kilometer at that time (Figure 96).

Beginning in 1973 the studies were broadened to include sections of river north of the Cayuga EGS as well as the section of river between Cayuga EGS and Wabash River EGS. The average catch rate then increased to three to four game fish per kilometer. During the entire period the average value of the index of well-being was usually less that 5.0 units.

It was the appearance in 1984 of significant numbers of game fishes, especially smallmouth bass, that first alerted us to the changes which were to occur. The catch rate of game fish that year exceeded 9\km including much higher numbers of flathead catfish, white bass, and smallmouth bass. This population boom was sustained for several years, but declined substantially after 1990. If freshwater drum were to be included in the game fish category the magnitude of the increase would have been far greater than is shown in Figure 96. A similar episode of population expansion appeared to be just underway by 1992, but the 1993 flood brought an end to that possibility. Another period of high water during midsummer of 1995 may have been responsible for delaying recovery of the fish populations until 1997.

The expansion of both game fish and nongame fish populations during this period was caused by enhanced reproductive success and survival throughout the Wabash River mainstem. The comparison of pre-1983 and post-1983 catch rates for some important taxa shown in Figure 97 indicate the positive increases for most species populations except for gizzard shad and carp. Successful year classes of fry led to healthy crops of yearlings the following year. Such an injection of young fish into the several populations should have had the effect of reducing the average size, but this did not occur for several species and the reduction in average size for many species was insignificant.

The enormous population increase of channel catfish did result in a 25% reduction in the mean size of fish following 1983 (Figure 98). Smaller reductions also occurred in populations of mooneye, white bass, and sauger. For many other expanding species populations, however, there was actually an increase in mean weight. The species populations of redhorse experienced increases in the mean weight as did sturgeon, gar, flathead catfish, drum, goldeye, and spotted bass.

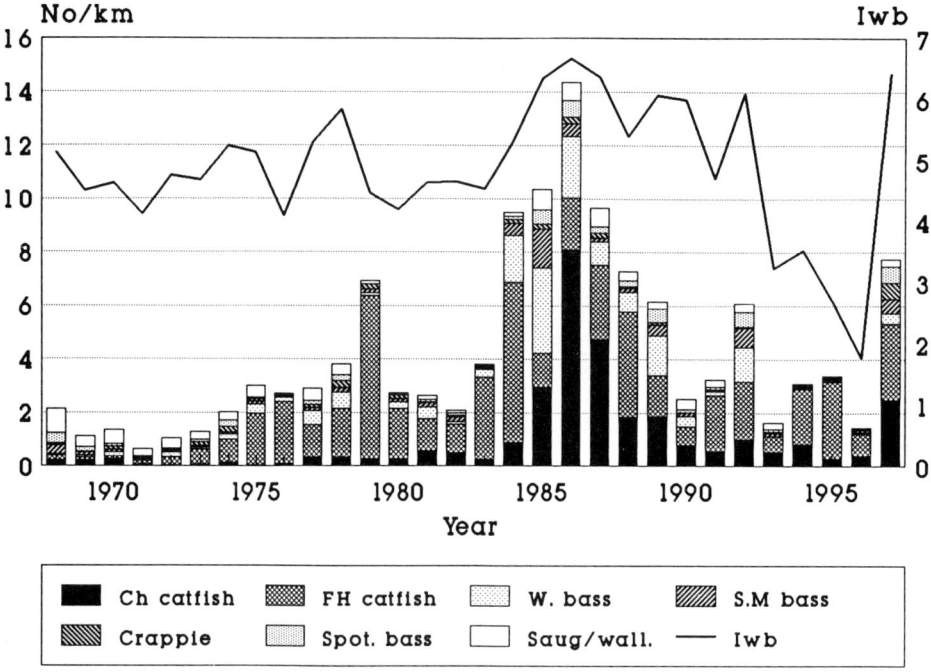

**Figure 96: Annual electrofishing catch rates (No/km) of some sport fishes
of the Wabash River 1968 - 1997.**

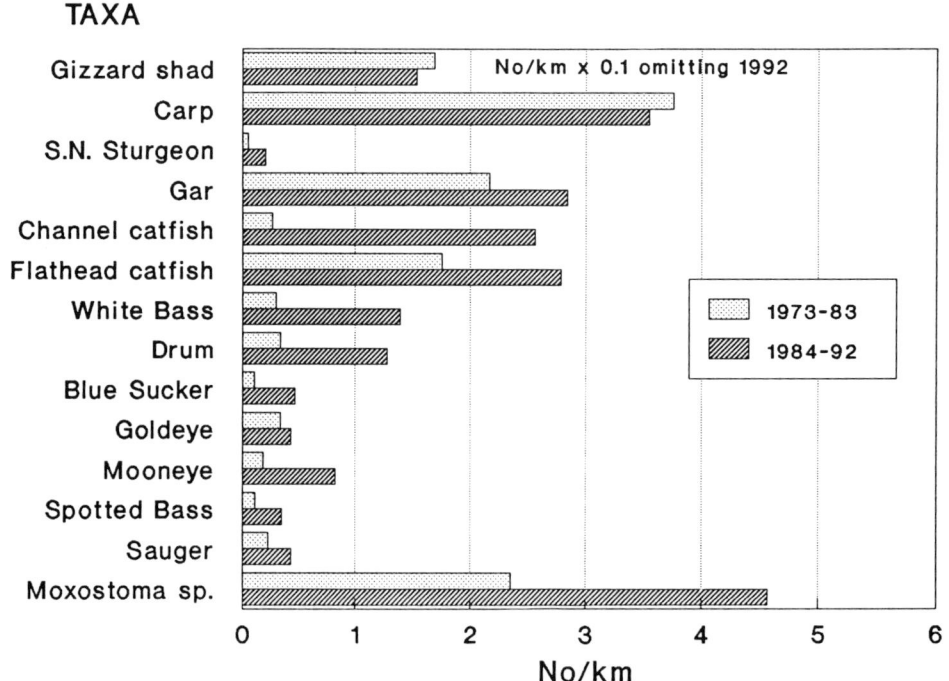

**Figure 97: Comparisons of electrofishing catch rates (No/km)
of some taxa of Wabash River fishes 1973-83 vs 1984-92.**

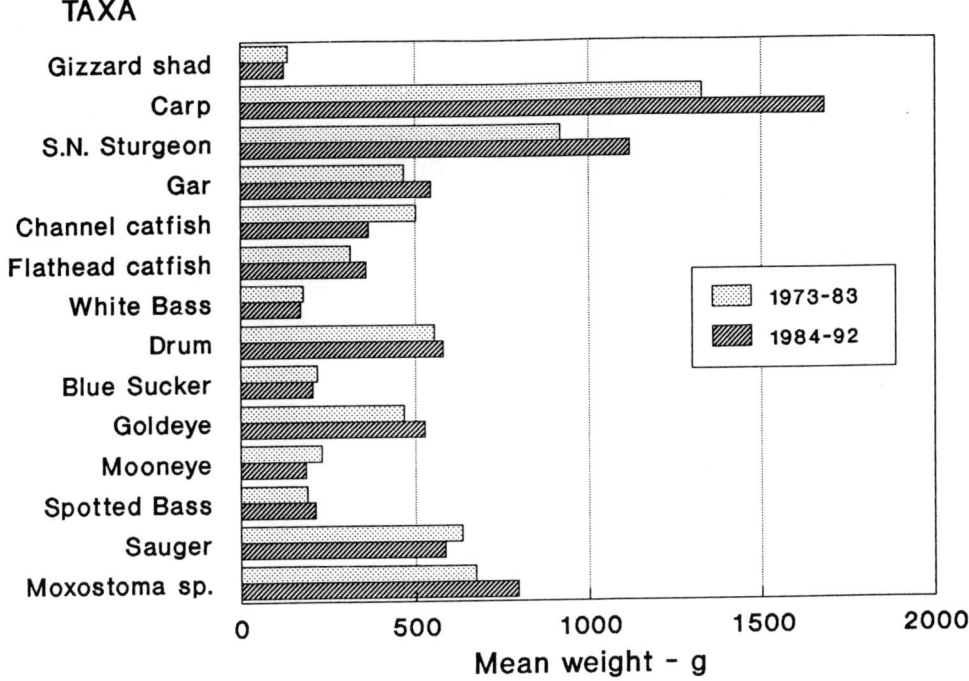

Figure 98: **Comparisons of the mean weight (g)**
of some taxa of Wabash River fishes 1973-1983 vs 1984-1992.

This unexpected phenomenon suggests that members of these species populations may be living longer and/or growing faster. If, indeed, these aspects of population biology have changed it is probably yet another indication of more favorable environmental conditions. These aspects need to be examined in considerably greater detail in the future.

All midwestern rivers have been in the past and are currently being affected by human activities to a greater or lesser degree, especially large rivers the size of the Wabash. An undisturbed river, if one existed today, would support a biotic community which could serve as a standard by which all other impacted rivers could be compared.

Since there is no unpolluted river for comparison, an alternative strategy is to locate a reference segment which has as little disturbance as possible. For the middle Wabash River the least disturbed segment is Reach 1 located between Delphi, Indiana and Lafayette, Indiana. Reach 1 usually supported the "best" fish community among all other Reaches. Even here, however, the fish community was clearly impacted negatively on at least one occasion during early summer 1983.

There is a distinct advantage in having studied the Wabash River over a long period of time during this period of environmental activity. The past 30 years embrace an era during which society focused its attention to an unprecedented degree on improving environmental quality in lakes and rivers. Furthermore, significant monetary and technological resources were devoted to the daunting task of reducing pollution and its effects. Underlying this enormous effort was the reasonable assumption that most surface waters were negatively affected by the by-products of man's diverse activities. It was further assumed that if pollutional loadings to our lakes and rivers were reduced then the biotic communities inhabiting those waters would benefit and improve in quality.

Some sections of the middle Wabash River have clearly suffered from pollutional effects for a long period of time. On the other hand, recent changes in the fish communities have been spectacularly positive. The range of quality in the fish communities of specific Reaches over the period of study makes it possible to extend the numeric Iwb evaluations to statements about community quality. Table 10 summarizes attributes of "excellent", "good", "fair", and "poor" fish communities in terms of community parameters and other qualities.

The very best fish community observed during the period of study was the community of Reach 1 during 1985-86 when the mean Iwb was 8.6. That example suggested, in the absence of a better standard, that when samples of the fish community yielded Iwb values equal to or greater than 8.5 they probably represent "excellent" communities.

The line of delineation for a "poor" community where Iwb was less than or equal to 5.5 seemed to be exemplified by the fish community found in Reach 8 during 1973-75 (mean Iwb = 4.85), although several other Reaches or locations would have been equally suitable. The intervening distance between these designations was bisected at Iwb = 7.0 to provide ranges for "fair" fish communities (Iwb value is between 5.5 and 7.0) and "good" fish communities (Iwb value is between 7.0 and 8.5).

The ranges of Iwb values indicated for each quality grouping reflect differences in most other measures of community quality such as sport fish abundance and trophic composition. Only the community index of evenness and the percent biomass as piscivores failed to vary over the range of fish communities found in the middle Wabash River from 1968 through 1997. Some other compositional attributes of the four quality groupings are summarized in Figure 99 in which pie size is proportional to abundance.

Table 10: Community parameters and quality of fish communities in the Wabash River.

PARAMETER	EXCELLENT	GOOD	FAIR	POOR
Community				
Iwb	> 8.5	7.0-8.5	5.5-7.0	< 5.5
Av. No. Species	> 15	8 - 15	5 - 8	< 5
No/km	> 100	60 - 100	25 - 60	< 25
Kg/km	> 50	25 - 50	15 - 25	< 15
Div. (no.)[*]	> 2.2	1.7 - 2.2	1.3 - 1.7	< 1.3
Div. (wt.)[**]	> 2.0	1.5 - 2.0	1.1 - 1.5	< 1.1
Even (no.)	0.75 - 0.90	0.75 - 0.90	0.75 - 0.90	0.75 - 0.90
Even (wt.)	0.7 - 0.8	0.7 - 0.8	0.7 - 0.8	0.7 - 0.8
Sport Fish[***]				
(No/km)	> 20	12 - 20	4 - 12	< 4
Trophic Composition				
% wt. Piscivores	15 - 30	15 - 30	15 - 30	15 - 30
% wt. Insectivores	> 30	15 - 30	5 - 15	< 5
% wt. Herbivores	< 10	10 - 20	10 - 20	> 20
% wt. Detritivores	> 5	2 - 5	1 - 5	< 1
% wt. Omnivores[****]	< 40	< 40	40 - 60	> 60

[*] Shannon diversity based on numbers
[**] Shannon diversity based on weight
[***] Centrarchid bass, white bass, catfish, sauger, walleye, sunfish, and crappie
[****] Carp exclusively in this study

154

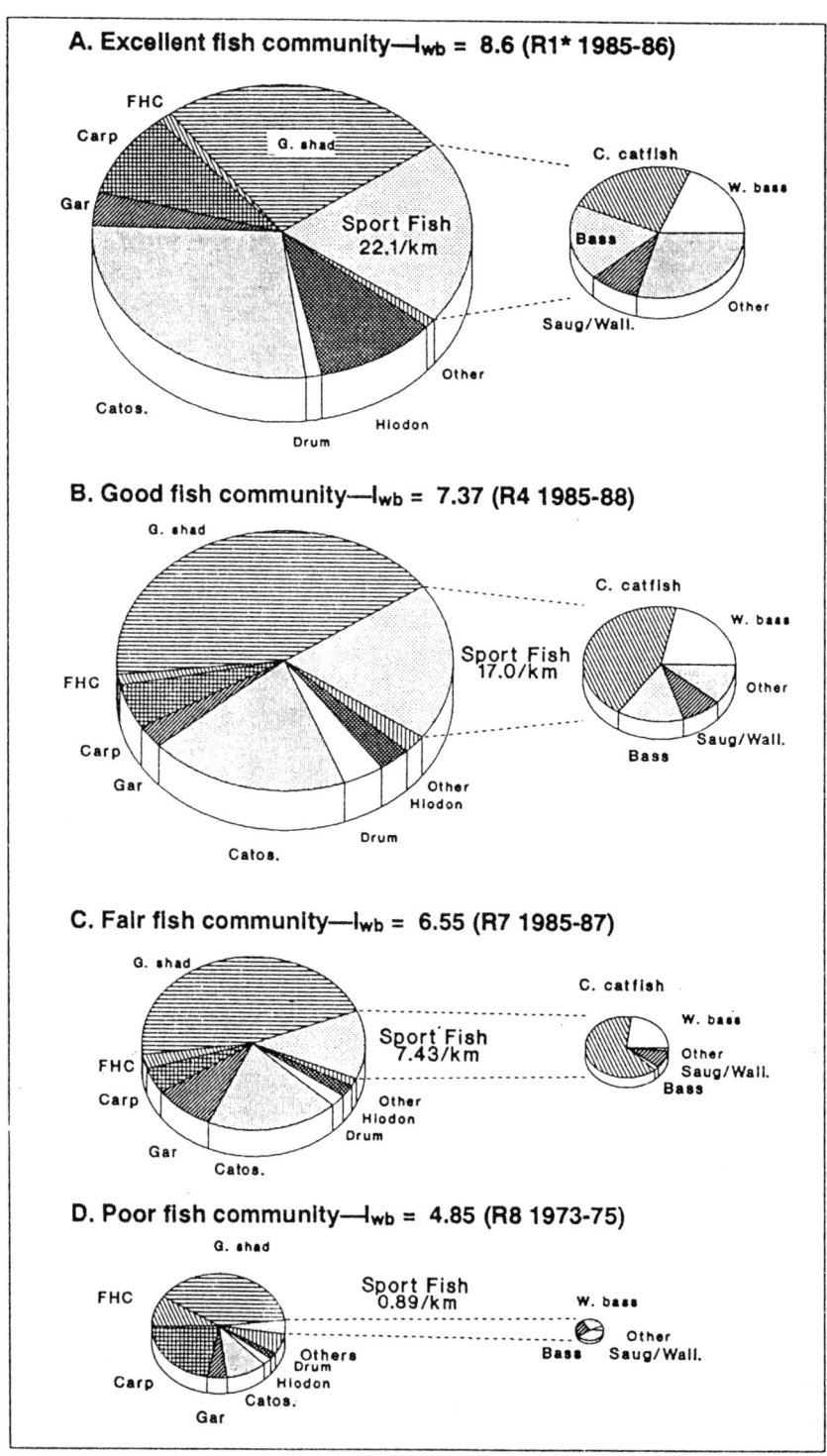

Figure 99: Examples of excellent, good, fair, and poor fish communities in the middle Wabash River.

155

The remarkable improvement of the overall fish community in the Wabash River from 1973-83 to 1984-92 lead to statistically significant increases in values of most community parameters (Table 11). Numeric and biomass catch rates increased markedly, Shannon-Weiner diversities and Iwb values increased substantially, and evenness values rose slightly. Furthermore, the extent of changes for other community parameters was even greater. The Iwb values and both Shannon-Weiner indices of diversity increased, as did the mean number of species captured per collection.

Most species of fish that reproduced and lived in the mainstem increased greatly in density throughout the river (e.g., channel catfish, flathead catfish, sauger, spotted bass, northern river carpsucker, blue sucker, and drum). Some species populations expanded their ranges into previously unoccupied areas of the river. For example, blue sucker catches tripled as fish moved into both the upper river (Reaches 1-4) and the lower section down-river from Terre Haute, Indiana (Reaches 9-12). Mooneye, sauger, smallmouth bass, and spotted bass became important components catches in Reaches 1 through 4. Goldeye and shortnose gar increased mainly in the lower Reaches.

Population increases also occurred for species entering the Wabash mainly from offriver reservoirs (white bass and walleye) and from clean tributaries (smallmouth bass and longear sunfish). During this period carp abundance was quite constant and gizzard shad populations declined notably because of increased predator pressure from a greatly expanded piscivore population.

The improvement extended into thermally influenced parts of the river at

Cayuga Electric Generating Station (Reach 6) and the Wabash River Electric Generating Station (Reach 9), changes which will be examined more closely in a later section.

The most remarkable aspect of the transformation was how quickly the entire community changed. The 1983 catches were dismal (Figure 100). Three years later the 1986 catches were outstanding and produced the highest Iwb values ever observed. The quality of the community after 1986 gradually diminished. A sharp recovery in 1992 was obliterated the following year by the disastrous 1993 flood which depressed the fish community for several years thereafter.

The 1993 flood depressed Iwb values in all Reaches proportionately and Iwb values in the lower Reaches were only marginally lower than those in the upper Reaches in 1994 (Figure 100). In short, the flood had a devastating effect on the entire river ecosystem. By comparison, the consequences of the 1983 and 1988 droughts were more severe in the lower Reaches, where water quality was lower, than in the upper reaches.

The improvements in the fish community in the 1980s probably resulted from a combination of events, including a long-term 50% reduction in BOD loading, the result of improvements in industrial and municipal waste treatment during the 1970's and early 1980's. The summer of 1983 was a low-flow summer, which has been shown statistically to facilitate good reproduction and survival through the first year of life for most mainstem species of fish. That particular year was also notable for the national Payment in Kind (PIK) program, for which farmers were paid not to grow crops. In Indiana this led potentially to a 25%

reduction in agricultural loadings to the river because 25% fewer acres of corn and soybeans were tilled and, presumably, 25% less fertilizer and herbicides applied.

Reach 8 was the only Reach to have communities reduced in quality after 1983. The 1993 flood reduced not only the community in Reach 8, but also to that of Reach 12. In retrospect, the fish communities and environmental problems of Reach 8 should have been investigated much more thoroughly. Unknown factors have negatively influenced this segment's fish community and possibly also the Reaches downriver.

Another way of gauging the degree to which fish communities of the individual Reaches have changed is to calculate a ratio of the composite index of well-being (Iwb) for each individual Reach divided by the Iwb value of Reach 1 and then compute a linear regression of this ratio over time. This regression examines the change over time of each Reach relative to the Iwb of Reach 1. One advantage of this technique is that it tends to neutralize the variable effects of large annual differences in weather, river flows, and collecting efficiency of the electrofishing crews.

Data from the years 1976, 1983, and 1993 was excluded from the regression analysis. The 1976 data was excluded because no collections were made in Reach 1 that year. Data from 1983 was exempted because of the forementioned anomalous and unexplained depression in Iwb values in Reach 1. As discussed previously, the 1993 data was biased because of high water and poor collecting conditions.

The catch data from 1994-96 yielded much lower ratios than normal, perhaps an indicating that the relatively healthy community in Reach 1 was recovering while the less robust communities in other Reaches were lagging.

The results of the regression analysis are summarized in Figure 101. Ratios were inordinately low during 1995-96, but rebounded again in 1997. The regression line was virtually flat for ratios at Reaches 2, 4, 6, and 7. Improving ratios over time characterized the lower five Reaches (Reaches 8 - 12). Reach 3 downriver from Lafayette/West Lafayette, Indiana exhibited a pronounced negative trend, while Reach 5 also trended downward somewhat.

Interpreting these overall trends with the data at hand is perplexing because the trends were all positive until the last three years, indicating fairly uniform improvement throughout the middle Wabash. Perhaps the post-1993 data is the anomalous result of this historic flood, something never again to be repeated. Perhaps recent activities in the Lafayette/West Lafayette, Indiana area which influence the river environment have increased.

Deviations from the norm also occurred during other years. The low-flow summers of 1977 and 1988 produced low ratios for most of the lower Reaches except for Reach 11. On the other hand, the year 1978 marked a time when the Iwb ratios for most Reaches was considerably higher than the trend line. The causes or causes for these departures are unknown.

157

Table 11: Statistical changes in the values of fish community
parameters (mean & S.E.) from 1973-83 to 1984-92.

COMMUNITY INDEX	1973-83	1984-92	T	% Change
No/km	29.93 (0.77)	51.09 (1.42)	1.83	+70.7
Kg/km	13.43 (0.34)	22.18 (0.46)	3.81**	+65.2
No. Species/collection	4.93 (0.07)	6.88 (0.09)	3.11**	+39.6
S-W Diversity(no)	1.097 (0.014)	1.399 (0.015)	3.33**	+27.5
S-W Diversity(wt)	1.002 (0.013)	1.314 (0.013)	3.90**	+31.1
Evenness(no)	0.71 (0.007)	0.76 (0.057)	1.41	+7.0
Evenness(wt)	0.65 (0.006)	0.71 (0.004)	3.79**	+9.2
Index of Well-being	4.64 (0.043)	5.86 (0.041)	4.01***	+26.1

* P<0.05   ** P<0.01   *** P<0.001

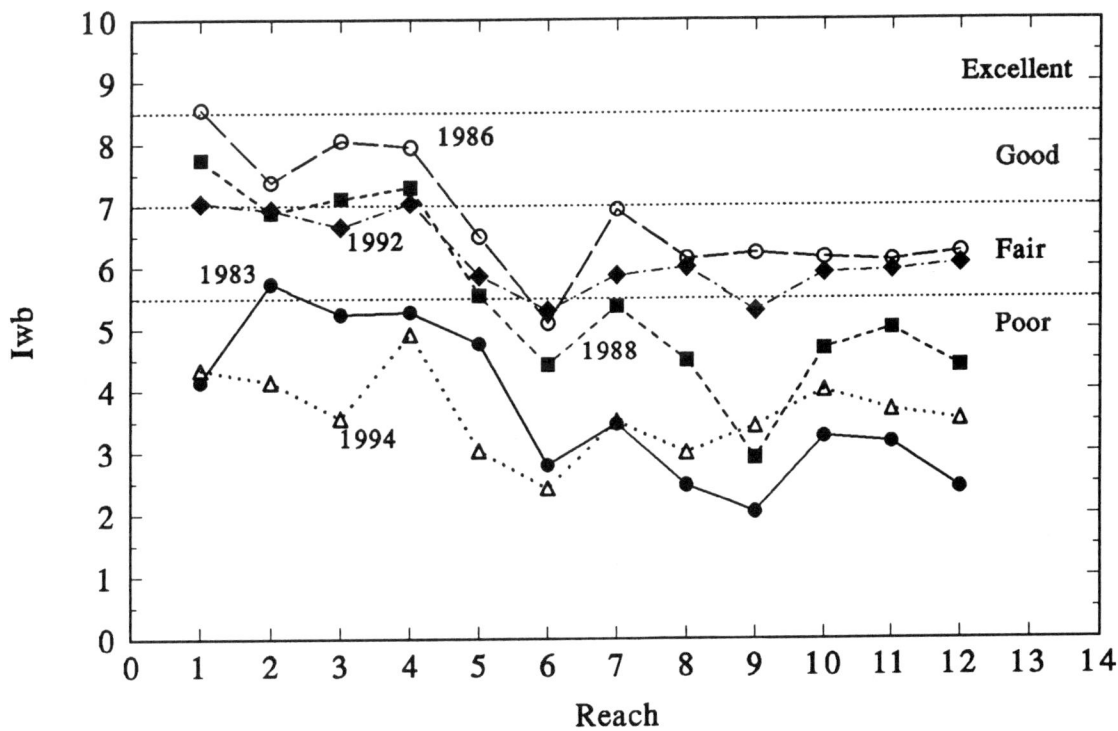

Figure 100: Longitudinal profiles of mean Iwb values in
1983, 1986, 1988, 1992, and 1994.

158

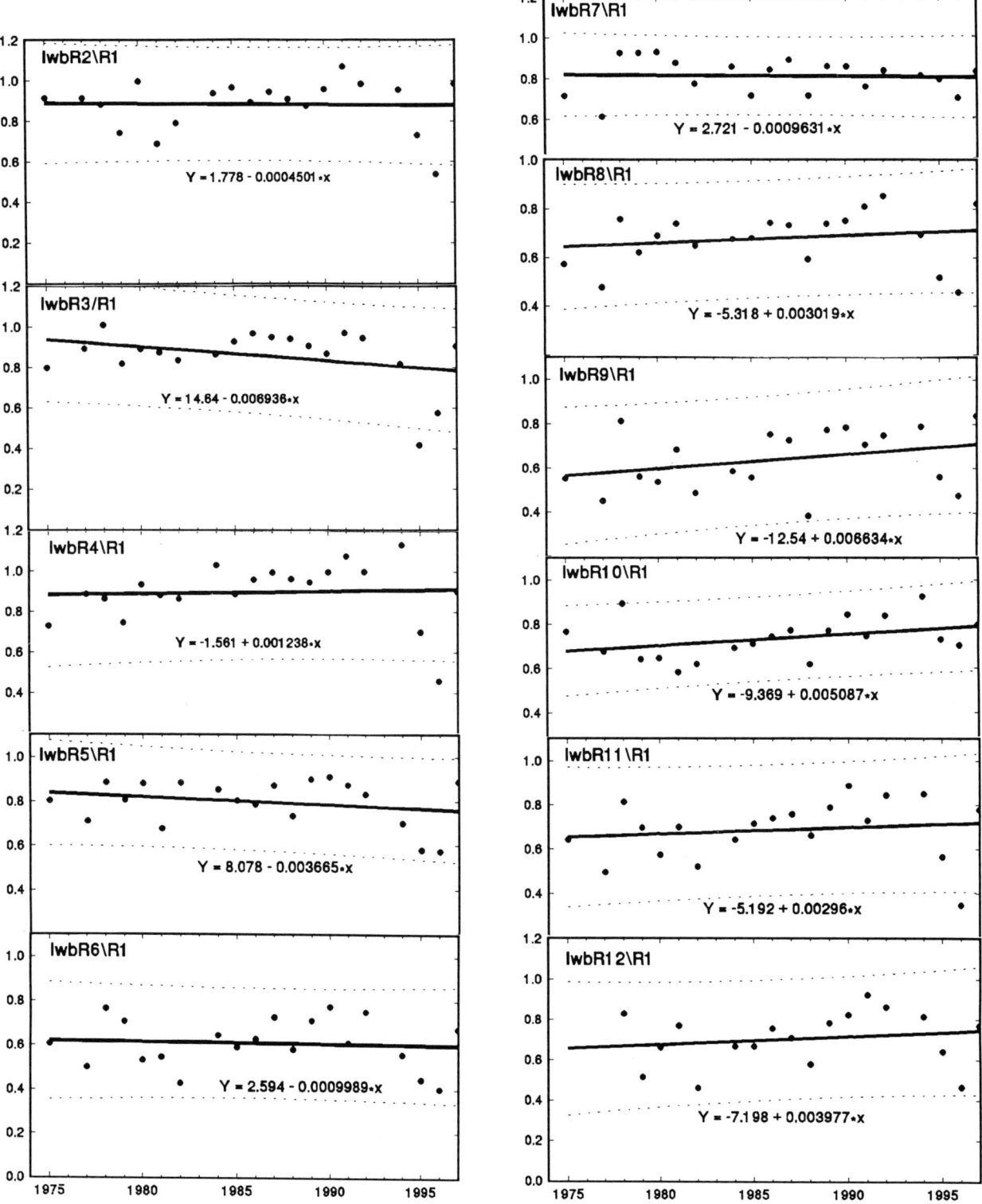

**Figure 101: Changes in the ratio IwbR$_x$\IwbR1 over time for Reaches 2 through 12.
The regression equation and line and the 95% confidence interval are indicated.**

DEVELOPMENT OF AN IBI FOR THE WABASH RIVER

Simon and Lyons (1995) defined great rivers as hydrological units with watersheds which are larger than 5700 km^2 (2200 mi^2) in area and possess a "large river faunal group" (Pflieger, 1971). According to that physical criterion the Wabash River undergoes a conversion from a "small" river to a "great" river below the mouth of the Mississinewa River near Peru, Indiana. In terms of the biological component of the criterion, however, the acquisition of a great river ichthyofauna may not yet have been achieved. Nevertheless, most great river fishes are found in the Wabash River at the level of Delphi, Indiana, where most of our studies begin.

Karr and his coworkers proposed, developed, refined, and extended methods for quantifying an Index of Biotic Integrity or IBI for small stream ecosystems (Karr 1987; Karr, Fausch, Angermieier, Yant, and Schlosser 1986; Karr, Yant, Fausch, and Schlosser 1987). Originally developed for small streams with a focus on the fish community, this approach has been expanded to include larger, more diverse aquatic systems (Miller and others 1988) and has been broadened to include macroinvertebrate and microbenthic communities as well.

The application of the IBI procedure to very large river ecosystems would seem to be a natural outgrowth of its use on smaller tributary streams. The primary reason for failing to develop such an index probably lies in the difficulty of comprehensively sampling all major habitats of large rivers. There is no single sampling device, short of extreme measures, which adequately samples fishes which live in such diverse habitats as shallow sand bars on the one hand, and deep, fast chutes on the other. Most studies of large rivers use methods which adequately sample some, but not all, components of the fish community. That being the case, the question arises as to whether fractional methods of sampling might, nevertheless, be adequate.

The extensive computerized data base available for the middle Wabash River system provides the basic raw material for the following exploration. As mentioned before, this long-term study began in 1967 at restricted sites in the vicinity of the Wabash River Electric Generating Station (EGS) located north of Terre Haute, Indiana. In 1968 a preoperational study was initiated at Cayuga EGS located about 90 km (35 mi) upriver. Collections of fish at multiple stations throughout the middle Wabash River began in 1973 and continue today from Delphi, Indiana to Merom, Indiana.

A variety of metrics suitable for large river fish communities were discussed with Dr. Tom Simon. It was decided to explore two forms of each metric, one based on numbers of fish and the other biomass.

We first examined the variability of each metric value as affected by sampling intensity. This feature was initially conducted on complete catches of all species, including gizzard shad (*Dorosoma cepedianum*). It was necessary to remove gizzard shad from the computations, however, because of the variability of its population size. Gizzard shad comprised approximately half of the total catch during many years, but in other years they were either scarce or overabundant.

Two data sets were used to establish the relationship between metric values and collecting effort: a) catches from the upper stations in Reach 1 which were most stable over the period of study and acted as an ambient collecting series, and b) 1970 catches from zone 3 of Reach 6 near the future discharge of the Cayuga Electric Generating Station. A departure from stability in Reach 1 occurred early in 1983 because of unknown factors, although the fish community progressively recovered as the summer wore on. The latter zone embodied the best example of a high quality riverine habitat, the type sought when the long distance sites were first established in 1973.

The catch data from some years were then used to establish scores for each metric. The years selected were 1977, 1983, 1985, 1986, 1987, together with the combined collections of 1993-94. The years 1977 and 1983 represented a long period when the Wabash River supported relatively mediocre fish populations, while the very best fish communities were found during 1984-87. The period 1993-94 was notable mainly for the populations of mostly adult fish decimated by the 1993 flood.

An IBI value was then determined for each sampling location for each period using the metric scores delineated in Table 12. IBI profiles together with the Iwb profile were then charted for each period.

We then examined the suitability of these IBI metrics for data from more comprehensive sampling programs such as a

combination of seining and electrofishing catches. Both of these sampling methodologies were used on the Wabash River in some years.

We also examined a set of data taken in September, 1993 using a different collecting methodology in order to examine the influence on the IBI when young-of-the-year fish were included in the computations.

The Effect of Collecting Effort on IBI Metrics

Two data sets were examined to determine individual metric responses to differences in collecting intensity: a) annual catches from Reach 1 over the period 1977 through 1994 and (b) one of the 1970 pre-operational catches in zone 3 of Reach 6 downstream from the future discharge of the Cayuga EGS into the Wabash River.

The first set of composite figures chart changes in each metric over a range of collecting efforts expressed as kilometers electrofished (Figure 102 A-M). The less intensive collecting efforts (0.5, 1.0, and 1.5 km) were obtained from the uppermost collecting station in Reach 1 (RM 329.0) by plotting metric values derived from the first collection series only (0.5km), collection series 1 and 2 combined (1.0km), and series 1, 2, and 3 combined (1.5km). Usually three series of collections were made each year, but occasionally only one (1979) or two series (1980) were possible and sometimes four series (1981 and 1982) were obtained.

The data points for more than 2.5 km effort represent pooled collections from sites in Reach 1. The metric values at the 2.5 km effort level consist of catches at all collecting sites combined (usually 5 sites) from the first series of annual collections. The first and second series combined yielded data points for the 5.0 km effort and all three series produced the points for 7.5 km effort. The composite thus obtained is not exactly equivalent throughout, but it does provide a fair representation of trends in metric values with increasing sampling effort.

The year 1983 was noteworthy for yielding the poorest catches of fish among all previous years, while the 1985 catches were spectacularly good because of the unusually high reproductive success in the previous two years. Thus, these two years bracket the high and low points within which the other profiles fall for most metrics.

Most metrics involving absolute numbers of species (total number of species, etc.) continually expanded with increasing effort to some asymptote well beyond the effort employed. On the other hand, most of the metrics based upon percentages of grouping stabilized quickly within 1.5 to 2.5km effort, as did the density (no/km).

The 1970 data was originally gathered in an effort to determine preoperational population size at the Cayuga EGS (Figure 103 A-J). Fifteen electrofishing collections over a distance of 1.4 km of shoreline were taken between June 30 and August 6, 1970. Metrics based on numbers of species were, again, subject to greater cumulative changes than metrics based on percentages, but even the latter varied considerably as cumulative effort increased. The different metric levels in early collections compared to later cumulative values may be the result of actual changes in community structure through time. The 1970 attempt failed because the degree of fish movement was too high and the population size too large.

Development of the Scoring Criteria

Methods for quantitatively ranking individual metrics range widely from quite subjective to totally objective. Theoretically, great river fish communities should not vary because of differences in drainage basin size. However, there is little information to either support or refute this assertion since all great rivers in the United States have been extensively modified by human activities. Physical modifications such as dams, channelization, dredging, and barge traffic all influence the character of fish communities in great rivers. Chemical modifications of great river tributaries and mainstems are also pervasive. The combined influence of industry, agriculture, and population centers on the biota of great rivers still remains to be ascertained.

Lacking the detailed, comparative information about uninfluenced great rivers, it is necessary to use the data at hand. The Wabash River has escaped the influence of dams for the most part and has not been used for transportation purposes for over a century. However, it has not avoided the early effects of canal building which drained associated wetlands and agricultural influences are pervasive, especially in the upper river. Industrial modification and a limited number of population centers exert their usual effects. Surface mining for coal is practiced in the lower part of the middle Wabash and there is considerable acid mine drainage as indicated earlier.

Fortunately for the Wabash River system, there is an abundance of data available including a series of years (1985-87) when the fish community demonstrated dramatic and unexpected improvement. The fish collections from those years together with two other years when unexceptional fish communities dominated the river (1977 and 1983) and one period following the devastating 1993 flood (1993-94) were used to determine maximum lines of expectation for each metric. Data taken during these representative periods were used to establish scoring levels.

Values for each metric at each collecting station (except for stations receiving direct heated effluents from EGSs) for all six periods are charted in Figure 104 A-U. Frequency analyses were made initially for each data set for the purpose of placing the maximum line of expectation as objectively as possible at a level which included 95% of the data points. This blindly objective approach proved to be unsatisfactory, however, since it often produced peculiar trisection values. Therefore, maximum lines of expectation were drawn subjectively so as to include all but approximately 10-15 data points. This was followed by trisection of the scattergram. The scattergrams of catch rates (no/km and kg/km) were converted to logarithms prior to trisection (Figures 104 T and U), but otherwise followed the same procedure. A line of minimum expectation was established in cases of the metrics for percentages of tolerant individuals and biomass (Figures 104 H and I).

The scoring indicated in Table 12 is the result of this effort. Catches of fish from each collecting station for each period were then used to determine individual metric scores and these scores were then summed for the overall IBI value. Computer software was developed to produce numbers of species and percentages directly from the catch data-base so as to minimize errors.

IBI Profiles for the Middle Wabash River

IBI profiles for six years based on calculations using the percent number and percent biomass and the number of species of roundbodied suckers, centrarchids, and sensitive species are shown in Figure 102. Actual IBI values for collecting sites are shown by circles and triangles while the lines are LOWESS plots (locally weighted sums of squares). Again gizzard shad were elimated from consideration prior to determining the metric values. Additionally, data from unheated zones immediately upriver from the heated discharges of the Cayuga EGS (RM250) and Wabash River EGS (RM215) were included in these profiles.

Although widely scattered, the IBI values consistently declined from upriver to downriver for each of the years. The fish communities in the Delphi-Tippecanoe River section (RM320-330) were generally the best communities with IBI values ranging from 40 to 50. IBI values declined as the Wabash River entered the Lafayette/West Lafayette metropolitan area (RM308-315) and then gradually increased again to a secondary high in the Attica/Independence area (RM280-290). Three relatively high quality streams enter this segment of the Wabash River: Big Pine Creek, Big Shawnee Creek, and Bear Creek, all of which may contribute positively to water quality. This section of river also lies more than 40 km (25 miles) downriver from Lafayette/West Lafayette, a sufficient distance to have permitted the river to recover somewhat from domestic and industrial influences. IBI values then enter another gradual decline over the next 50 km (30 mi) from Attica to the Cayuga EGS where the local thermal effect is apparent during only four of the six periods.

Table 12: Metrics used to compute Index of Biotic Integrity scores for Wabash River sites.

Metric Category / Metric	Scoring Classification 5	3	1
Species richness and composition			
1. Total number of species	15+	7 -14	< 7
2. a. Number of round-bodied sucker species	4+	2 - 3	0 - 1
b. RBS as a % of total species	>26%	13%-26%	<13%
3. a. Number of centrarchid species	4+	2 - 3	0 - 1
b. Centrarcid sp. as a % of total species	>20%	10%-20%	<10%
4. a. % Great River individuals(no.)	>60%	30%-60%	<30%
4. b. % " " " (wt.)	"	"	"
Species tolerance			
5. a. Number of sensitive species	7+	4 - 6	0 - 3
b. No. Sensi. sp. as % of total species	>36%	18%-36%	<18%
6. a. % tolerant individuals (no)	<50%	50%-75%	>75%
b. " " " (wt)	"	"	"
Trophic composition			
7. a. % macrovorous individuals(no)	>50%	25%-50%	<25%
b. % " " (wt)	>60%	30%-60%	<30%
8. a. % insectivorous individuals(no)	>40%	20%-40%	<20%
b. % " " (wt)	>44%	22%-44%	<22%
9. a. % omnivorous individuals(no)	<20%	20%-40%	>40%
b. % " " (wt)	<30%	30%-60%	>60%
Reproductive guild			
10. a. % simple lithophils(no)	>32%	16%-32%	<16%
b. % " " (wt)	>40%	20%-40%	<20%
Fish abundance			
11. a. Catch per unit effort - (no/km)	>30	12 - 30	<12
b. " " " " - (kg/km)	>25	9 - 25	< 8
Fish Condition			
12. Percentage DELT anomalies (deformities, fin erosion, lesions, tumors.)	<0.1%	0.1%-1.3%	>1.3%

164

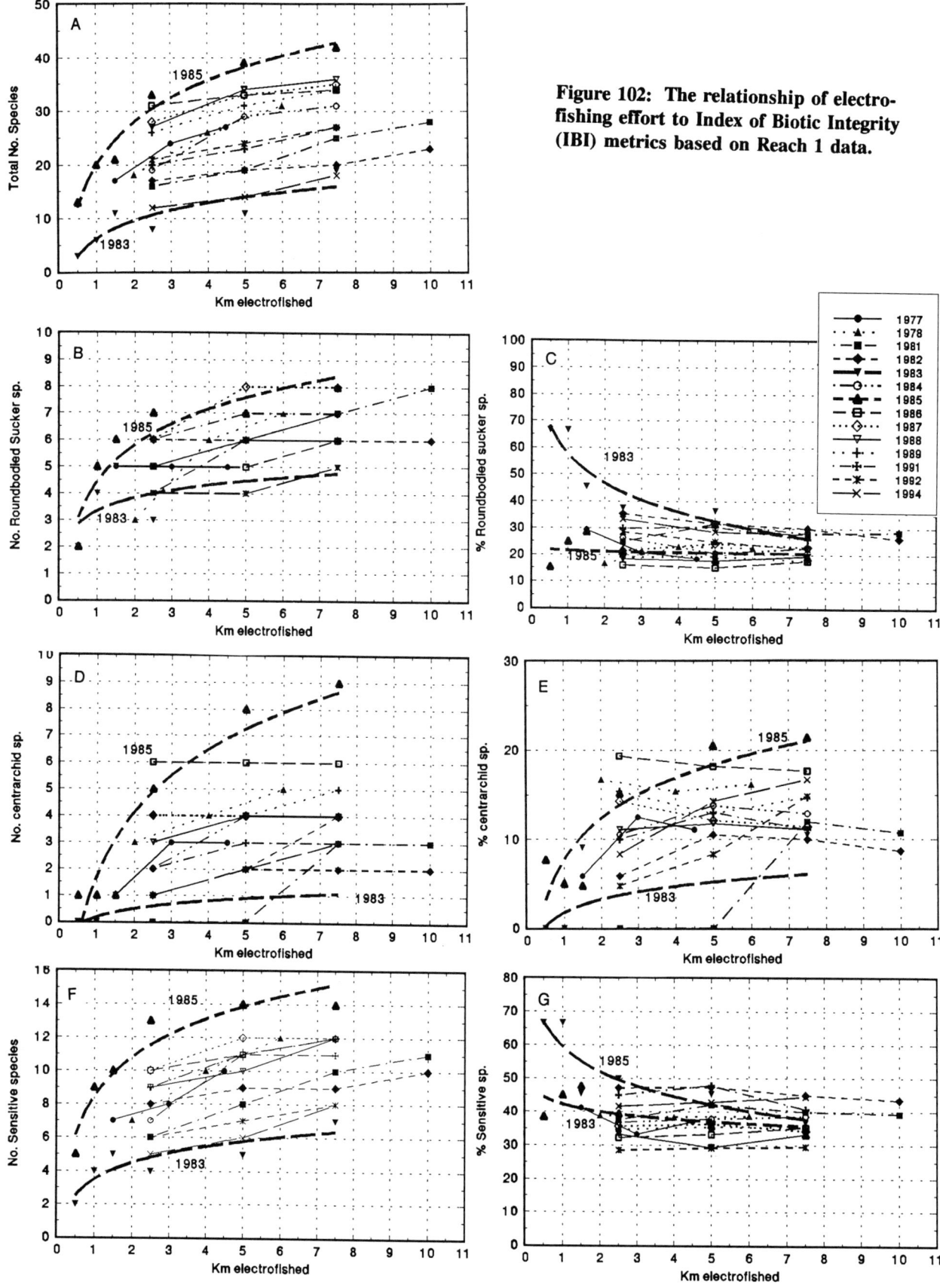

Figure 102: The relationship of electrofishing effort to Index of Biotic Integrity
(IBI) metrics based on Reach 1 data.

Figure 102: (con't.)

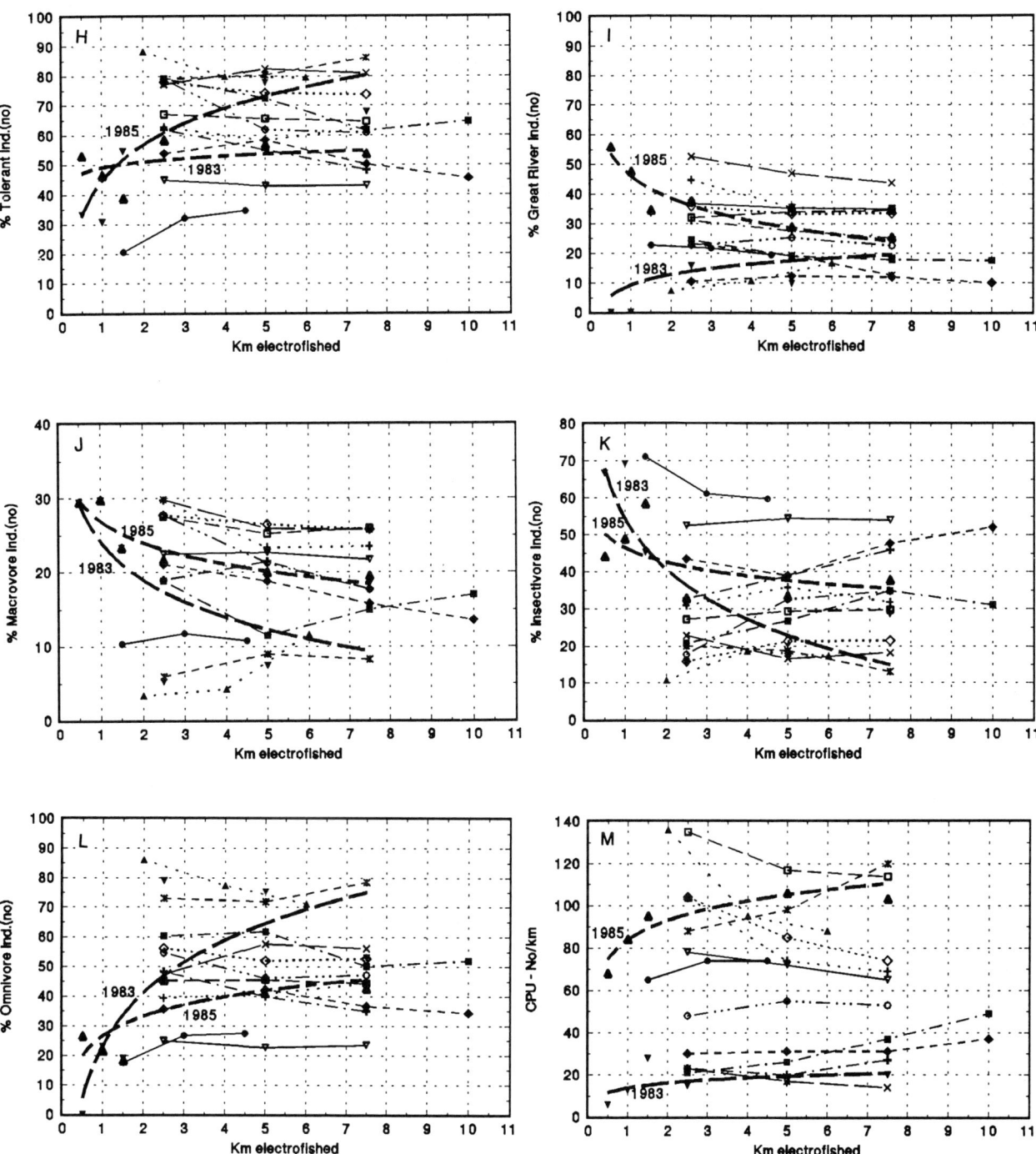

Figure 103: Cumulative changes in Index of Biotic Integrity (IBI) metrics based on 15 electrofishing collections from June 30 to August 6, 1970 at zone 3 of preoperational Cayuga EGS.

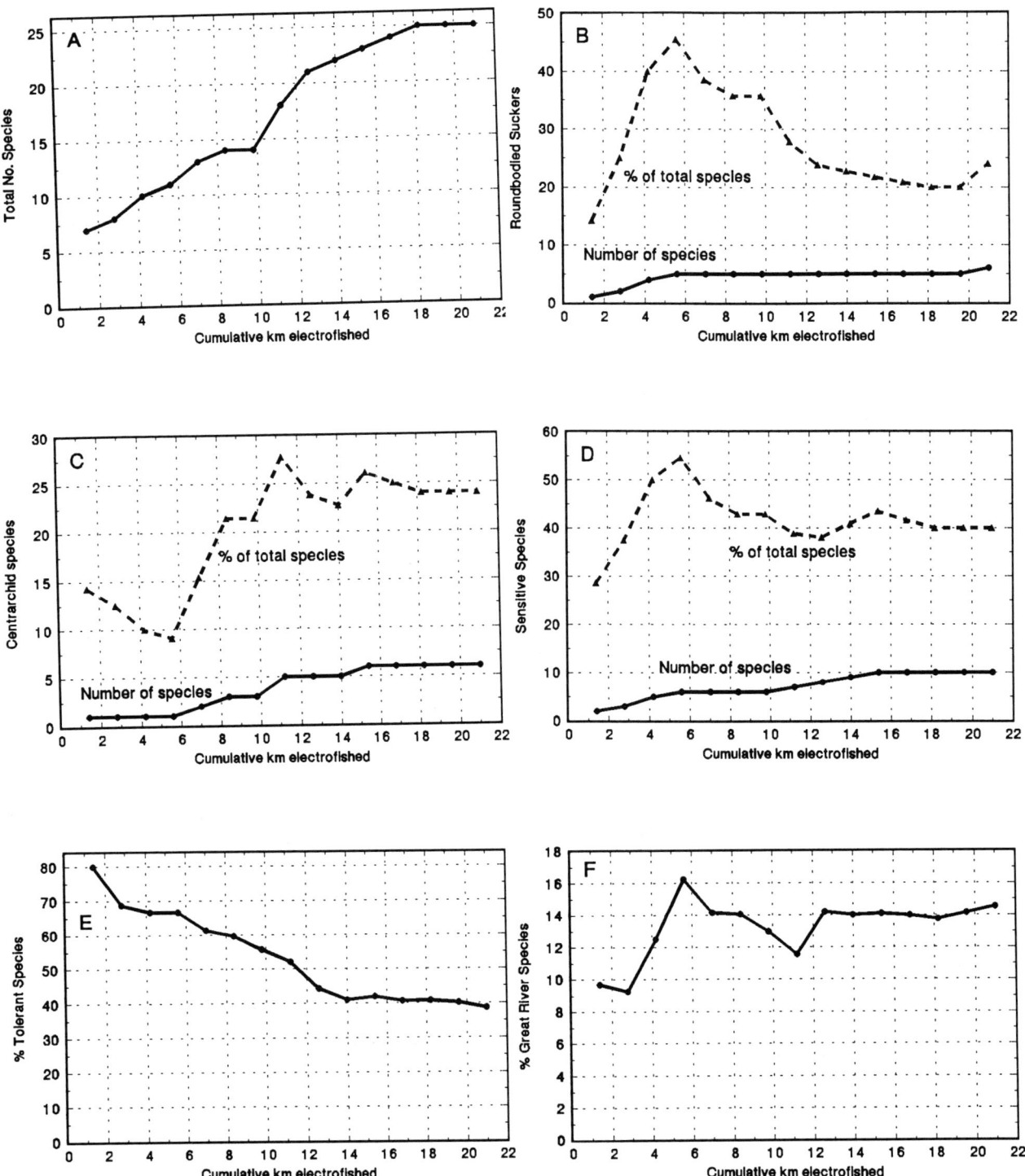

Figure 103: (con't.)

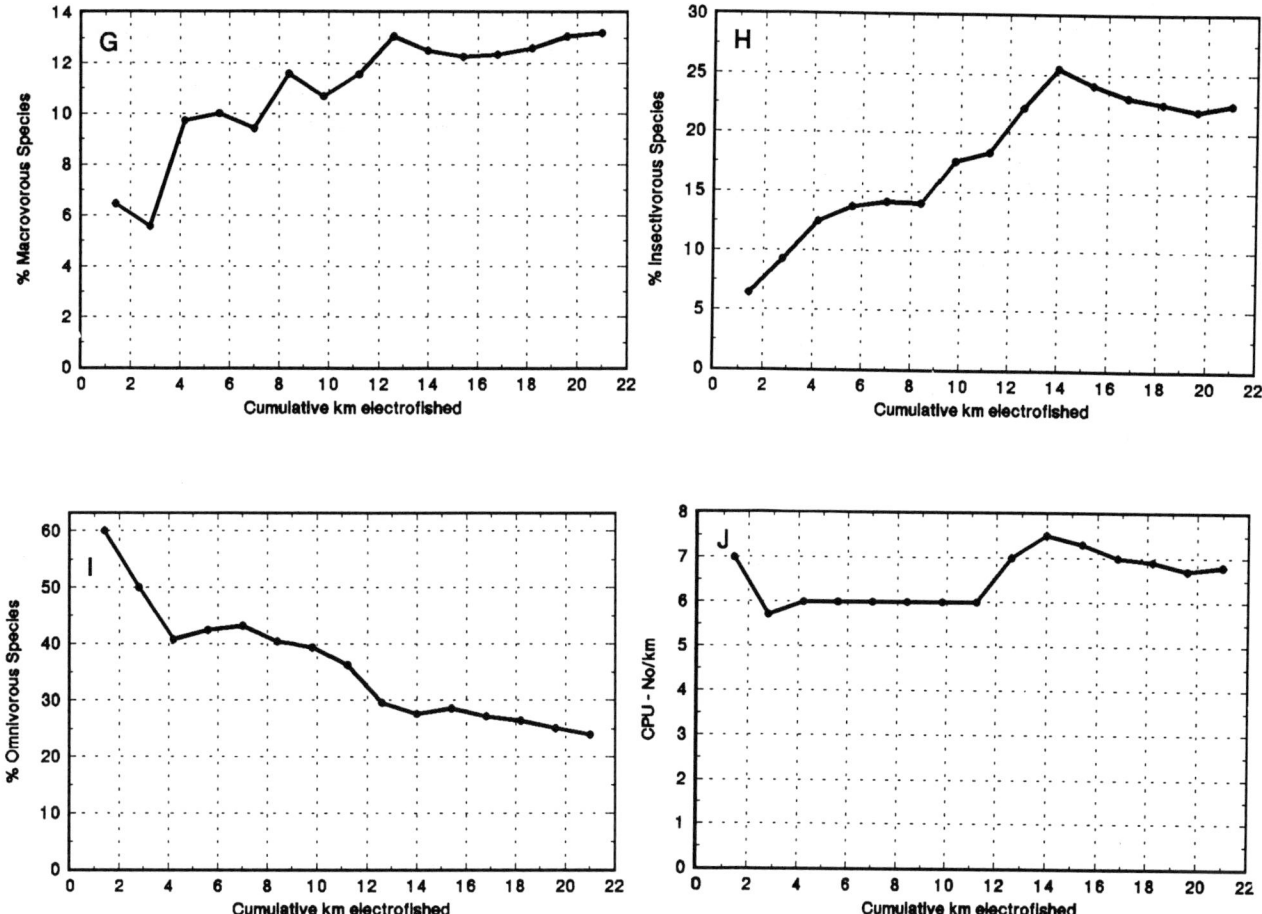

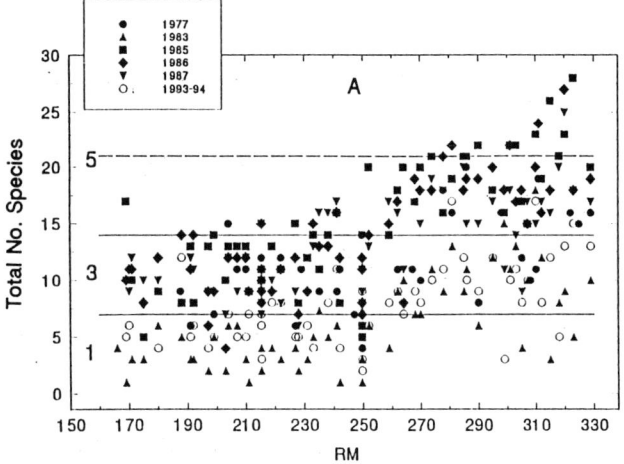

Figure 104: Scattergrams and scoring criteria for metrics of a Great River Index of Biotic Integrity (IBI) based on electrofishing catches from the Wabash River.

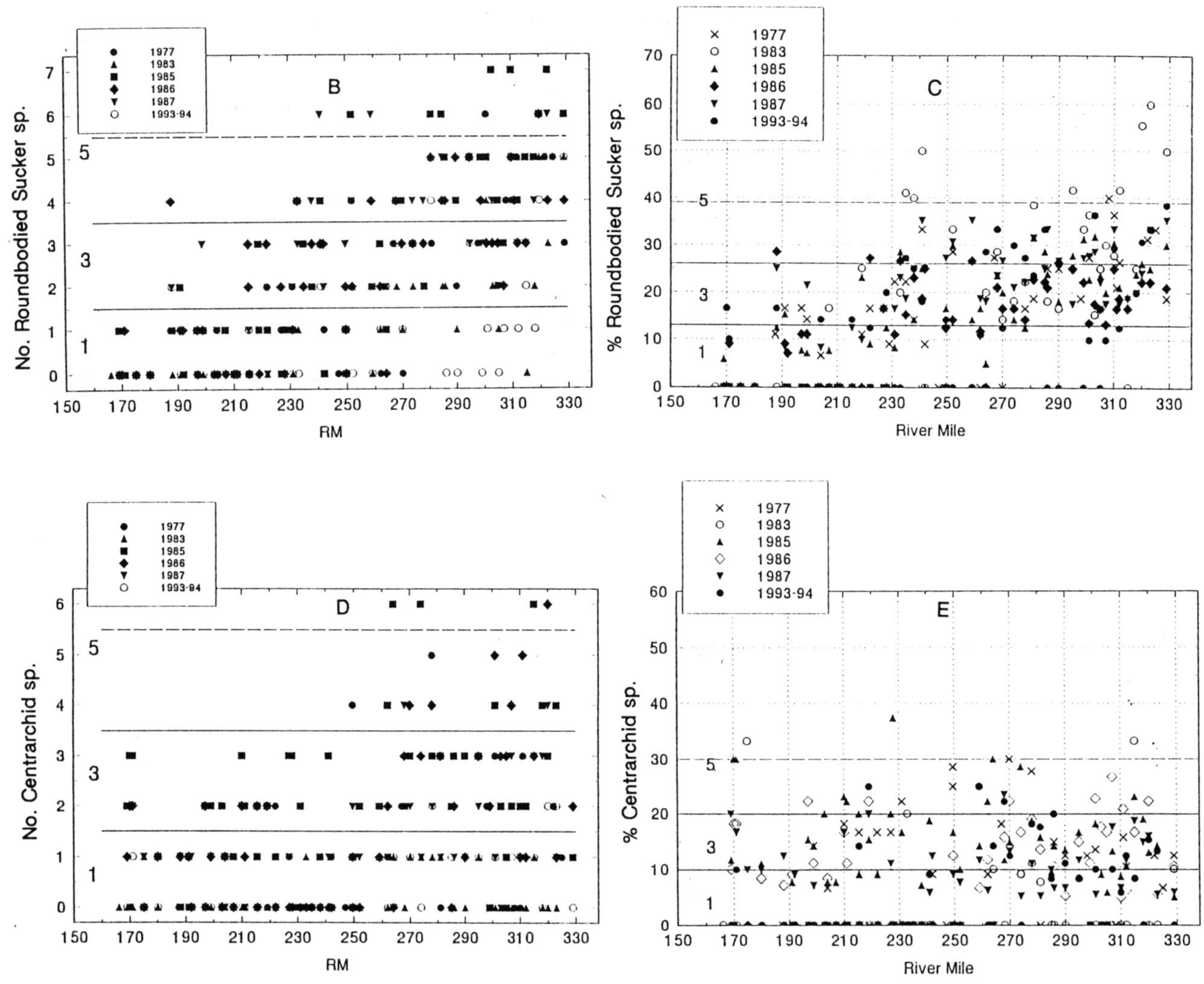

Figure 104: (con't.)

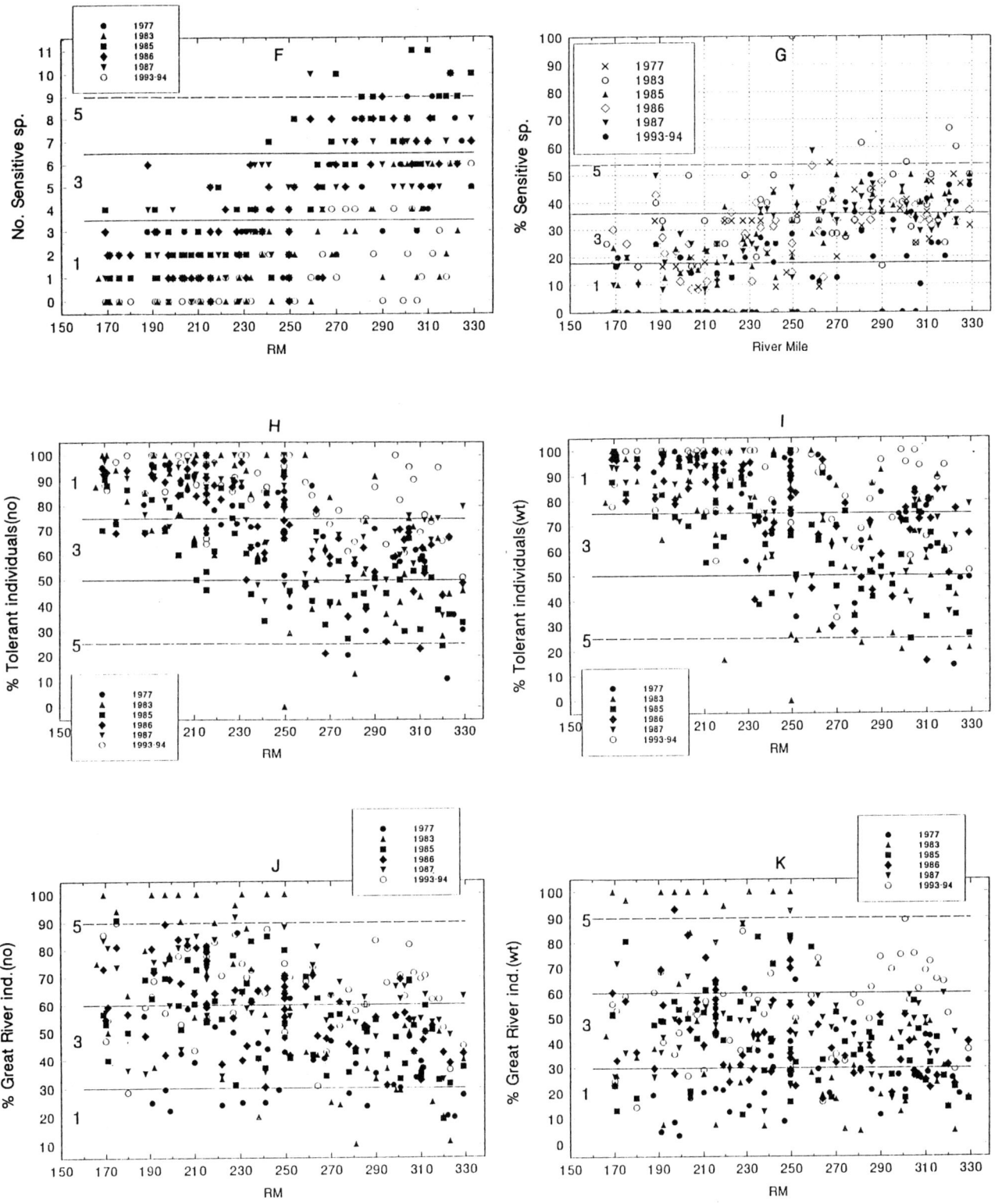

170

Figure 104: (con't.)

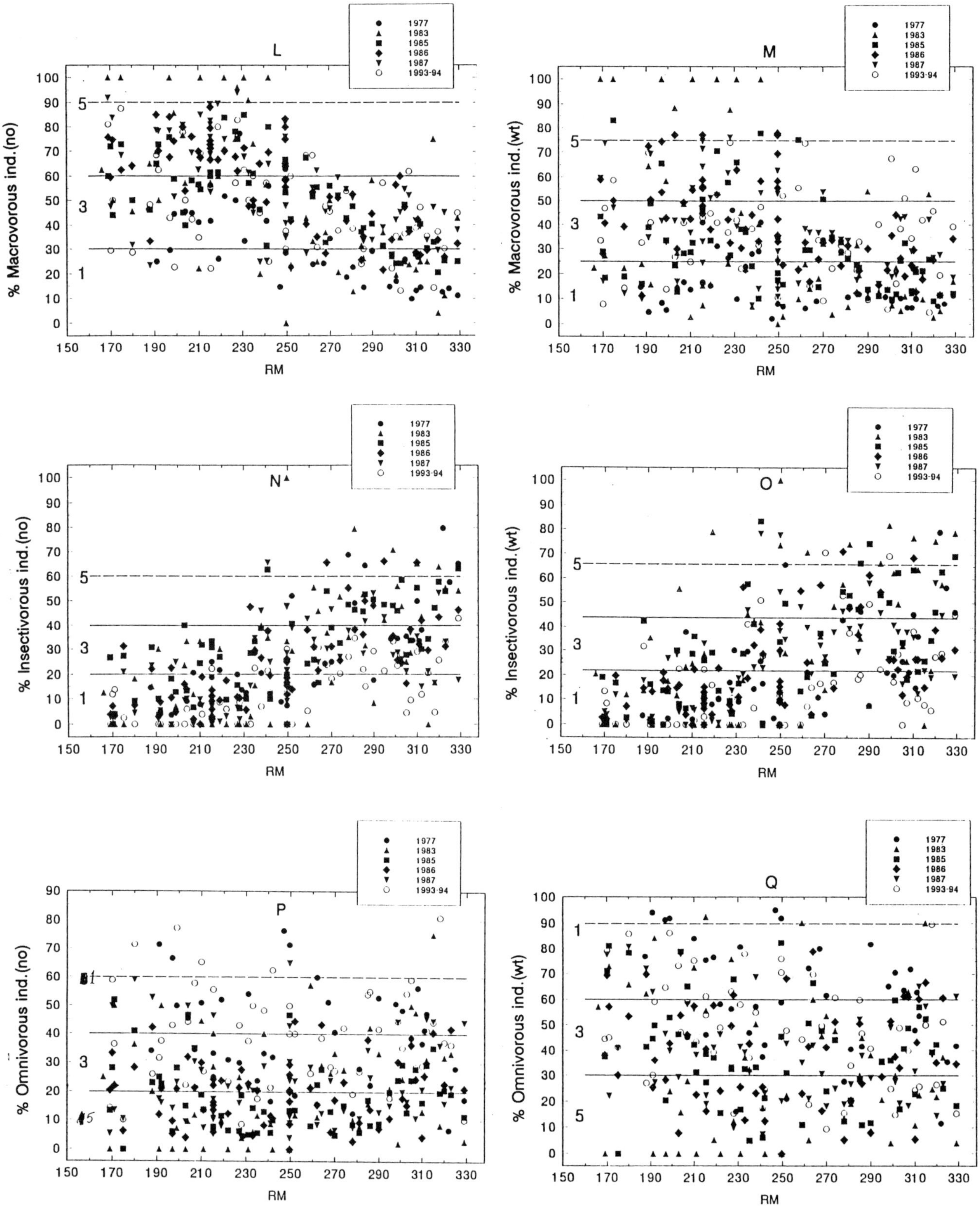

Figure 104: (con't.)

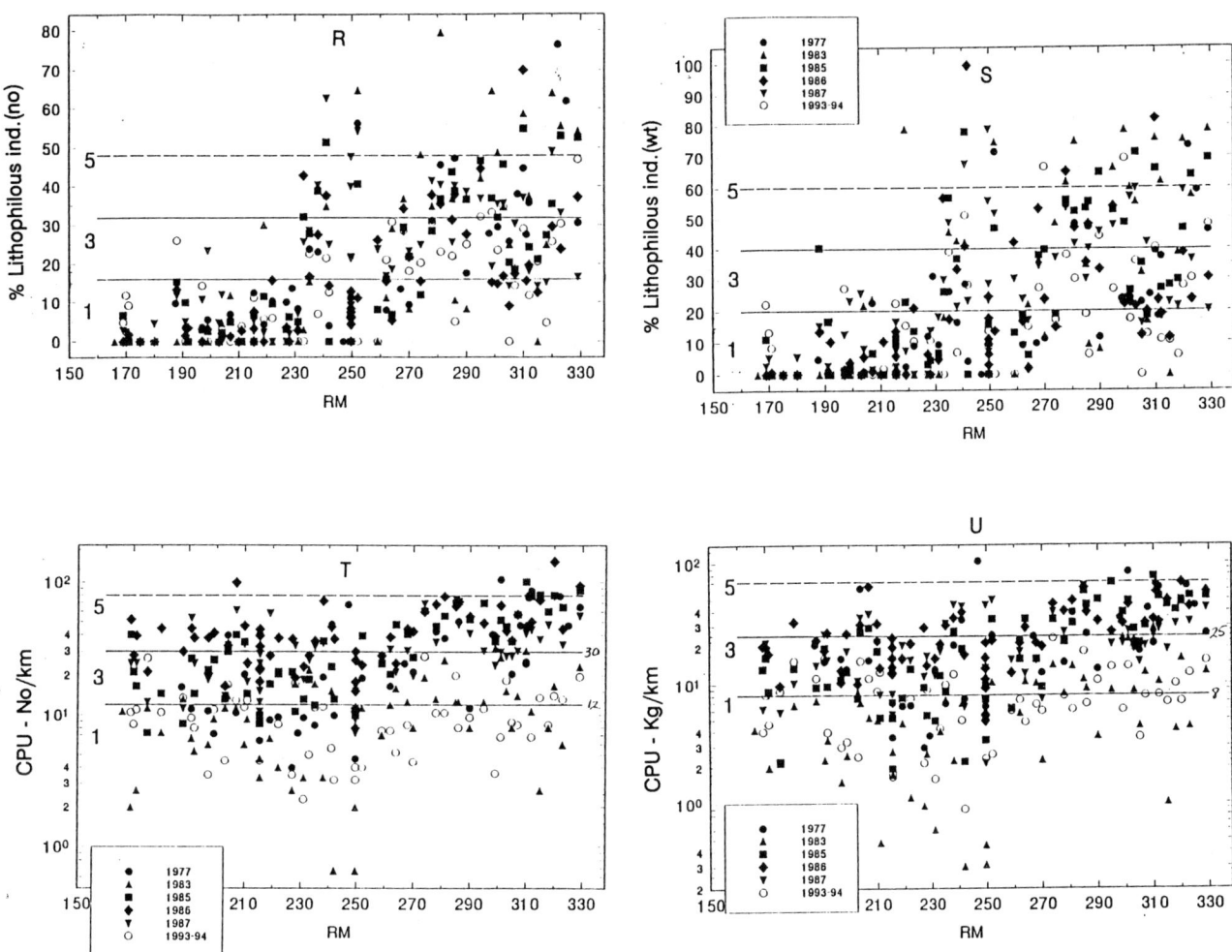

The IBI also peaked slightly at the point where Sugar Creek enters the Wabash River 13 km (9 mi) downriver from the Cayuga EGS, but then gradually declined downriver. During most years there was an obvious depression below the Terre Haute metropolitan area (RM 204-212). However, the river usually recovered by the time it reached the Hutsonville, Ill. area (RM168).

The IBI profiles based on numbers were almost always higher than biomass calculations, except for 1977. However, both profiles were very similar in configuration with corresponding peaks and valleys.

Modifications of IBI Criteria for Low Catch Rates

When catch rates were very low and consisted of only a very few species, as was the situation at many sites during 1983 and 1993-94, it was necessary to modify procedures for calculating IBI scores. The justification for altering the process is shown in Figure 106, which plots unmodified IBI values versus Iwb values for the six years. This curvilinear plot of is based on No/km data from 1977, 1983, 1985, 1986, 1987, and 1993-94. A plot of IBI based on biomass (IBIwt) versus Iwb was very similar and is not presented here. The relationship is approximately linear for IBI values greater than about 25 and Iwb values of about 3.0, but when Iwb values are lower than 3.0 the relationship becomes distinctly curvilinear.

The best way to achieve a linear relationship when the IBI scores are low is to lower them still more. It was obvious when the overall dataset was examined that catches yielding low Iwb scores possessed higher than desirable IBI scores because of high scores in two of the metrics involving the percent great river species and percent macrovores. Furthermore, these problems were generally correlated with low catch rates of only one or two species. Therefore, for all catch rates less than 10 fish/km each of those two metrics was scored as "1" regardless of the usual scoring criteria.

The results of modifying the IBI based on numeric data is shown in Figure 107. Regression slopes were very similar for all years and the scatter was also sufficiently consistent so that all of the years were combined as shown in Figure 107. The equation (Modified IBI = 12.29 + 4.362 Iwb) indicates an excellent correspondence of the lowest IBI and Iwb values, IBI = 12 and Iwb = 0. The overall correlation of these two indices is also quite good ($r = 0.72$).

Comparable results were obtained when analogous modifications were made whenever the biomass catch rate was lower than 3.0 kg/km in which case both of percent great river biomass and percent macrovore biomass were scored "1". A regression of this relationship is not included here.

Longitudinal profiles of modified IBI values were slightly higher than unmodified IBI profiles. High modified IBI values were identical with or very similar to high unmodified IBI values, but the low values were even lower. These effects were particularly evident for IBI profiles of years of low population abundance (Figure 108).

The modified IBI profile of 1977 is particularly interesting because of a late summer fishkill which occurred from about RM225 to RM240, the section shown in Figure 108 to be most depressed by the modified IBI. The depressed area is even better defined by the Iwb profile. This low dissolved oxygen episode is discussed in greater detail by Gammon and Reidy (1981).

In 1983 the modified IBI profile was especially depressed downriver from Terre Haute, Indiana in comparison with the unmodified profile. However, in 1993-94 the greatest difference between modified and unmodified profiles occurred in the RM 205-270 section of the river. The modified IBI scores correlated better with Iwb values than the unmodified scores.

We also examined alternative forms for three of the metrics based on numbers of species, specifically 1) the number of species of roundbodied suckers, 2) the number of species of centrarchids, and 3) the number of sensitive species. All of these metrics are directly correlated with sampling effort. In order to minimize the effects of effort we converted the raw species numbers to percentages of the total number of species captured. Scattergrams of these values for the Wabash River studies were presented earlier in Figures 104C, E, and G.

Incorporating these changes into the modified IBI produced only a slight effect and made the modified-modified IBI values (mod-mod IBI) more similar to the unmodified IBI values. All three versions of IBI were correlated positively with the Iwb values (Table 13 and Figure 109) as well as with each other. Only IBI scores based on numeric catch rates (No/km) were examined, but it is likely that similar results would have been obtained using biomass data (kg/km).

Table 13: Results of regressions of three versions of IBI computation on Iwb values.

IBI Version	r^2	Regression Equation
Unmodified IBI	0.62	IBI = 17.47 + 3.57 Iwb
Modified IBI	0.72	IBIm = 12.29 + 4.36 Iwb
Modified modified IBI	0.65	IBImm = 15.97 + 3.87 Iwb

174

Figure 105: Index of Biotic Integrity (IBI) profiles using numbers and biomass based on electrofishing catches in 1977, 1983, 1985-87, and 1993-4 using criteria for numbers of species where appropriate.

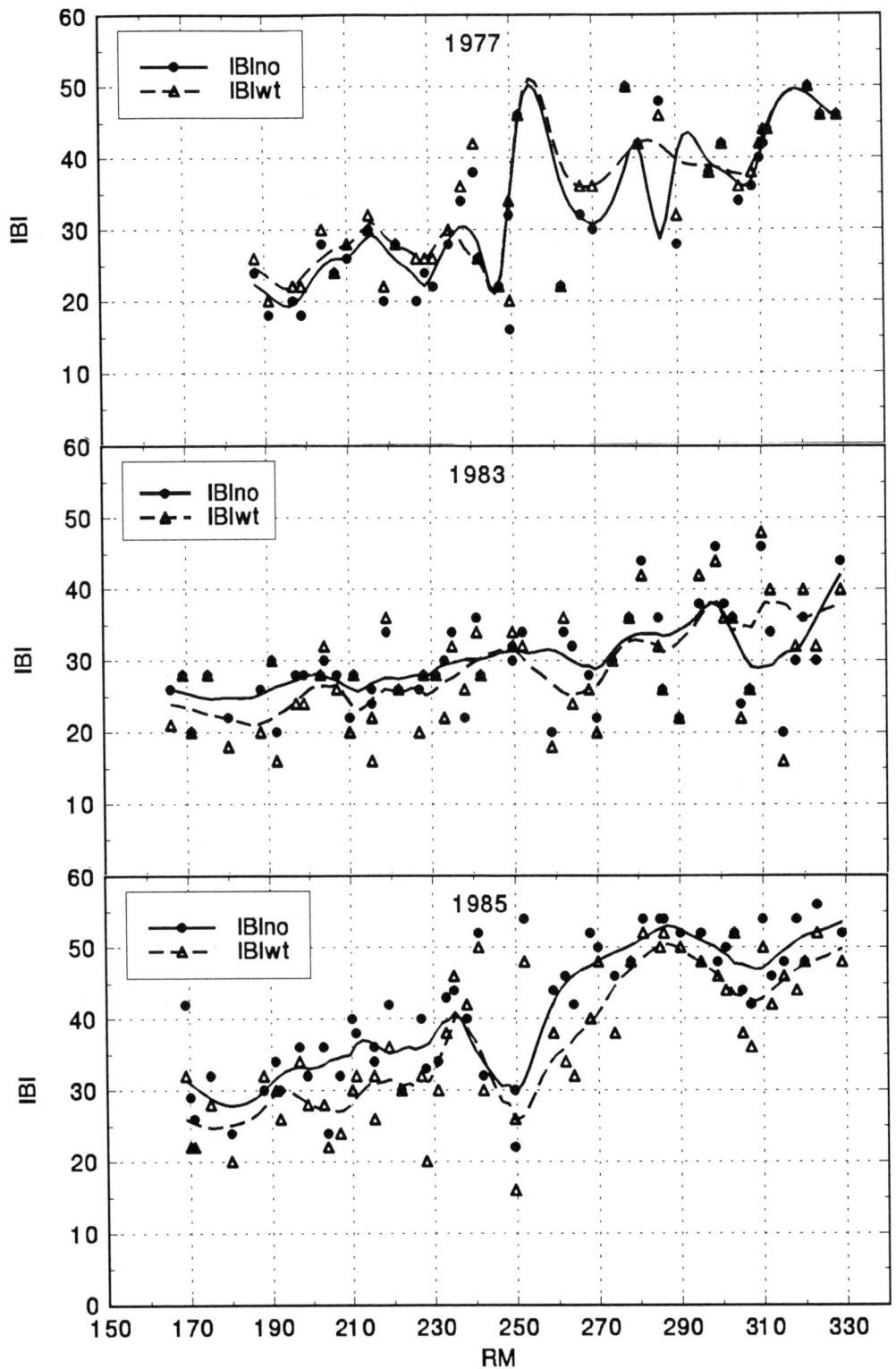

175

Figure 105: (con't.)

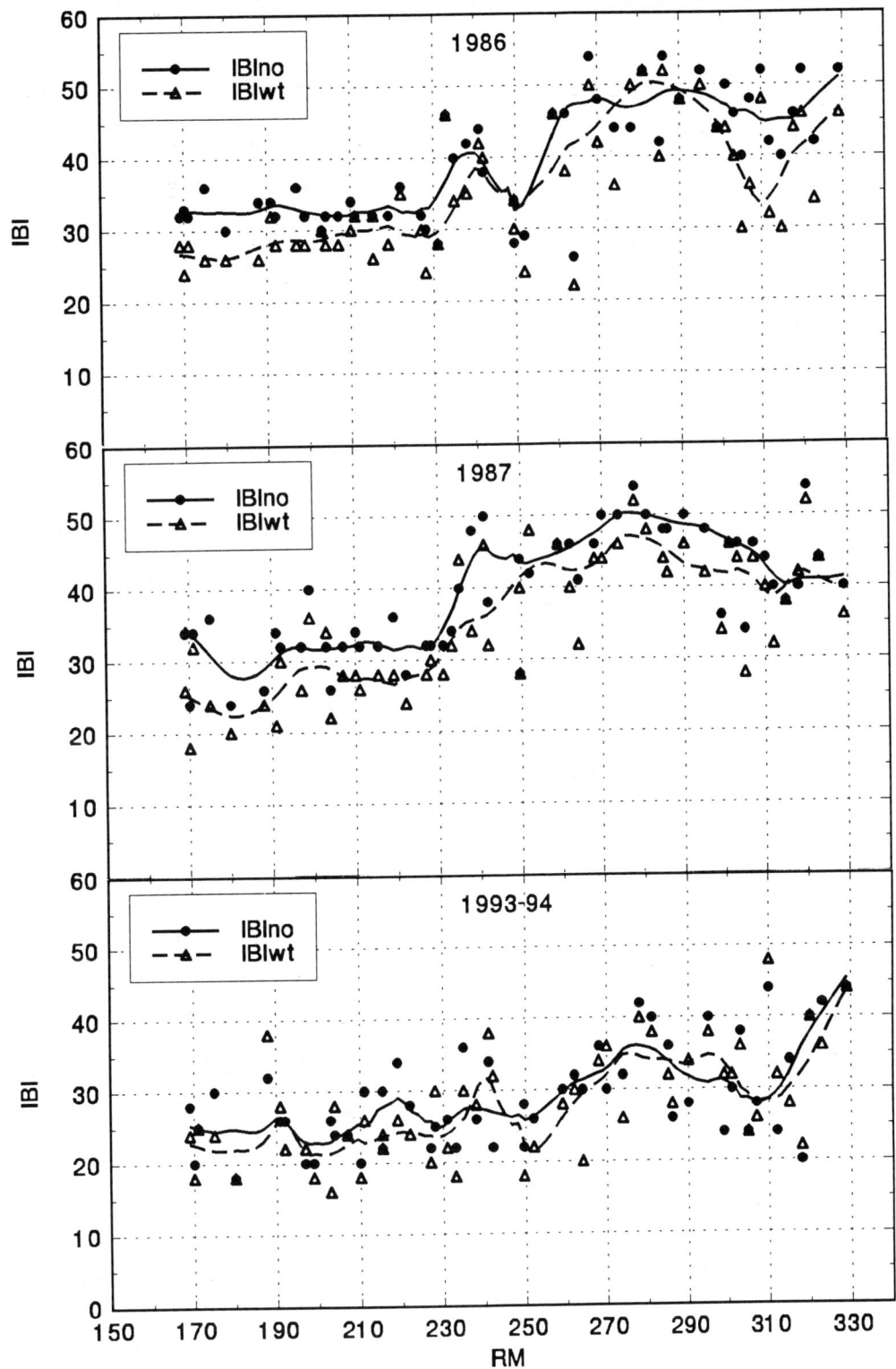

Figure 106: Curvilinear regression of unmodified IBI scores on Iwb values.

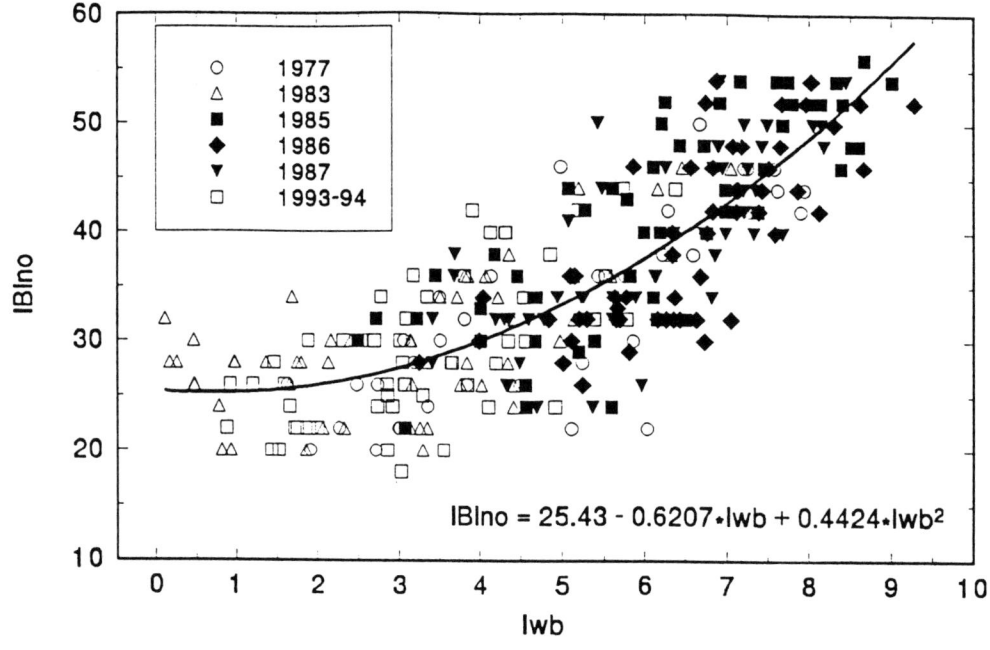

$$IBIno = 25.43 - 0.6207*Iwb + 0.4424*Iwb^2$$

Figure 107: Linear regression of modified IBI scores on Iwb values.

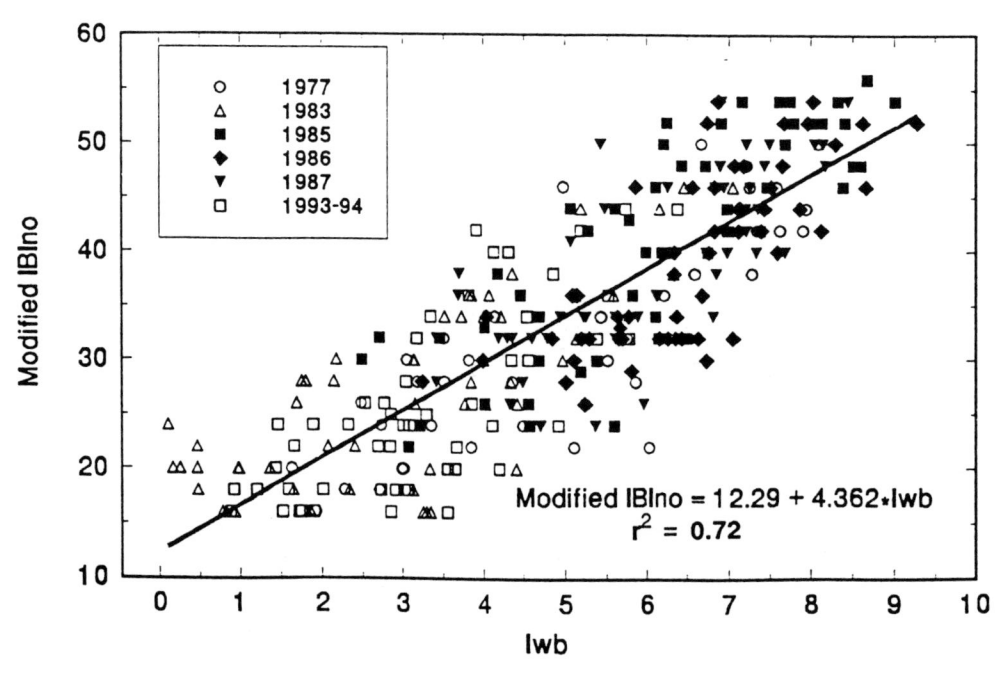

$$Modified\ IBIno = 12.29 + 4.362*Iwb$$
$$r^2 = 0.72$$

Figure 108: Comparison of modified IBI profiles with unmodified IBI and Iwb profiles for the years 1977, 1983, and 1993-94.

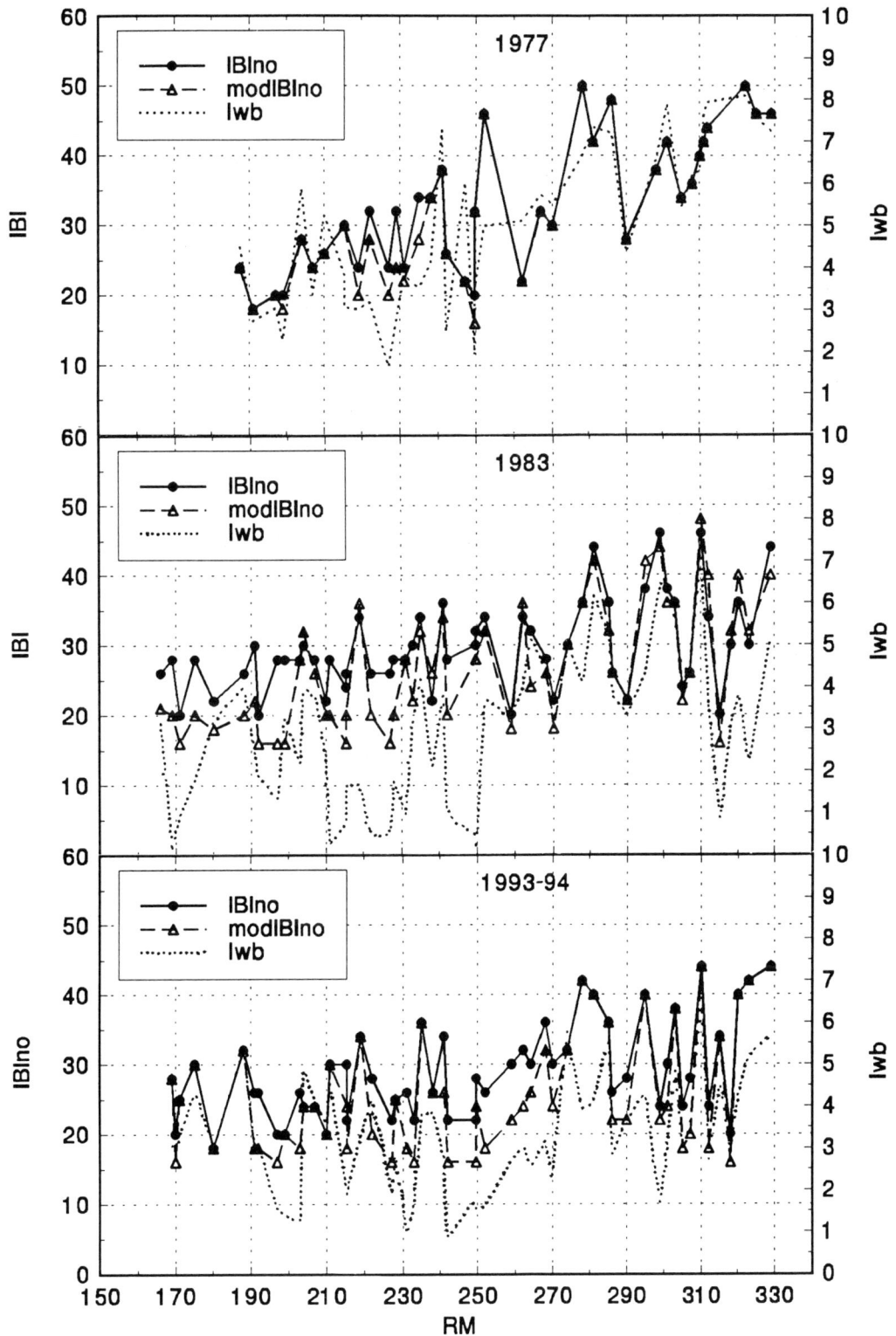

Figure 109: Regressions of mod-mod IBI, modified, and unmodified IBI values on Iwb values for all test years, using numbers of fish.

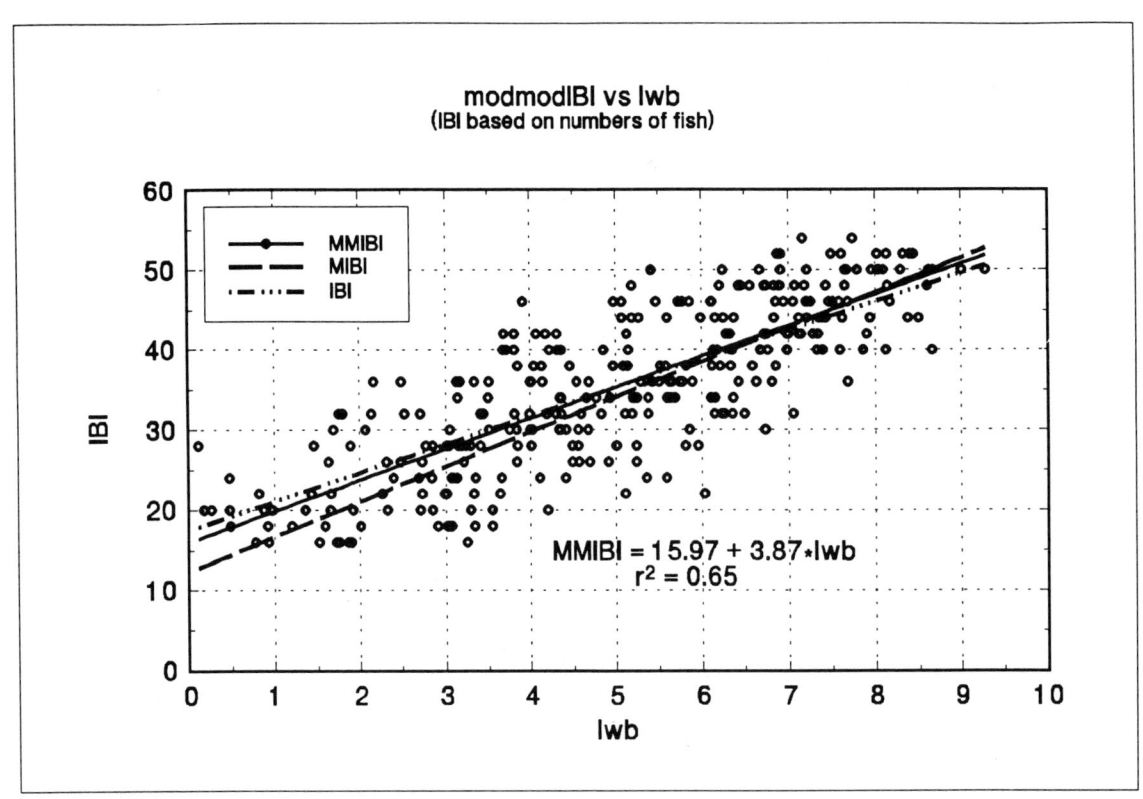

THE WABASH RIVER'S RENEWABLE
COMMERCIAL RESOURCES

The Wabash River produces only two economically valuable resources, other than water itself: 1) fish and 2) mussels or clams.

The Commercial Fishery

In Indiana, the commercial fishery using nets is limited to about 900 km (560 miles) of the Wabash, White, and Patoka rivers and 580 km of the Ohio River (Stevanavage 1990; Blackwell 1991). The lower 325 km (200 miles) of the Wabash River forms the boundary between Indiana and Illinois and may be fished without limits of the numbers of nets or seines. Indiana waters, however, may be fished by a maximum of four hoop or trap nets with stretched mesh not less than 5 cm. The license is free, but each of the required metal net tags costs $4.

Most of the commercial harvest is used directly by the fishermen and their families. Few fishermen, if any, derive their entire income from the river, although some fish may be sold locally.

The number of commercial licenses issued annually from 1974 to 1990 fluctuated between 800 and 400. In 1977 a policy was adopted which mandated reporting the commercial harvest. This led to a 40% decline in licensed fishermen in 1978.

In January 1985 a fish consumption advisory was issued for the Wabash River from Lafayette, Indiana to Darwin, Illinois and for the West Fork of the White River from Broad Ripple in northern Indianapolis, Indiana to its junction with the East Fork of the White River near Petersburg, Indiana. This may have been the reason for a 20% decline in commercial fishing licenses from 605 in 1984 to 447 in 1986 (Glander 1987). It may also explain the 20% decline in overall harvest. During this period, however, there was an increase in catches of channel catfish and flathead catfish, the most important species in the commercial catch. The decrease was mainly because of reduced catches of carp, suckers, and drum. Catfish species currently constitute about 70% of the harvest.

Mollusca of the Wabash River

Most studies of Wabash River mollusca have focused on commercially valuable species of clams. No comprehensive studies of snails have been conducted. The shells of many species of clams were used in making buttons from the late 1890's through mid 1900's, when plastic replaced shell. More recently, the cultured pearl industry mills sections of shells into round "seeds," which are then implanted into pearl oysters to create cultured pearls.

The Wabash River was once occupied by over 75 species of unionids (Call 1900; Blatchley and Daniels 1903; Daniels 1903, 1914; Goodrich and van der Schalie 1944).

During 1966 and 1967, 30 species of Unionidae were collected at 63 sites from the lower sections of the Wabash River and East

Fork White River (Krumholz et al. 1970). The 10 species most important to the commercial market made up 77.1% of the total catch (*Quadrula quadrula, Obovaria olivaria, Q. pustulosa, Actinonaias ligamentina, Amblema plicata, Fusconaia ebena, Fusconaia flava, Megalonaias nervosa, Q. metanevra*, and *Tritogonia verrucosa*) (Figure 110). Over-harvesting was considered to be the main cause for depleted commercial species. However, the negative effects of past increases in pollution on the overall fauna, including extirpated species of mussels which are not commercially valuable, has probably been significant (Anderson, personal communication).

Cummings et al. (1987, 1988, 1991) examined the abundance and distribution of clams of the Wabash, White, and Tippecanoe rivers. They found a drastic reduction in range and abundance of many species formerly common and widespread on the upper and middle Wabash River. The three most abundant species were *Obovaria olivaria, Leptodea fragilis*, and *Quadrula quadrula*, which made up 61% of the live mussels collected (Figure 111).

Fourteen of 26 mussel collection sites were located within our 12 Reaches with other sites located further up-river and down-river. Twenty-eight live species were found. The hickorynut (*Obovaria olivaria*) was common throughout the middle Wabash River except between the mouth of Big Vermilion River and Terre Haute. It constituted 30% of the live clams collected and appeared to be more common in recent years than in the past. Nevertheless, this species is growing scarce in other rivers including the Mississippi and upper Ohio (Cummings and Mayer 1992). Fragile papershell (*Leptodea fragilis*) accounted for

22% of all live mussels collected by Cummings et al. (1988). Its distribution was virtually identical to that of the hickorynut.

The mapleleaf (*Quadrula quadrula*) accounted for 9% of live mussels in the 1988 study and was common except at sites between Big Vermilion River and Terre Haute, Indiana where it was rare.

Three species of mussels considered to be endangered or threatened (Williams et al. 1993) were also found in 1988: (1) fanshell (*Cyprogenia stegaria*), (2) rabbitsfoot (*Quadrula cylindrica*), and (3) sheepnose (*Plethobasus cyphyus*). A single live fanshell was found below Hutsonville. One live rabbitsfoot was found downriver from Lafayette. Six live sheepnose were found near Delphi. Only the shells of 14 other rare, endangered, or extinct species were found.

Lewis (personal communication) believes that the dearth of species noted by Cummins et al. between the mouth of the Big Vermilion River and Terre Haute, Indiana may be the result of poor clam habitat or low densities near the three sites where they collected. In this same reach, Lewis et al. (1997) found 24 live species of clams at established collecting stations near the Cayuga and Wabash River EGSs including the commonly collected hickorynut (*Obovaria alivaria*), mapleleaf (*Quadrula quadrula*), and fragile papershell (*Leptodea fragilis*).

Bingham (1968) found that the reproductive seasons of eight resident mussels differed greatly. The pistolgrip (*Tritongonia verrucosa*) was thought to reproduce during winter. The hickorynut (*Obovaria olivaria*) and mucket (*Actinonaias ligamentina*) are long-term breeders with the females becoming gravid in late summer and fall and shedding their glochidia early the following summer. In most other species, including all Ambleminae, the females became gravid in late spring or early summer and shed glochidia in late summer or fall.

Restrictions on commercial collecting methods were first instituted in April 1967 by prohibiting the use of mechanical dredges and diving with auxiliary air supplies. Acceptable, but less efficient, methods of collection include handpicking, short forks, tongs, and crowfoot bar (brail) (Figure 112). The mussel season traditionally extends from April 15 through October 31, but most mussels are taken in June, July, and August. A minimum size limit of 64 mm protects smaller mussels. A $15 fee is charged for a license. Mussel harvest reports were collected by Indiana Department of Natural Resources (IDNR), but were not summarized or tabulated prior to 1988 (Henschen 1989, Stevanage 1992).

Sales of Indiana shells totaled 2,000 tons in 1965, 4,200 tons in 1966, 1,080 tons in 1967, and less than 250 tons in 1968. In 1991, 690 tons valued at $1.6-3.0 million were bought by mussel buyers. The East Fork of the White River and the Wabash River accounted for 76% of the total harvest in 1991.

A combination of extremely low river flows which increased clam vulnerability and high prices for clam shells led to a serious overharvest of clams in 1988 and 1991 (Flatt et al. 1992). A total of 950 mussel harvesters purchased licenses in 1991, nearly quadruple the average sold from 1982 to 1990. Clammers combed exposed areas in the Wabash and White rivers and then extended

their harvest efforts into tributaries. The 1991 collecting season was abruptly halted in September by the Indiana DNR on exclusively Indiana waters. The season was officially closed in 1992 for an indefinite period, a move favored by most legal musselers. Past harvesting regulations were obviously inadequate and over-collection was having a negative impact on the mussel community. However, the extent to which point-source and nonpoint-source pollution affected mussel populations was not at all apparent.

From 1992 through 1994 the section of the Wabash River from Lafayette, Indiana to the mouth of the Mississinewa River was studied by IDNR biologists, as well as the Tippecanoe River and East Fork of the White River (Ball and Schoenung 1996).

The study's three objectives were to: 1) determine the location, species composition, and relative abundance of beds in the core mussel-producing river reaches, 2) evaluate population density, age structure, and growth characteristics of commercially valuable and endangered species at representative mussel beds, and 3) estimate sustainable yield levels for each commercially valuable species.

Mussels were collected by brail and quadrat sampling (10mx10m) at five Wabash River stations. Densities averaged 1.19 total mussels/m^2 and 0.84 mussels/m^2 of 12 commercial species. Mussels were more common in the other two rivers, but all three Indiana rivers supported fewer mussels than the Mississippi River.

Mussel growth was determined for 112 specimens by microsectioning shells and examining the sections microscopically.

Some individual clams were older than 50 years in age (elephant-ear, pimpleback, purple wartyback, threeridge, and washboard). Growth of an additional five species was determined by measuring the external growth rings (clubshell, fanshell, rabbitsfoot, round hickorynut, and round pigtoe).

Seven of nine commercially valuable species for which age frequency data were available failed to recruit small indivuals into the six-year-plus age category in sufficient numbers to maintain population density. There was no evidence of recruitment whatsoever for the Ohio pigtoe, rough pigtoe, butterfly, and ebonyshell and their population densities were so low that it was felt they may be on the verge of disappearing from the Wabash River and the East Fork of the White River. It was recommended that the collection ban on commercial mussels be continued to permit recovery of their populations.

Because water quality was a concern total nitrates, ammonia, turbidity, and a group of herbicides were measured from April to July or August 1992-94 at five Wabash River stations. The average nitrate concentrations was highest in 1992 at 5.3 mg/l, but lower in 1993 (3.5 mg/l) and 1994 (3.4 mg/l). Nitrate concentrations were lower at the other two rivers. Ammonia concentrations were usually 0.1 at the Wabash stations, but higher in the other two rivers. Mussel die-offs were observed in mid-summer of 1992 and 1993 in the East Fork of the White River, but not in the Wabash or Tippecanoe Rivers, for reasons apparently not associated with a pollution event.

Aldridge et al. (1987) exposed three species of unionid mussels to 600 to 750 ppm diatomaceous earth and measured reduced rates of oxygen consumption, food clearance, and ammonia nitrogen excretion. However, Spacey and Chaney (1993) found no detectable changes in those physiological parameters when two species of unionid mussels were exposed to 500 and 1000 mg/l bentonite clay.

Cummings et al. (1988) found the Asian clam (*Corbicula fluminea*) at all Wabash River collecting sites from Wabash, Indiana down to the White River. However, it was not as abundant in the Wabash River as it was in the Ohio River and other southeastern United States rivers.

Ball and Schoenung (1996) also found *Corbicula*, but not in problem densities. They found no evidence that zebra mussels (*Dreissena polymorpha*) had become established in the Wabash River, but they have become abundant in the Ohio River in recent years. However, there is little doubt that they will soon be present because in 1996 they were found in the reservoirs of the Tippecanoe River and resident mussels may well face yet another potentially damaging factor.

183

Quadrula pustulosa
(**Pimpleback**)

Fusconaia ebena
(**Ebonyshell**)

Fusconaia flava
(**Wabash Pigtoe**)

Megalonaias nervosus
(**Washboard**)

Quadrula metavevra
(**Monkeyface**)

Tritogonia verrucosa
(**Pistolgrip**)

Figure 110: Some common commercially valuable mussels of the Wabash River.

184

Obovaria olivaria
(Hickorynut)

Leptodea fragilis
(Fragile Papershell)

Quadrula quadrula
(Mapleleaf)

Figure 111: The most common species of clams in the Wabash River.

185

Figure 112: Wabash River clam boats with stored crowfoot bars and hooks.

PHYTOPLANKTON
AND THE WABASH RIVER ECOSYSTEM

Interactions between the biota and the physical and chemical environment are sporadic, complex, and obscure in the highly variable environment of most rivers. The intricate linkage between the producer phytoplankton and the fish community is especially intriguing. It is believed that the entire Wabash River ecosystem is strongly influenced by phytoplankton, especially during the summer months, but there are few data to support that contention. Our knowledge about this particular group is disparate to its ecological importance.

Phytoplankton constitutes by far the most important producer group in the middle Wabash River. Among other producer groups beds of water willow (*Dianthera*) are well represented in the upper river below Huntington Dam, but they become scarce in the lower two-thirds. Periphyton is severely limited because of the turbidity of the water, but is no doubt important in shallow areas which receive light. Submerged aquatic vegetation only appears during extended periods of low discharge.

The primary source of information for this group is the Indiana Department of Environmental Management (IDEM), formerly a Division of the Indiana State Board of Health (ISBH). Collections of phytoplankton from various river stations were made on a monthly schedule beginning in 1971 (ISBH 1957-85; IDEM 1986-90), but ceased a few years ago because of budgetary and manpower constraints.

The basic spatial pattern of phytoplankton density and the influence of discharge is shown best at eight stations along the Wabash River during the low-flow summers of 1977 and 1988 (Figure 113). Summer density in the upper river is low initially, but increases sharply as the river flows southward, reaching its highest density level downriver from Lafayette, Indiana.

Phytoplankton density is typically high from May through October. Low densities occur during the winter months because of low temperatures and also during the spring because of high discharge rates. The summer of 1977 followed an unusually dry winter and spring and algal densities were very high in June and July. However, densities decreased substantially thereafter when the river discharge increased in August and September. Usually the phytoplankton community was dominated by diatoms, but green algae were as abundant as diatoms in 1988.

During the summers of 1981 and 1982 an interdisciplinary team studied the dissolved oxygen (DO) status of the middle Wabash River (Bridges et al. 1986). Phytoplankton densities at this time ranged from 20,000 to 120,000 cells/Ml, chlorophyll $\underline{a}$ ranged from 18 to 247 ug/L, pheophytin $\underline{a}$ was 6.5 to 51 ug/L, gross primary productivity (GPP) ranged from 70 to 850 mg $C/m^2/h$, and net primary productivity was 90 to 784 mg $C/m^2/h$. Turbidity was high and the euphotic zone was only 0.2 to 1.2 m.

Estimates were also made of point-source biological oxygen demand (BOD), sediment oxygen demand (SOD), and chemical oxygen demand (COD). All of this information was used to estimate the dissolved oxygen deficit (DOD) using a version of the DIURNAL computer model (HydroQual, Inc. 1984). The DOD was calculated as the saturation concentration at the prevailing temperature less the actual dissolved oxygen concentration (Smith et al. 1987). This model projected low-flow DOD values of 2.0 to 2.5 mg O/L from Delphi, Indiana to Montezuma, Indiana and then increasing values to about 4.0 mg O/L from Terre Haute, Indiana to Merom, Indiana.

Phytoplankton respiration was estimated to be responsible for 50-60% of the dissolved oxygen deficit in the upper study section and about 70% downriver from Terre Haute, Indiana. The second largest source, carbonaceous BOD or CBOD which entered the river from multiple point sources, accounted for only about 10% of the DOD in the upper reaches and over 15% downriver from Terre Haute, Indiana.

Sediment oxygen demand (SOD) was estimated to be the third largest oxygen sink, and nitrogenous BOD and other substances also contributed to a minor degree.

Phytoplankton and other organic substances created stress conditions during low flows in some parts of the river by reducing DO concentrations (Parke and Gammon 1986). For example, the inter-action of river morphology, large algal populations sustained by high nutrient inputs, and thermal loading from an electric generating station produced low DO concen-trations in a 9.6 km long, lake-like section of river dammed by gravel from Sugar Creek.

When river discharge diminished to about 1,500 cfs there was a sharp increase in phytoplankton density with chlorophyll a increasing from about 160 ug/L to nearly 230 ug/L. Significant amounts of suspended solids settled to the bottom as the water passed through the ponded segment. Chloro-phyll a decreased to less than 150 ug/L and total suspended nonfilterable solids decreased from 80 mg/L to about 50 mg/L. Secchi disk transparency increased as the suspended materials settled out and SOD increased. Fish avoided this area of the river in 1977 when the dissolved oxygen concen-tration was depressed (Gammon and Reidy 1981). Fish were killed here in 1983 and 1986 when the DO problem was even more severe.

The dissolved oxygen sag curve was reexamined in 1988, both with and without the Cayuga Electric Generating Station in operation (EA Engineering 1990). It was concluded that a DO sag occurs naturally in the "Sugar Creek Pool" because of its morphology and high summer temperatures.

Such events probably occur irregularly elsewhere in the river, causing stress if not mortality of fish and other aquatic organisms. If an understanding of the underlying dynamics of the Wabash River's ecosystem is to be attained it will be absolutely essential to study the phytoplankton population and its relation to nutrient inputs in far greater detail.

The minimal efforts of IDEM in tracking the population density of phytoplankton appear to have been terminated at this time. This is most unfortunate and short-sighted in the extreme since the problems of most large rivers and reservoirs are related to enrichment.

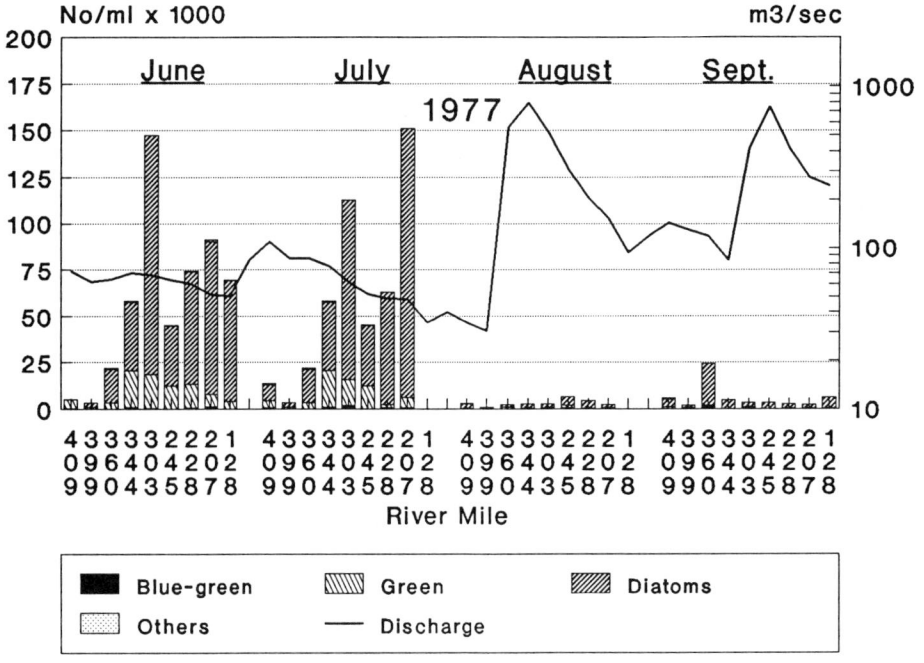

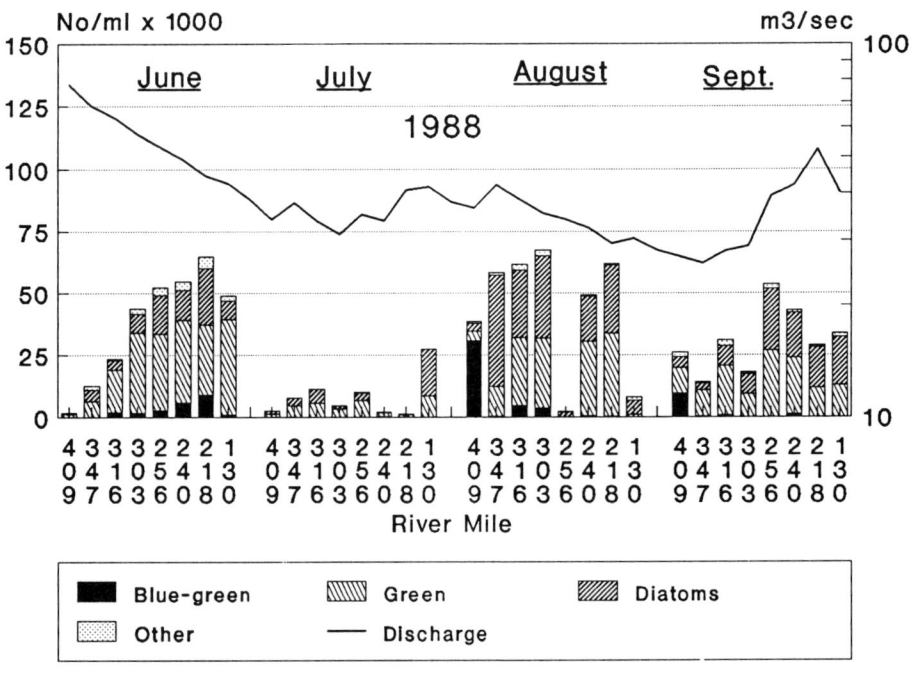

Figure 113: Spatial pattern of phytoplankton density and their relationship to river discharge during the summers of 1977 and 1988.

MACROINVERTEBRATES OF THE WABASH RIVER

The most extensive studies of benthic macroinvertebrate populations in the middle Wabash River have been conducted by the various electric power companies at the sites of generating plants. WAPORA, Inc. examined benthic communities from 1971 through 1980 at the Breed Electric Generating Station (EGS) west of Fairbanks, Indiana. This station is located approximately 48 river kilometers (30 mi.) south of Terre Haute, Indiana near the mouth of Prairie Creek. The Indiana and Michigan Electric Company operated Breed until 1994, when it was terminated. The purpose of this 10-year study was to determine the effects of the Breed EGS's thermal effluent on macroinvertebrate populations. Sampling stations were sited upriver from Prairie Creek, between Prairie Creek and Breed EGS, within the EGS discharge canal, at the ash pond discharge, and above the first riffle downriver from Breed EGS.

Macroinvertebrates were sampled at five locations using both a ponar grab sampler and Hester-Dendy multiple-plate samplers suspended one foot below the river's surface for six to eight weeks.

The bottom substrate was thoroughly surveyed in 1974 because of its influence on the benthic community. Subsequent comparisons were based on similar bottom substrates. A majority of substrate types consisted of mud, sand, and gravel with sand predominating. Among these substrate types sand was the poorest in terms of benthic density. For Ponar samples in 1980, for example, the mean macroinverte-brate density in sand was only $518/m^2$, compared to $1,848/m^2$ in mud and $3,135/m^2$ in gravel.

In the ambient upriver station, oligochaetes predominated in mud substrates from July through September. They were also common in sand and about as abundant as nematodes, but only during the month of July. During August and September chironomids were most common in sand. Hydropsychidae dominated in gravel substrates in all months, although oligochaetes and Ephemeroptera were almost as numerous in the July samples. A generalized relationship between substrate, current, and macroinvertebrate complement follows and the species collected are listed in Table 14.

Current Velocity
  Slow ————————————————————————————————————> Fast

Substrate Size
  Small ————————————————————————————————————> Large
  Mud ————> Sand ————> Coarse Sand ————> Gravel ————> Rubble

Predominant Invertebrate Complex

Tubificidae	Tubificidae	Hydropsychid caddisflies
Chironomidae	Chironomidae	Chironomidae
Burrowing mayflies	*Potamyia flava*	*Corbicula manilensis*

Cinergy Corp. (formerly PSI-Energy) sampled macroinvertebrates as well as fish at six collecting sites near their Cayuga Electric Generating Station with three sites upriver from the thermal effluent and three below (Lewis et al 1997). Approximately 100 samples were taken during each summer over a five-year period as part of their NPDES discharge permit requirements.

Hester-Dendy plate samplers remained in place for a minimum of a six weeks before lifting and two such samples were usually obtained during the period of warm weather. The H-D plate samplers consisted of eight pieces of 3" x 3" x 1/8" tempered hardboard separated by one-inch square spacers. Five plate samplers were tied to each concrete construction block which were placed in river runs with some current at each station.

Other macroinvertebrates were obtained using 30-mesh dipnets in all available natural habitats over a minimum of a 30-minute period at those times when the plate samplers were recovered.

The material from all five plate samplers was combined into a single sample. A Folsom sample splitter was used for subsampling both the plate and qualitative samples. Approximately 250 organisms from each sample were identified to genus or a higher level of classification whenever possible. A variety of taxonomic experts were used for confirmation of difficult taxa.

Community indices such as taxa richness, enumeration, community similarity, Hilsenhoff biotic index, species diversity, and Invertebrate Community Index (ICI) were also estimated for each sampling location during each sampling period. The Ephemeroptera, Plecoptera, and Trichoptera (EPT) taxa list was derived from the qualitative data. Mussels were also collected in this assessment.

The taxa from the natural substrates in the Wabash River were generally much more diverse than taxa which occupied the Hester-Dendy plate samplers. Densities of macroinvertebrates on the plate samplers generally increased from early summer to late summer, while ICI values remained fairly steady until September when they fell to lower levels.

Coefficients of Similarity values for all six collecting stations were usually high, an indication of strong similarities among the stations and little or no influence on community structure by the Cayuga Electric Generating Station.

On several occasions, especially in late summer during low-flows, fine organic sediments were deposited in significant amounts along the shoreline shallows creating ideal conditions for the production of hugh concentrations of chironomids and snails. These, in turn, attracted shore birds in large numbers.

191

Table 14: Macroinvertebrates collected from the Wabash River near two electric generating stations.

Taxon (common name)	Breed EGS[1]	Cayuga EGS[2]	
		H-D Plate	Substrate
TURBELLARIA (flatworms)	X		
TRICLADIDA			
Planariidae			
Dugesia tigrina		X	X
NEMATODA (roundworms)	X		
ANNELIDA (segmented worms)			
OLIGOCHAETA			
Naididae	X	X	
Tubificidae			X
Branchiura sowerbyi	X		
Limnodrilus cervix	X		
L. claparedianus	X		
L. hoffmeisteri	X		
L. udekemianus	X		
Lumbricidae			X
ARTHROPODA			
CRUSTACEA			
Amphipoda (scuds)			
Gammarus spp.	X		
Isopoda (sow bugs)			
Asellidae - *Caecidotea sp.*			X
Decapoda (crayfish)			
Cambaridae - *Orconectes virilis*			X
INSECTA			
Ephemeroptera (mayflies)			
Baetidae			
Baetis spp.	X		
Baetis intercalaris		X	X

Table 14: Continued

Taxon (common name)	Breed EGS[1]	Cayuga EGS[2] H-D Plate	Substrate
Labiobaetis longipalpus		X	X
Caenidae			
Branchycercus spp.	X		
Caenis spp.	X		
Caenis hilaris			X
Ephemeridae			
Hexagenia spp.	X		
H. limbata			X
Pentagenia spp.	X		
P. vittigera			X
Tortopus primus			X
Heptageniidae			
Heptagenia flavescens		X	X
Stenocron spp.	X		
Stenocron interpunctatum			X
Stenoma spp.	X		
Stenonema exiguum		X	X
S. integrum		X	X
S. mediopunctatum			X
S. pulchellum		X	X
S. terminatum		X	X
Oligoneuriidae			
Isonychia spp.	X	X	X
Potamanthidae			
Anthopotamus spp.			X
Potamanthus spp.	X		
Tricorythidae			
Tricorythodes spp.	X	X	X

Table 14: Continued

Taxon (common name)	Breed EGS[1]	Cayuga EGS[2] H-D Plate	Substrate
Odonata (dragonflies and damselflies)	X		
Anisoptera			
Gomphidae	X		
Gomphus crassus			X
Ophiogomphus spp.			X
Stylurus notatus			X
S. spiniceps			X
Zygoptera			
Calopterygidae			
Haeterina spp.			X
Coenagroinidae			
Argia spp.	X	X	X
Corduliidae			
Neurocordulia molesta		X	X
Libellulidae			
Erythemis simplicicollis		X	X
Aeschnidae			
Boyeria spp.			X
Hemiptera (water bugs)			
Belostomatidae			
Belostoma fluminea			X
Corixidae			
Sigara spp.			X
Trichocorixa spp.			X
Gerridae			
Metrobates spp.			X
Trepobates spp.			X
Notonectidae			
Notonecta raleighi			X

Table 14: Continued

Taxon (common name)	Breed EGS[1]	Cayuga EGS[2] H-D Plate	Substrate
Megaloptera (dobsonflies)			
Corydalidae			
Corydalus cornutus			X
Sialidae			
Sialis spp.			X
Plecoptera (stoneflies)			
Pteronarcyidae			
Pteronarcys dorsata		X	X
Trichoptera (caddisflies)			
Hydropsychidae			
Cheumatopsyche spp.	X	X	X
Hydropsyche bidens		X	
H. orris	X	X	X
H. simulans	X	X	X
Potamyia flava	X	X	X
Leptoceridae			
Nectopsyche spp.			X
Psychomyiidae			
Cyrnellus fraternus		X	X
Polycentropus spp.	X		
Coleoptera (water beetles)			
Dytiscidae			
Copelatus spp.			X
Laccophilus maculolosus			X
Dryopidae			
Helichus spp.			X

Table 14: Continued

Taxon (common name)	Breed EGS[1]	Cayuga EGS[2] H-D Plate	Substrate
Coleoptera (continued)			
Elmidae			
Ancyronx variegata		X	X
Dubiraphia spp.	X		
Dubiraphia vittata	X		X
Macronychus glabratus		X	X
Neoelmis spp.	X		
Stenelmis spp.	X		
Stenelmis humerosa-sinuata		X	X
Gyrinidae			
Dineutus assimilis		X	X
Gyrinus analis			X
Haliplidae			
Haliplus immaculosus			X
Peltodytes lengi			X
P. muticus			X
Hydrophilidae			
Tropisternus blatchleyi			X
Diptera (midges, mosquitos, gnats & flies)			
Chaoboridae			
Chaoborus spp.	X		
Chironomidae	X		
Ablabesmyia spp.			X
Chironomus spp.			X
Clinotanypus spp.			X
Coelotanypus spp.			X
Corynoneura spp.		X	
Cryptochironomus spp.	X		X

Table 14: Continued

Taxon (common name)	Breed EGS[1]	Cayuga EGS[2] H-D Plate	Substrate
Diptera - Chironomidae (continued)			
Encochironomus spp.	X		
Glyptotendipes spp.	X	X	X
Nanocladius spp.		X	X
Parachironomus spp.			X
Phaenopsectra spp.	X	X	
Polypedilum convictum	X	X	X
P. illinoense		X	X
P. scalaenum		X	
Procladius spp.			X
Robackia demeijerei	X		
Rheotanytarsus spp.		X	X
Stenochironomus spp.		X	
Stictochironomus spp.	X		
Tanyarsus spp.	X		
Thienemannimyia group		X	X
Tvetenia vitracies		X	X
Xenochironomus spp.	X		
Empididae	X		
Hemerodromia spp.		X	
Simuliidae	X		
Simulium spp.		X	
Tipulidae			
Brachypremna spp.	X		
Hexatoma spp.			X

[1] 1980 data condensed from Table 5-29 of Finni (1981)
[2] 1995 data from Table 16 and 17 of Lewis et al (1997)

THE EFFECT OF ELECTRIC GENERATING PLANTS ON FISH COMMUNITIES

Fish must adapt to the ever changing environment of the river physiologically and/or behaviourally in order to survive. Natural environmental conditions may change suddenly such as altered water velocity and turbidity and decreased temperature and dissolved oxygen (DO) following a heavy rain. Man's activities also produce sudden changes for a passing organism as the result of the entry of industrial or domestic wastes. Longer term environmental alterations of the river ensue when corn and soybeans replace the natural vegetation of a drainage basin or when large areas are mined.

Describing the effects on fish of two electric generating plants (EGS) has been a central objective of the monitoring effort since 1967. Here we used multivariate statistical analysis of fish species catch data and changes over space and time as indicated by the composite index of well-being (Iwb) to describe basic changes over time.

A detrended correspondence analysis (DCA) (Ter Braak 1987) was used to examine the modifications in fish community structure. The sections of river examined included (a) the Wabash River Electric Generating Station (Reach 9), and (b) the Cayuga Electric Generating Station (Reach 6), (c) the Eli Lilly plant located north of Clinton, Indiana (located at the boundary of Reaches 7 and 8), (d) the Lafayette-West Lafayette metropolitan area (Reach 2), and (e) the Terre Haute metropolitan area (Reach 10) (Figure 114).

In all of the following cases the data consisted of raw species abundance (no/km) of the 33 most common and/or important species of fish in the middle Wabash River, as shown in Table 15. Most of the selected species are found throughout the middle Wabash River although some may be more abundant in the upper section and others more common the lower section. The species list originally included gizzard shad,

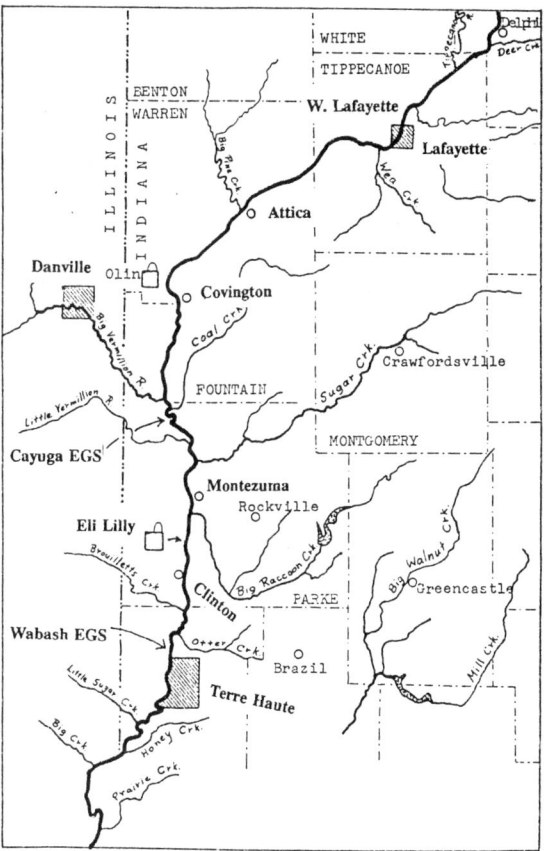

Figure 114: Location of point-sources evaluated with detrended correspondence analysis (DCA).

the most common species taken by electrofishing. However, the ordination pattern was much clearer when gizzard shad was excluded from the analyses. The basic cause of the problem was the erratic fluctuations in population abundance from year to year for a variety of reasons. Sometimes weather patterns and local hydrological conditions were influential. From 1984 through 1991 predator fishes were sufficiently abundant to keep the gizzard shad from achieving high population densities.

The detrended correspondence analysis patterns produced by catch data transformed to natural logarithms were essentially similar to patterns produced using untransformed data. Therefore, all DCA analyses were performed using raw relative density values (No/km).

Five electric generating stations (EGS) were located directly on the Wabash River between Delphi, Indiana and Merom, Indiana during the period of study, all of them coal-fired and employing once-through cooling. Only three EGSs remain operational at the present time.

Cinergy Corp. operates two electric generating stations in the middle Wabash River, (1) the Cayuga EGS which is located between Newport, IN and Cayuga, IN at RKm 402 (RM 250) in Vermillion County and (2) the Wabash River EGS which is located upriver from Terre Haute, IN located at RKm 346 (RM 215) in Vigo County. Further downriver is the Hutsonville EGS (Central Illinois Public Service), a four-unit plant rated at 214 MW which is located about 3 km (2 miles) north of Hutsonville, Illinois.

Cinergy Corp.'s Dresser EGS, rated at 149 Megawatts (MW) and located at Rkm 330 (RM 205) south of Terre Haute, was operational until 1975. Further downriver at Rkm 295 (RM 183) the Breed EGS (420 MW) was operated by the Indiana and Michigan Electric Company. It discharged water elevated about 5°C above ambient temperatures until it, too, closed in 1994.

The Wabash River EGS is rated at 970 Megawatts (MW) with a maximum heat rejection of about 5.0×10^9 Btu/h (average monthly heat rate = 10,063 Btu/kWh). The maximum pumping rate is approximately 30 m^3/s. Operating since 1956, it increased its capacity in 1968 and recently began operating one unit under a coal gasification process.

The Wabash River is about 120 m wide and 2.7 m deep near the Wabash EGS during normal summer flows. Near-shore substrates consist of mud mixed with sand changing to sand and gravel toward the middle of the river. The discharge is located immediately downriver from the intake embayments and is separated from the river water by a 120 m long concrete "cell structure" (Figure 115).

The discharge water is heated 7 - 9°C maximum over ambient intake water. When river discharge is low (< 230 m^3/s or 8,000 cfs) the effluent is separated from river water by the cell structure. After clearing the cell structure, the heated plume moves diagonally to the east shore and quickly mixes with river water facilitated by the location of a shallow riffle about 1.2 km (0.75 m) downriver. At higher flows the river water passes over the top of the cell structure and mixes quickly with the heated discharge. The heated plume then moves downriver along the west shore for some distance.

The Cayuga EGS began operating in October 1970 with a 500 MW unit. A second 500 MW unit started up in May 1972. About 37.3 m³/s of cooling water is pumped to remove about 4.88 x 10⁹ Btu/h heat (average monthly heat rate = 9600 Btu/kWh). Operating at full capacity this EGS is theoretically capable of using 100% of the river. The intake is located near the head of a large horseshoe bend in the river which subsequently receives heated discharge water nearly 3.7 km (2.3 m) downriver (Figure 116). The heated discharge is then returned to a canal located close to the intake structure and flows 1.2 km (0.75 m) to finally enter the Wabash River at the lower end of the horseshoe bend. A low wier located midway along the canal aids in pumping water into two small cooling towers.

The river at the Cayuga EGS is about 117 m wide and offers a wide variety of aquatic habitats from slow-moving, sandy-bottom shelves to fast, deep chutes. Woody cover is common, but aquatic vegetation is absent. The thermal plume adheres closely to the west bank for about three kilometers before mixing with river water (Bartolucci, Hoffer, and Gammon 1973). Both the Wabash River EGS and the Cayuga EGS use intermittent chlorination to control fouling organisms.

The river studies at the Wabash and Cayuga EGSs since 1967 have been summarized in numerous special and annual reports, 11 published documents, and 8 M.A. theses which are listed in the Literature Cited section. The following summarizes some of the major activities and findings.

Early studies included field determinations of temperature preference in the immediate vicinity of the Wabash River EGS using several collecting methods (Gammon 1973, 1982). Hoop-nets with 3.8 cm bar-mesh fished without wings or leaders and D.C. electrofishing collections were made in three sections of river. These sites were selected because of their distinct thermal regimes ranging from a cool ambient upriver section to hot effluent in the discharge canal with an intermediate mixed thermal zone. A Smith-Root Type VI electrofisher producing 400 to 600 V DC at 5.5 amps was mounted on a 5 m Jon-boat and used for the electrofishing activities.

Raw catch data were converted to relative catch rates (No/km) to remove variations in annual abundance and gear efficiency by dividing each lower catch value by the highest catch value. These relative catch values were then plotted as a function of temperature (Gammon 1973). Provided with free access to a range of thermal conditions, fish will normally congregate in a "preferred" thermal zone which is dependent upon acclimation temperature and species attributes. Increasing the acclimation temperature will elevate the preferred temperature, but only as high as an *upper preferred temperature* or *final preferendum* (Fry 1969). This final preferred temperature appears to be physiologically optimal for the species in terms of growth, movement, food conversion, etc. Avoidance follows further elevations in temperature or, if cooler water is not found, death at a thermal level called the *upper ultimate incipient lethal temperature* or critical thermal maximum (TMax).

The preferred field temperatures of resident species in the Wabash River and Ohio River were compared to published upper preferenda and upper ultimate incipient lethal temperatures as determined from laboratory studies (Gammon 1982).

There was generally good agreement of values determined through field and laboratory studies. No mortality was observed due to thermally lethal temperatures because behavioral avoidance effectively operated and induced fish to move out of thermally lethal water.

Preoperational assessments of the fish community at the Cayuga electric generating station began in 1968, three years before the first 500 MW unit began generating. A river segment 5.2 km (3.25 miles) long was subdivided into 8 collecting zones. After Cayuga began operating three of these zones were strongly influenced by the heated effluent and two other zones were sporadically affected. Catch statistics for common species and community attributes before (1968-71) and after (1972-81) full start-up of Cayuga EGS were also examined (Gammon 1983). The population responses generally followed predictions based on thermal preferenda with population decreases in thermally sensitive species such as smallmouth bass, redhorse, sauger, mooneye, and goldeye and population increases for thermally tolerant species such as drum, carpsucker, gizzard shad, carp, and flathead catfish.

The calculated Iwb values at both Wabash EGS and Cayuga EGS have fluctuated considerably since 1968, as have the Iwb values at Reach 1, the "ambient" section of river located between Delphi, Indiana and Lafayette, Indiana, (Figure 117). All of the collecting zones at both electric generating stations were included in this example, heated zones as well as unheated zones.

The changes in the fish community following start-up at the Cayuga EGS resulted in a substantial local decrease in the Iwb values. The fluctuations were similar at both electric generating stations after 1970, but the amplitude was more accentuated at the Wabash River EGS than at Cayuga EGS. Iwb values were especially low during the low-flow summers of 1983, 1988, and 1991 and after the 1993 flood.

The overall pattern of Iwb change in Reach 1 was generally similar to that of the electric generating stations (Reaches 6 and 9). Iwb values here were also notably depressed in 1983, 1991, 1993, and 1994, but not during 1988.

The input for the DCA results included the catch data from all 8 zones at Cayuga (Reach 6) combined, together with data from Reach 5 upriver, Reach 7 downriver, and Reach 1. Catches in Reach 1, extending from Delphi, Indiana to Lafayette, Indiana were included as a reference or ambient data set. However, collections of fish from Reach 1 date back only to 1975, hence, the time periods do not exactly coincide.

The fish populations at the Cayuga EGS construction site during 1968, 1969, and 1970 were quite homogeneous as shown by the DCA pattern (Figure 118). They changed only slightly in 1971 after the first 500 MW unit began operating. There is also a surprisingly good correspondence at this time and place with the fish populations in Reach 1 of 1975-93 despite the different time periods. By 1972, however, when both Cayuga EGS generating units were operational, the pattern shifts downward and to the right as indicated by the arrow. The period 1973-85 is marked by great annual variability with the extremely poor catches of 1983 displaying the greatest deviation.

Improvements in the fish community throughout the river, first noticed in 1984, crested in 1985-86. Prior to 1987 the Cayuga EGS discharge water, elevated about 8° C (14° F), was discharged directly into the Wabash River, although the cooling towers were used sporadically to moderate the temperature of discharge water.

Beginning in 1987, however, the mode of operation at the Cayuga EGS was altered by agreement with the Indiana Department of Environmental Management. When the water temperature of the Wabash River increased to the level of 24.5° C (78° F), the heated discharge water was pumped from the discharge canal into the cooling towers. This operational mode was continuous through the hot summer months, but was discontinued when the ambient water temperatures declined again in the fall. Beginning in 1991 the cooling towers started operating when the water temperature reached 24.5°C (78°F) and the river flow decreased less than 4000 cfs.

One interesting question is whether this change in operation had any beneficial effect on the fish community. One important indication of improvement was noted in 1987, the very first year of the modified operation, when the Iwb value of Reach 6 alone increased while in all other Reaches it declined. The increased Iwb was induced in part by the return in small numbers of several thermally sensitive species after a long absence.

The DCA indicates clearly that there was, indeed, a significant improvement in the fish community following 1987. This recovery is indicated by a shift of the fish community toward its preoperational composition for the years 1987 through 1992. As indicated previously, the 1993 data was severely limited by high discharge rates and limited collecting effort which may account for the displacement of that data point. Data from 1994 through 1996 was not included in the analysis because of continued depression by additional periods of high water during mid-summer.

Table 16 examines the changes in catch rates at the Cayuga EGS for three periods. The preoperational period (1968-71) includes the first post-operational year 1971 because the resident fish community was similar to those of the previous three years. The postoperational period extends from 1972 through 1986. The modified post-operational period includes 1987 through 1992, but not 1993 data because of its limitations. It should be emphasized that this period includes two drought years.

The fish listed in Table 16 include the more common and important species of the community which are divided into four groups separated by dashed lines. Carp and flathead catfish both increased numerically after Cayuga EGS started operating. However, the catch rates of both species have decreased since 1987 under conditions of modified operation.

The second group consists of seven species of fish whose catch rates increased after the Cayuga EGS became operational and increased still further when the operation was modified in 1987. They include blue sucker, drum, gizzard shad, channel catfish, shovelnose sturgeon, buffalofish, and white bass. Some of these species, but not all, increased greatly in abundance throughout most of the river during the late 1980s. Nevertheless, it is possible that the modified operation was advantageous to them in the thermally elevated sections of river.

The third group includes smallmouth bass, shorthead redhorse, silver redhorse, spotted bass, mooneye, goldeye, shortnose gar, longnose gar, and northern river carpsucker. Most of these species, but not all, are less thermally tolerant than the preceding groups. All of these species declined in abundance after Cayuga EGS began operating full-time. Additionally, all of them increased in numeric abundance after 1986, indicating improved environmental conditions in the vicinity of the Cayuga EGS as well as throughout the Wabash River.

The last group includes the most thermally sensitive species of fish: longear sunfish, black and white crappie, skipjack herring, golden redhorse, and sauger. All of these species declined in abundance after Cayuga EGS began operating and the modified operating procedures did nothing to enhance their numbers in the vicinity of the power plant. Sauger and skipjack herring populations throughout the Wabash River appear to have grown smaller in recent years.

One consequence of the altered operation at the Cayuga EGS is that the fish community is now much more similar to the community in Reach 5, the segment of river upriver from Cayuga EGS. In addition, the change in operation at Cayuga EGS has also improved the fish populations in Reach 7 downriver where the recent fish communities approach preoperational status.

The Wabash EGS (Reach 9) was already operational when we began studying the river in 1967. Fish collections using DC electrofishing date from 1968. Regular electrofishing collections upriver at Reach 8 and downriver at Reach 10 (Terre Haute, Indiana) began in 1973. The results of the

DCA are based on catch data from Reaches 8, 9, and 10, again with data from Reach 1 serving as a reference or ambient community. The DCA indicates that the fish communities in R8, R9, and R10 are very similar to one another, but that as a group they are quite different from the community of Reach 1 (Figure 119). The fish community in R8 was more variable over time than the other two Reaches. Reach 10, the Terre Haute, Indiana area, exhibited the least temporal variability among the fish populations in these three Reaches. Nevertheless, the fish populations in R10 differed most from those of R1.

Figure 120 contains the same DCA plots as the previous figure, but groups the data points for Reach 9, the Wabash River EGS, into two time periods, a) the operational 1968-88 period and b) the modified operational 1989-92 period.

The Wabash River EGS lacks cooling facilities, but it began operating under a maximum thermal ceiling of 32.2°C (90°F) beginning in September 1988, as measured at the Indiana Highway 63 bridge located 3 km (2 miles) downriver from the plant. When the river temperature reaches that limit, the EGS reduces its generating capacity and, therefore, the amount of heat delivered to the Wabash River.

The fish communities near the Wabash River EGS after initiating the modified operating procedure were very similar to those which preceded them, as shown by the mostly overlapping patterns in Figure 120. From the DCA alone it appears that this change in plant operation had little benefical impact on the fish community, unlike the situation at Cayuga EGS. However, an examination of catch rates of fish before and after the operational change indicates that most species populations were much more common after 1988 (Table 17). However, carp, buffalofish, and crappie abundance was unchanged, while goldeye, skipjack herring, golden redhorse, and sauger populations declined.

Table 15: Species included in the detrended correspondence analysis (DCA).

Shovelnose sturgeon - *Scaphirhynchus platorynchus* (Rafinesque)
Longnose gar - *Lepisosteus osseus* (Linnaeus)
Shortnose gar - *Lepisosteus platostomus* Rafinesque
Goldeye - *Hiodon alosoides* (Rafinesque)
Mooneye - *Hiodon tergisus* Lesueur
Skipjack Herring - *Alosa chrysochloris* (Rafinesque)
Gizzard shad - *Dorosoma cepedianum* (Lesueur)
Carp - *Cyprinus carpio* Linnaeus
Silver chub - *Macrhybopsis storeriana* (Kirtland)
Emerald shiner - *Notropis atherinoides* Rafinesque
Northern river carpsucker - *Carpiodes carpio* (Rafinesque)
Quillback carpsucker - *Carpiodes cyprinus* (Lesueur)
Highfin carpsucker - *Carpiodes velifer* (Rafinesque)
Blue sucker - *Cycleptus elongatus* (Lesueur)
Northern hog sucker - *Hypentelium nigricans* (Lesueur)
Smallmouth buffalo - *Ictiobus bubalus* (Rafinesque)
Bigmouth buffalo - *Ictiobus cyprinellus* (Valenciennes)
Silver redhorse - *Moxostoma anisurum* (Rafinesque)
River redhorse - *Moxostoma carinatum* (Cope)
Black redhorse - *Moxostoma dequesnei* (Lesueur)
Golden redhorse - *Moxostoma erythrurum* (Rafinesque)
Shorthead redhorse - *Moxostoma macrolepidotum* (Lesueur)
Channel catfish - *Ictalurus punctatus* (Rafinesque)
Flathead catfish - *Pylodictis olivaris* (Rafinesque)
White bass - *Morone chrysops* (Rafinesque)
Longear sunfish - *Lepomis megalotis* (Rafinesque)
Smallmouth bass - *Micropterus dolomieu* Lacepede
Spotted bass - *Micropterus punctulatus* (Rafinesque)
Largemouth bass - *Micropterus salmoides* (Lacepede)
White crappie - *Pomoxis annularis* Rafinesque
Black crappie - *Pomoxis nigromaculatus* (Lesueur)
Sauger - *Stizostedion canadense* (Smith)
Walleye - *Stizostedion vitreum* (Mitchill)
Freshwater drum - *Aplodinotus grunniens* Rafinesque

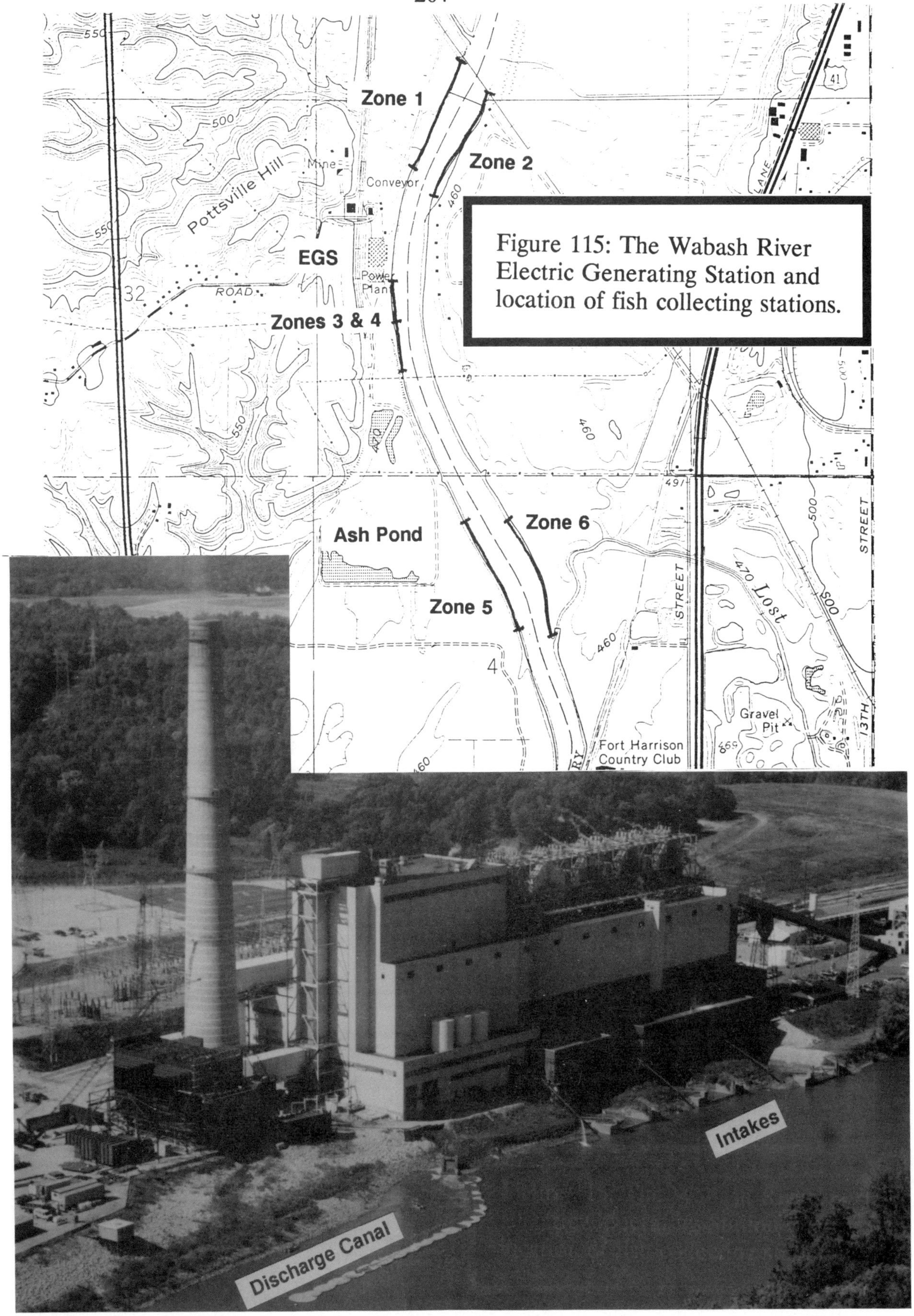

204

Figure 115: The Wabash River Electric Generating Station and location of fish collecting stations.

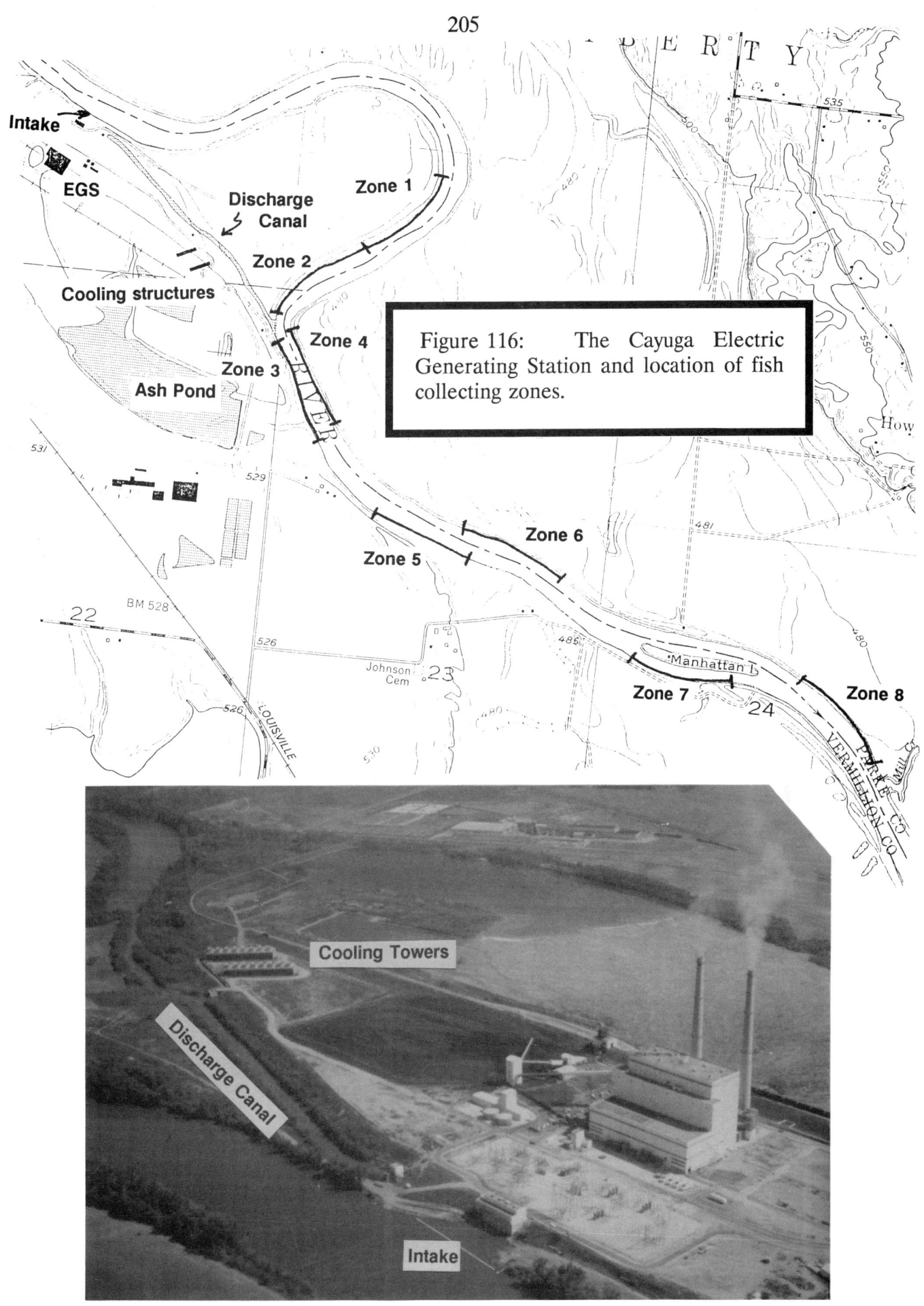

Figure 116: The Cayuga Electric Generating Station and location of fish collecting zones.

206

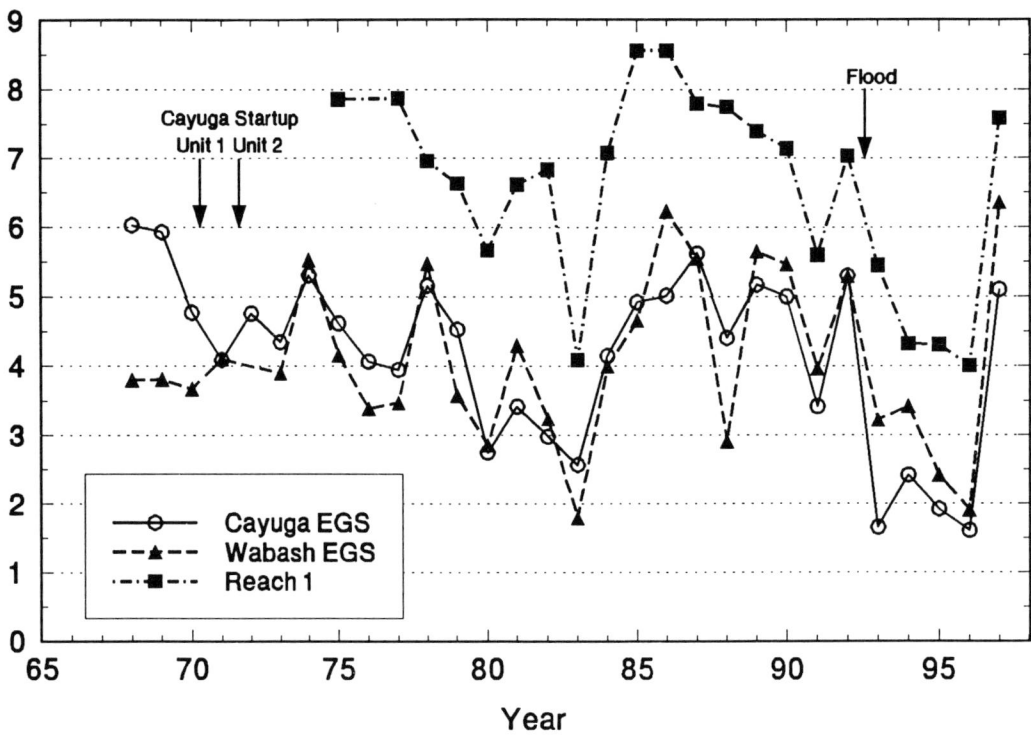

Figure 117: Changes in mean composite index of well-being (Iwb) values at Wabash EGS, Cayuga EGS, and Reach 1.

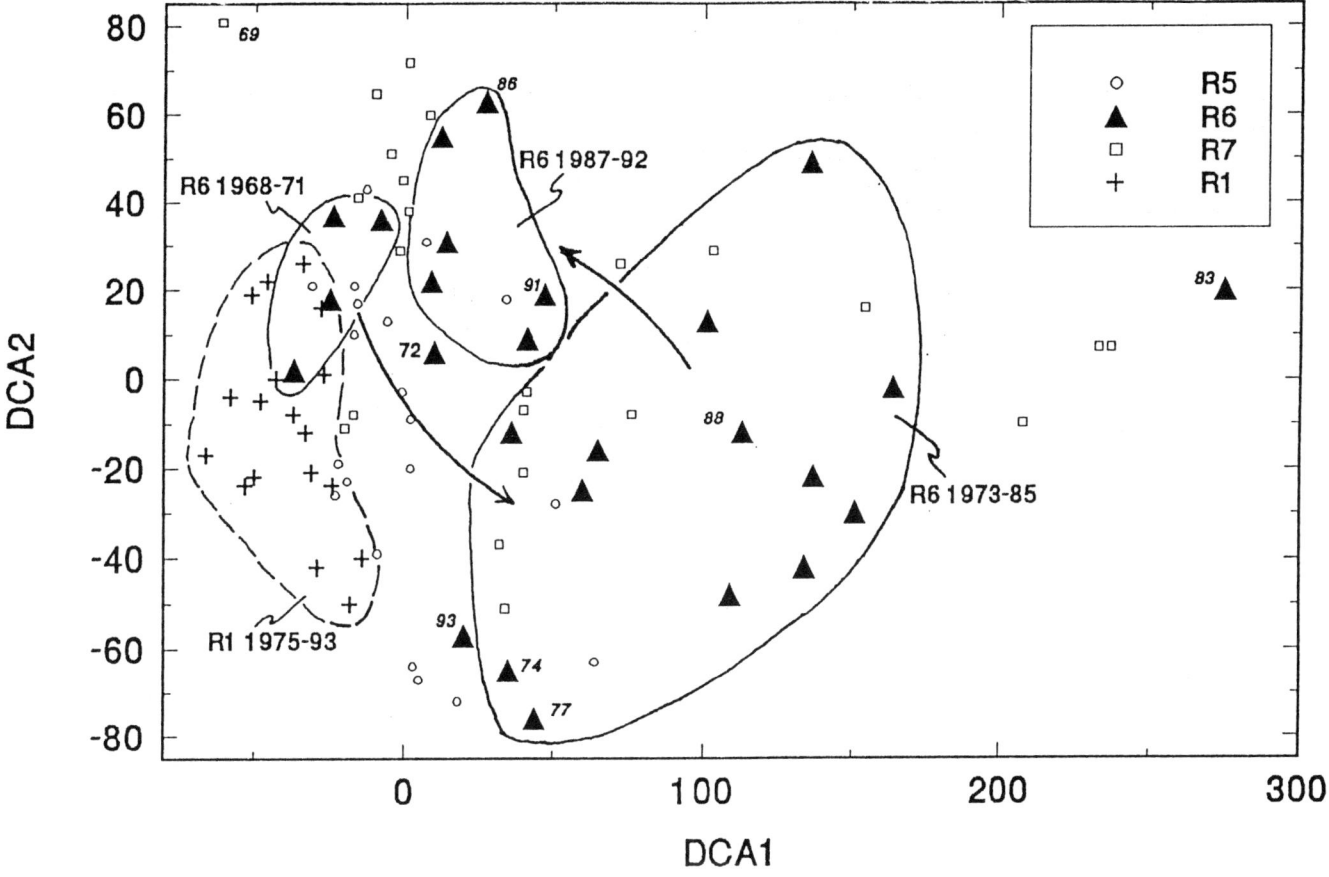

Figure 118: Plots of the first two axes of detrended correspondence analysis (DCA) of species abundance (no/km) at the Cayuga Electric Generating Station (Reach 6) and data from Reaches 1, 5, and 7.

Table 16: A comparison of catch rates of fish (No/km) at the Cayuga Electric Generating Station during the following periods: a) preoperational (1968-71), b) postoperational (1972-86), and c) modified postoperational (1987-92).

Species	1968-71	1972-86	1987-92
Carp	0.546	2.026	1.471
Flathead catfish	0.083	1.634	1.426
Blue sucker	0.025	0.046	0.235
Freshwater drum	0.105	0.127	0.662
Gizzard shad	12.494	30.885	41.515
Channel catfish	0.148	0.198	0.382
Shovelnose sturgeon	0.018	0.046	0.059
Buffalofish	0.116	0.138	0.544
White bass	0.166	0.320	0.515
Smallmouth bass	0.166	0.023	0.103
Shorthead redhorse	1.037	0.277	0.529
Silver redhorse	0.434	0.148	0.338
Spotted bass	0.155	0.090	0.147
Mooneye	0.123	0.071	0.088
Goldeye	0.206	0.164	0.206
Shortnose gar	0.752	0.551	1.103
Longnose gar	0.614	0.565	1.250
Northern river carpsucker	0.932	0.666	2.132
Longear sunfish	0.163	0.046	0.044
Crappie	0.120	0.109	0.029
Skipjack herring	0.155	0.150	0.118
Golden redhorse	0.857	0.364	0.176
Sauger	0.647	0.189	0.147

Total number of catches =	215	514	136
Total km electrofished =	277	434	68

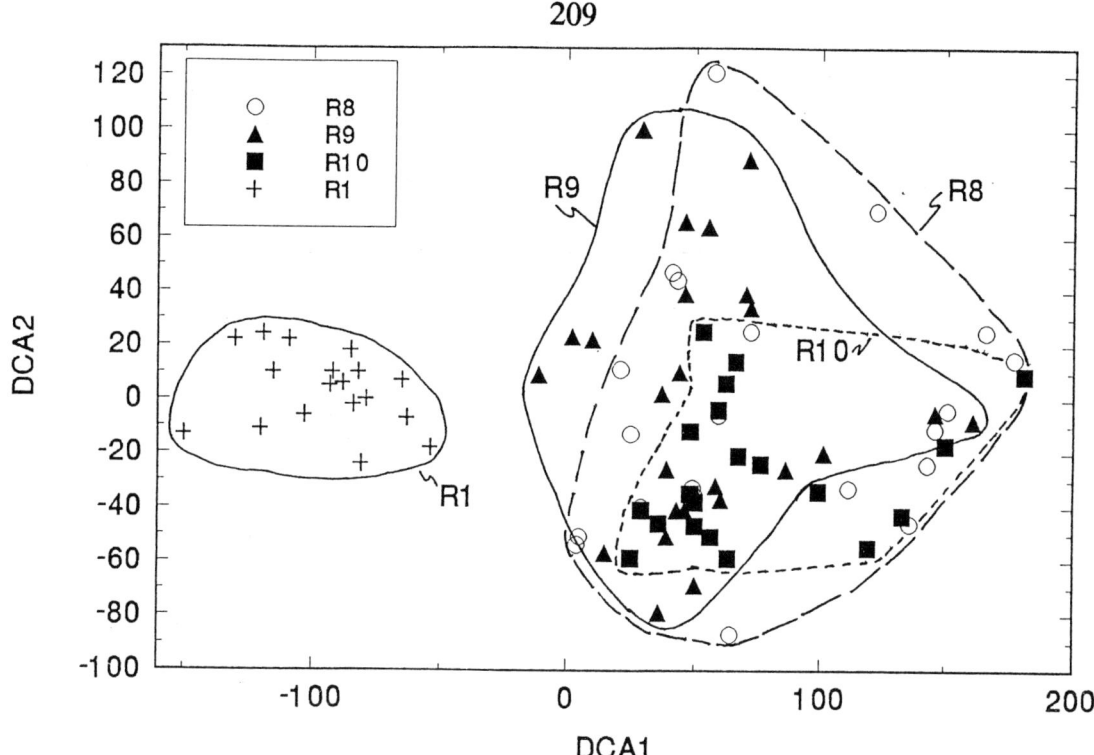

Figure 119: Plots of the first two axes of detrended correspondence analysis (DCA) of species abundance (no/km) at the Wabash EGS (R9) together with flanking Reaches R8 and R10, and R1 for comparison.

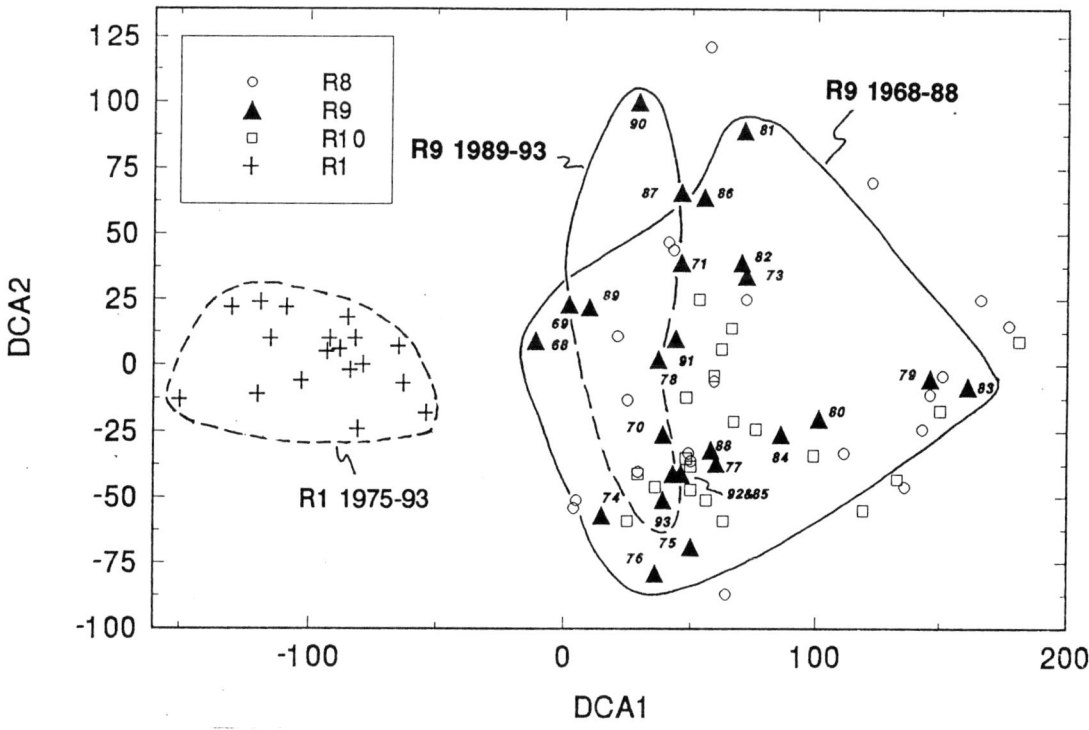

Figure 120: Same DCA plot as the previous figure, but grouping the data at the Wabash EGS (Reach 9) into two time periods: 1968-88 and 1989-93.

210

Table 17: A comparison of catch rates of fish (No/km) at the Wabash River EGS during two time periods: a) 1968-88, and b) modified operation 1989-92.

Species	1968-88	1989-92
Carp	1.891	2.121
Flathead catfish	0.770	2.121
Blue sucker	0.029	0.394
Freshwater drum	0.170	0.818
Gizzard shad	15.658	38.697
Channel catfish	0.510	1.000
Shovelnose sturgeon	0.024	0.450
Buffalofish	0.178	0.182
White bass	0.244	0.758
Smallmouth bass	0.008	0.091
Shorthead redhorse	0.058	0.273
Silver redhorse	0.029	0.121
Spotted bass	0.088	0.182
Mooneye	0.088	0.364
Goldeye	0.475	0.303
Shortnose gar	1.054	1.879
Longnose gar	0.906	2.333
Northern river carpsucker	0.783	1.818
Longear sunfish	0.027	0.273
Crappie	0.056	0.061
Skipjack herring	0.337	0.212
Golden redhorse	0.037	0.030
Sauger	0.149	0.091

Total number of catches = 418 66
Total km electrofished = 377 33

THE EFFECT OF THE ELI LILLY AND COMPANY PLANT AT CLINTON, INDIANA

Eli Lilly and Company operates two large pharmaceutical facilities on the middle Wabash River, the Tippecanoe Plant located on the south edge of Lafayette, Indiana at Rkm 497 (RM309) and the Clinton Plant located between Montezuma and Clinton, Indiana at Rkm 374 (RM232). Analyses of data collected near the Clinton Plant are presented here. Data collected near the Tippecanoe Plant will be considered as part of the Lafayette/West Lafayette area municipal/industrial complex.

Two data sets are available to evaluate the effect of the effluent from Eli Lilly's Clinton facility on the Wabash River fish community: 1) the regular annual electrofishing collections of fish in adjacent upriver and downriver zones which flank the effluent canal and 2) the results of special electrofishing studies conducted in 1984 and 1991. An additional special series of collections was made when the effluent system was altered and moved upriver, but that data is not considered here.

The locations of the collecting sites at the Clinton Plant are shown in Figure 121. The same sites were used both in 1984 and 1991 except for the addition in 1991 of site E where the new effluent canal was relocated during the summer of 1994.

Data from regular electrofishing collections at sites immediately above and below the effluent canal were compared for the years 1973 through 1992 using a Detrended Correspondence Analysis (DCA).

Data from the years 1979 and 1993 were excluded because of insufficient collecting effort. The length of arrows shown in Figure 122 indicate the relative differences in the fish populations above and below the plant. The longer the arrow the greater the compositional differences between upstream and downstream populations.

Assuming that effluent volume and toxicity remained fairly constant over time, any negative impacts to the fish community should be greater during low-flow periods than during periods when river discharge was "normal" or high. Therefore, population differences between upriver and downriver stations should be greater during drought years of 1977, 1988, and 1991.

In 1977 and 1988 the populations above and below the effluent canal differed substantially. In 1991, however, the differences were no greater than for other high or normal flow summers. In 1977, the first upriver station electrofished was located at RM335 (A) immediately above the mouth of Big Raccoon Creek. Catches downriver from the effluent canal at Summit Grove (RM231 or G) included longear sunfish, spotted bass, smallmouth buffalo, blue sucker, longnose and shortnose gar while those upriver included bowfin and goldeye in addition to other species common to both locations (Table 18). Therefore, the community downriver appeared subjectively to be somewhat "better" than the community upriver. Values of various community parameters, however, were very similar.

212

In 1988 stations D and G constituted the flanking collecting sites. Catches upriver at D were considerably more diverse than those downriver at G and included channel catfish, white bass, two species of *Moxostoma*, and particularly large numbers of northern river carpsuckers in addition to species common to both sites. Community values were substantially higher upriver than downriver as a result.

Fish populations above and below the plant were quite similar in 1991 when the river flow was also extremely low. However, all community values were higher downriver from the effluent canal than upriver.

Thus, the DCA and compositional examination yielded inconsistent results. In the absence of preoperational data there is no convincing evidence that the effluent from the Eli Lilly Clinton plant consistently exerts a negative impact on the fish community of the adjacent Wabash River.

Next we considered the second series of data. Two special electrofishing collections were made during the periods August 6-15, 1984 and July 18 to August 14, 1991 in order to examine more closely the effect of Eli Lilly - Clinton effluents on the Wabash River fish community. As indicated previously, 10 sites were sampled by electrofishing 3 times each in 1984 and 11 sites were electrofished 3 times each during 1991. Catch data (no/km) for all sites for both years were analyzed by DCA. The results of these studies which compare upriver catches with downriver catches are summarized in Figure 123.

Variability among stations was considerably greater in 1991 than it was in

1984. In 1984 the fish communities downriver from the Eli Lilly Clinton plant were very similar to those upriver as indicated by the intersecting hypervolumes. In 1991 there was less similarity, but still an intersection of the above-below hypervolumes. Station G, immediately downriver from the effluent canal, and station I deviated most from populations upriver. The catch data indicates somewhat fewer species at these two sites than elsewhere, but nothing unusual otherwise.

Composite index of well-being (Iwb) profiles for 1984 and 1991 are shown in Figure 124. The higher Iwb values for the 1991 catch clearly reflect the improved fish community found in the Wabash River in recent years. Locally weighted least-squares linear regression (Lowess) lines suggest that the fish communities downriver from the effluent canal were depressed in 1984, but not in 1991. The summer of 1984 was not a particularly low-flow summer and the effluents for Eli Lilly - Clinton would probably have been dispersed and diluted more effectively that year than during 1991 when river discharge was lower. Included in Figure 124 are the Index of Biotic Integrity (IBI) values using the unmodified criteria discussed in an earlier chapter.

The overall results are inconsistent and inconclusive. Considering all of the evidence at hand, there is no convincing evidence that the Eli Lilly-Clinton plant has had any negative impact on the nearby fish community of the Wabash River. The IBI patterns are very similar to the Iwb patterns for both years, thereby reinforcing the conclusion that effluent from the plant has little or no effect on the fish communities downnriver.

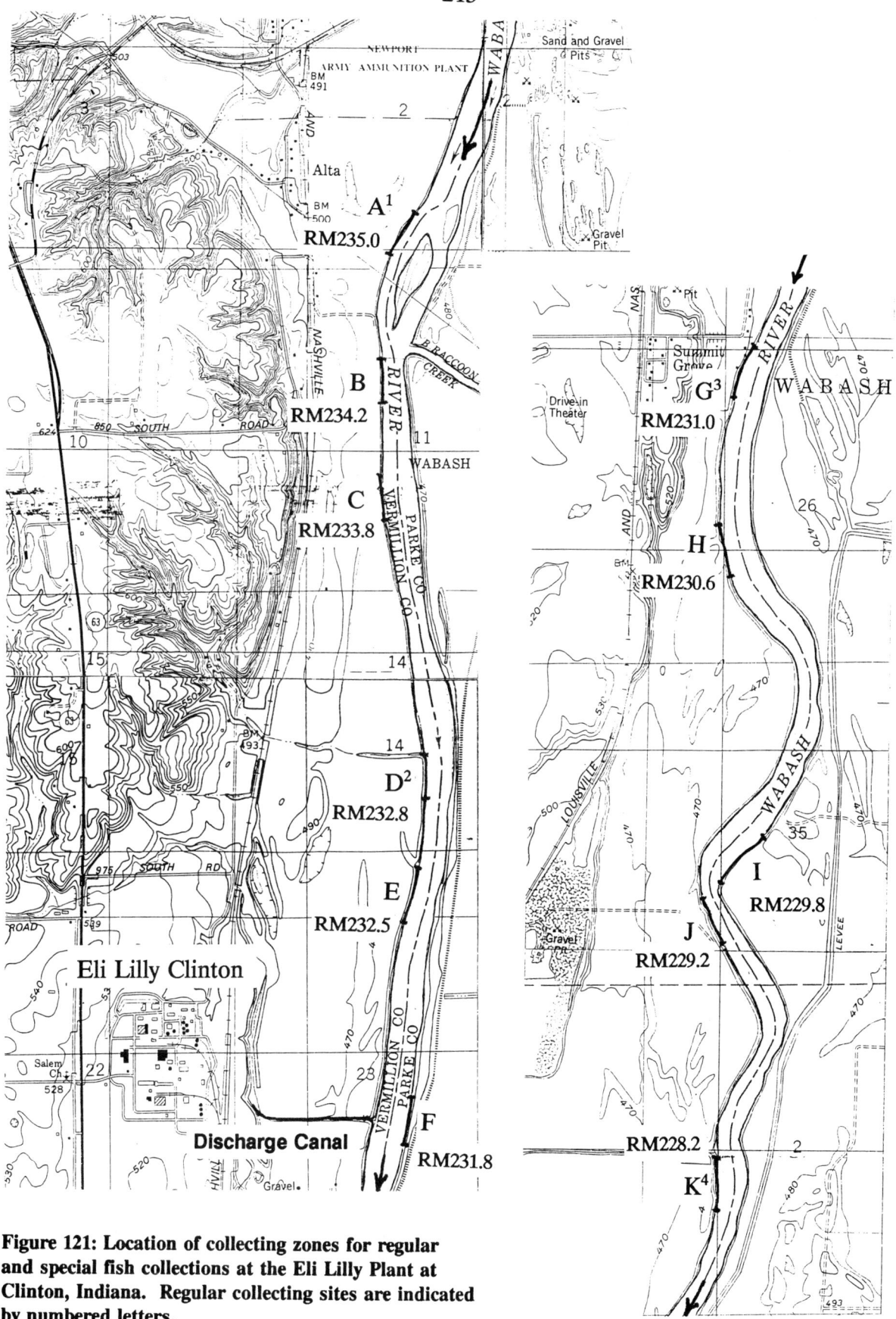

Figure 121: Location of collecting zones for regular and special fish collections at the Eli Lilly Plant at Clinton, Indiana. Regular collecting sites are indicated by numbered letters.

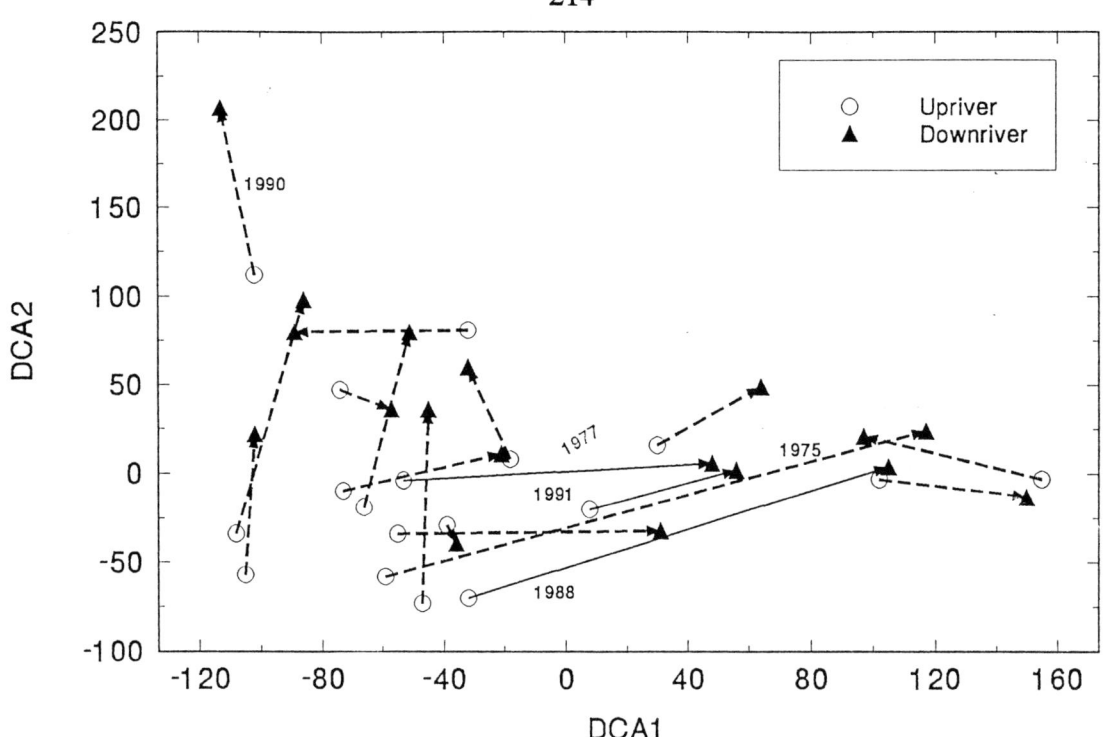

Figure 122: Plots of the first two axes of detrended correspondence analysis (DCA) of species abundance (no/km) at regular upstream and downstream collecting stations closest to the Eli Lilly Clinton effluent canal.

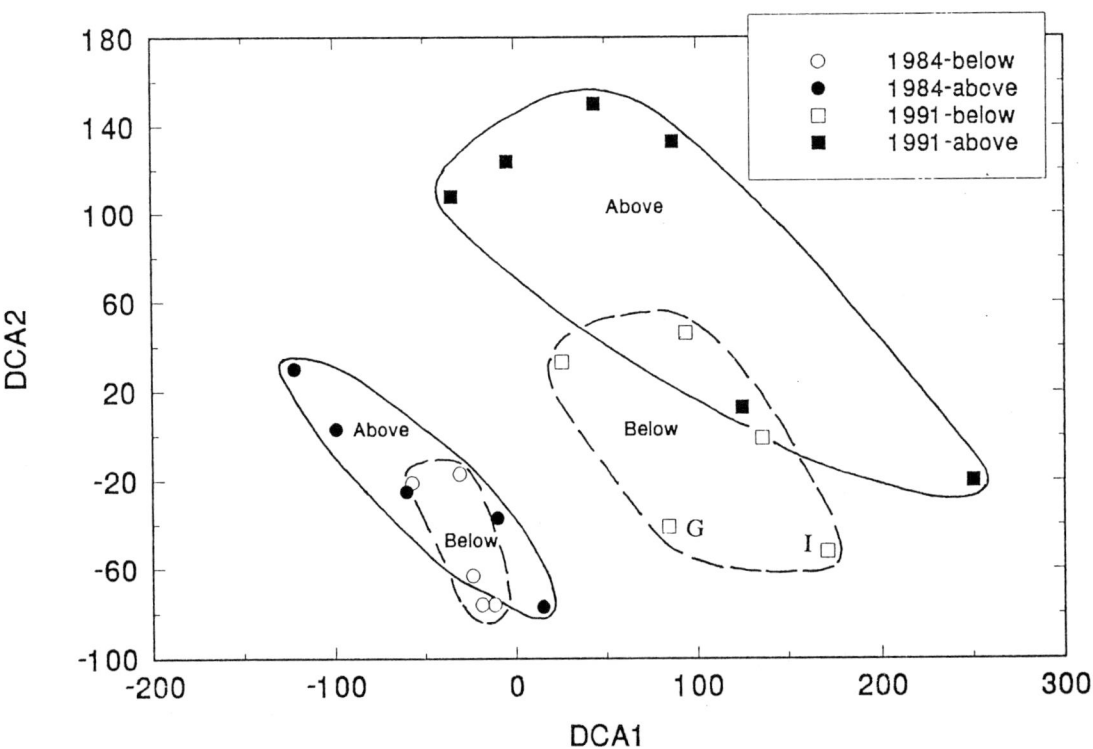

Figure 123: Plots of the first two axes of detrended correspondence analysis (DCA) of catch data taken in special studies in 1984 and 1991.

Table 18: Electrofishing catch rates (No/km) above and below the Eli Lilly - Clinton effluent canal during the drought summers of 1977, 1988, and 1991.

Species or Community Parameter	1977 Above (A)	1977 Below (G)	1988 Above (E)	1988 Below (G)	1991 Above (E)	1991 Below (G)
Longnose gar	-	1.333	-	-	-	-
Shortnose gar	-	0.333	0.667	0.667	-	0.667
Gizzard shad	29.333	19.333	2.000	3.333	4.667	1.333
Goldeye	0.333	-	-	-	-	0.667
Channel catfish	-	-	1.333	-	-	-
Flathead catfish	2.000	0.667	4.000	8.000	3.333	9.333
White bass	-	-	1.333	-	-	1.333
Freshwater drum	1.000	-	0.667	-	0.667	0.667
Carp	1.333	3.667	5.333	3.333	3.333	4.000
Silver redhorse	0.667	-	0.667	-	0.667	-
Golden redhorse	-	-	0.667	-	-	-
Shorthead redhorse	1.667	0.333	-	-	-	-
Northern river carpsucker	0.667	0.667	11.333	-	1.333	-
Blue sucker	-	0.333	-	-	-	-
Spotted bass	-	0.333	-	-	-	-
Number per kilometer	37.667	27.667	28.000	15.333	14.667	21.333
Kilograms per kilometer	13.094	15.080	22.842	10.654	10.309	18.593
Ave. no. species per collection	3.667	3.667	5.667	2.667	4.000	5.000
S-W Diversity(no.)	0.749	0.830	1.390	0.713	1.277	1.418
Composite Index of Well-being	4.747	4.468	5.340	3.436	4.724	5.412

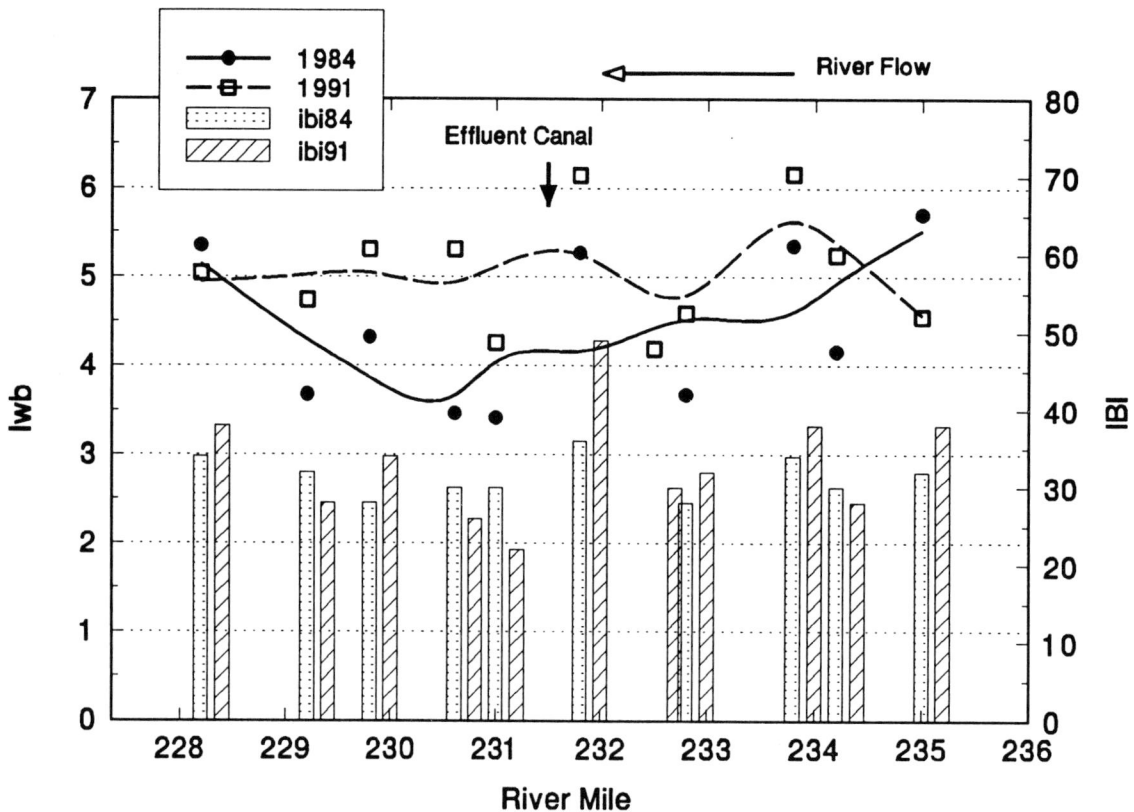

Figure 124: Composite index of well-being (Iwb) profiles derived from
electrofishing catches of fish during species studies of 1984 and 1991
together with IBI values obtained from unmodified criteria.

THE EFFECT OF THE LAFAYETTE/WEST LAFAYETTE METROPOLITAN AREA

Evaluation of the Lafayette/West Lafayette section of the middle Wabash River is a major challenge because of the close proximity of multiple influences, all of which potentially affect water quality negatively to some extent. The upriver boundary of this section of river is the US 52 bridge (Rkm 504 = RM 313) and the lower boundary is Ouiatenon (Rkm 493 = RM 305), but the urban influence extends downriver at least as far as Granville Bridge.

In addition to general urban nonpoint-source pollution which periodically and diffusely enters along 10 km (6 miles) of river, there are 4 major point sources (Figure 129). In order from upriver to downriver, they are: 1) the West Lafayette, Indiana Sewage Treatment Plant (STP) effluent, 2) the Lafayette, Indiana STP, 3) the Eli Lilly Tippecanoe Laboratory STP, and 4) Wea Creek. These point sources enter the river at approximately 1.5 to 2.0 km (1 mile) intervals. The Elliot Ditch tributary of Wea Creek, as well as lower Wea Creek, have been major sources of industrial pollutants for decades, but may be less polluted in recent years (Bridges, personal communication). An additional effluent will soon enter the Wabash River between the Lafayette and West Lafayette STPs.

A minimum of five fish collecting stations were established between the US 52 and Granville bridges, including these point sources, since 1977. Figure 125 shows the recent locations of some of these sites. Rather than examine individual years, the data has been grouped into 4 periods of time: 1) 1975-80, 2) 1981-85, 3) 1986-90, 4) 1991-93. Each period includes one low-flow summer (1977, 1983, 1988, and 1991.

The composite index of well-being (Iwb) profile through this section of river is shown in Figure 126. The fish community with the highest Iwb is located immediately downriver from the West Lafayette STP before the effluent thoroughly mixes with river water. The fish communities deteriorate progressively as mixing occurs downriver and as the other effluents are added. During the period 1975-80, the cumulative depression of water quality and the fish community was conveyed well downriver from Ouiatenon. During the last three periods, however, the recovery appeared to be more rapid. The depression was very mild during 1986-90, a period which was characterized by the "best" fish communities throughout the river.

The DCA analysis indicates the same basic pattern of changes through the Lafayette\West Lafayette section of the river (Figure 127). The fish communities of 1975-80 were qualitatively different from those in the other three periods so that the population changes which occurred as the river flowed through Lafayette/West Lafayette area are indicated by open circles and arrows. Major changes take place between Mascouten and the West Lafayette STP, after which some slight degree of recovery occurs. However, the populations continue to deteriorate downriver from Ouiatenon.

218

The patterns of population changes which the last three periods exhibit are similar to each other, but differ from the 1975-80 pattern. During the period 1975-80 the populations were severely altered between Mascouten and the West Lafayette STP, after which they recovered slightly before continuing to change in the region downriver from Wea Creek. The populations recovered by the time the river reached Granville Bridge after 1980.

During the drought years of 1977, 1983, 1988, and 1991 the quality of the fish communities downriver from the Lafayette and West Lafayette Sewage Treatment outfalls declined substantially. (Figure 126). The lowest Iwb values occurred below Wea Creek and Ouitenon, as was the case with other years, but the communities recovered by the time the river reached Granville Bridge.

The various community indices and catch rates of species populations were also examined at Mascouten, the Ouiatenon area, and the first two stations downriver from Granville bridge ("The Hills" and Collier Island) by combining all of the catches from 1975 through 1993 (Table 19).

All community index values were lower in the Ouiatenon section than in flanking areas. Relative biomass (kg/km) was particularly depressed, because the age/size of most species of fish residing in that section was lower than in flanking areas. Goldeye, mooneye, spotted bass, and blue suckers were much less abundant in the Ouiotenon section and the first three species remained scarce downriver from that point.

Catch rates of catfish, carp, white bass, silver redhorse, shorthead redhorse, and carpsuckers were also depressed at Ouiatenon, but then recovered downriver.

Several other species of fish were at least as abundant at Ouatenon as they were in flanking areas, including shortnose gar, gizzard shad, drum, golden redhorse, longear sunfish, and smallmouth bass.

It is impossible to distinguish individual effluent effects in the Lafayette\West Lafayette area because no effects were instantaneously exerted, with the possible exception of chlorination products. Nor could point-source impacts be separated from the general urban nonpoint-source effects. Rather, it appears that the depression in the fish community between Wea Creek and Granville Bridge was probably the result of the cumulative impact of all of these effluents combined. However, the negative effects that Lafayette/West Lafayette do exert on the riverine fish community appear to be limited to a relatively short distance after which the direct effects fade.

This is not to say that urban contributions of nutrients do not have some measureable influence further downriver on such biotic components as phytoplankton, for example. This important and poorly understood aspect needs to be addressed by future research.

Iwb profiles of low-flow years (1977, 1983, 1988, and 1991) were similar to those of the five-year groupings (Figure 132). The summers of those years were particularly dry with less than normal nonpoint-source pollution from city streets. Therefore, the impacts exhibited probably reflect the sum of point-source pollutants, ie. the STPs at Lafayette, West Lafayette, Eli Lilly's

219

Tippecanoe Laboratory, and other industries. The recovery of fish populations is fairly complete by the time the river reaches Granville bridge. The profiles of the composite index of well-being (Iwb) indicate that the "best" fish populations were usually found immediately downriver from the West Lafayette STP. A similar concentration of fish occurred at the Terre Haute, Indiana STP.

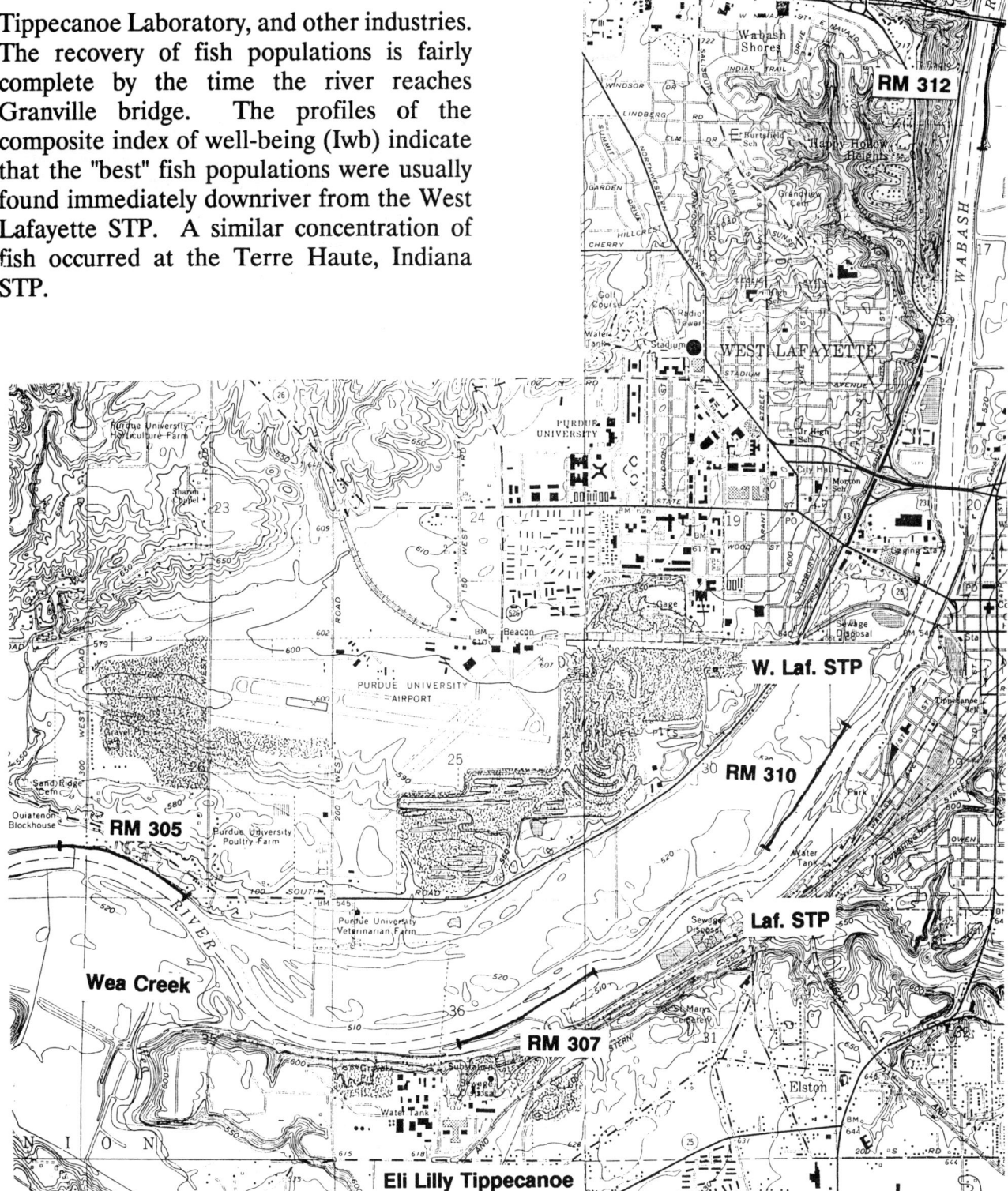

Figure 129: Map of the Lafayette\West Lafayette\Eli Lilly - Tippecanoe Laboratory area showing the location of fish collecting sites.

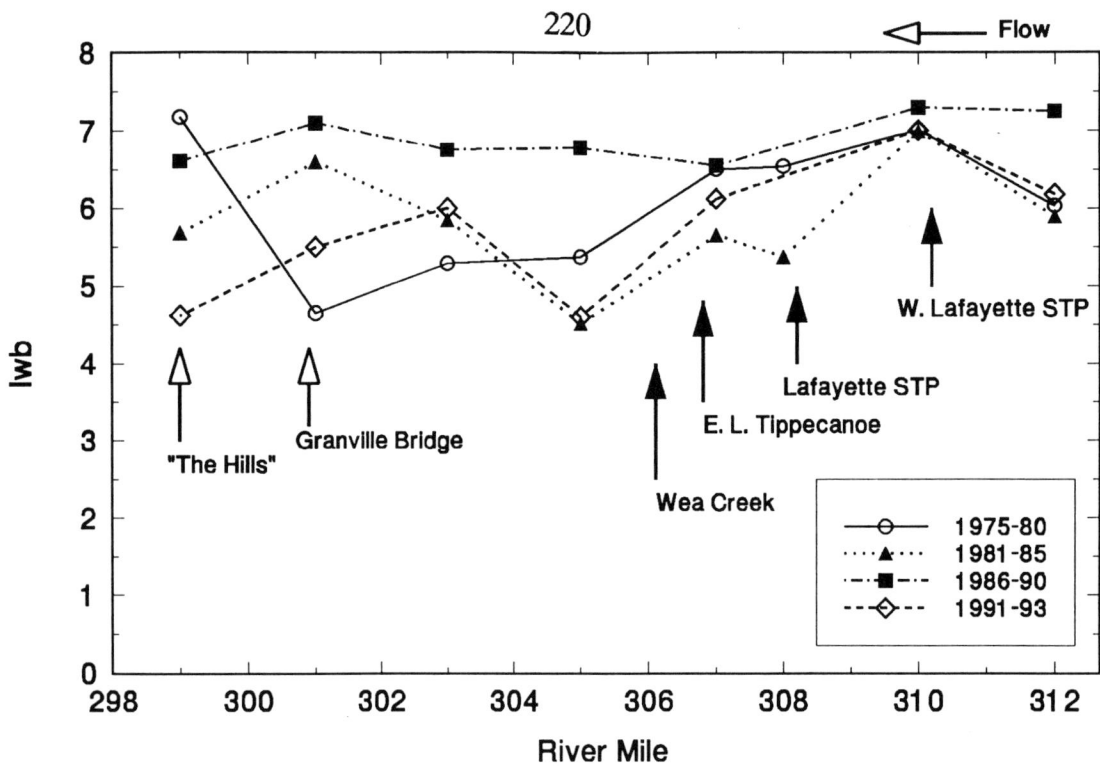

Figure 126: Composite index of well-being (Iwb) profiles for the Wabash River at Lafayette\West Lafayette grouping the data into five year blocks. Each block contains one low-flow summer.

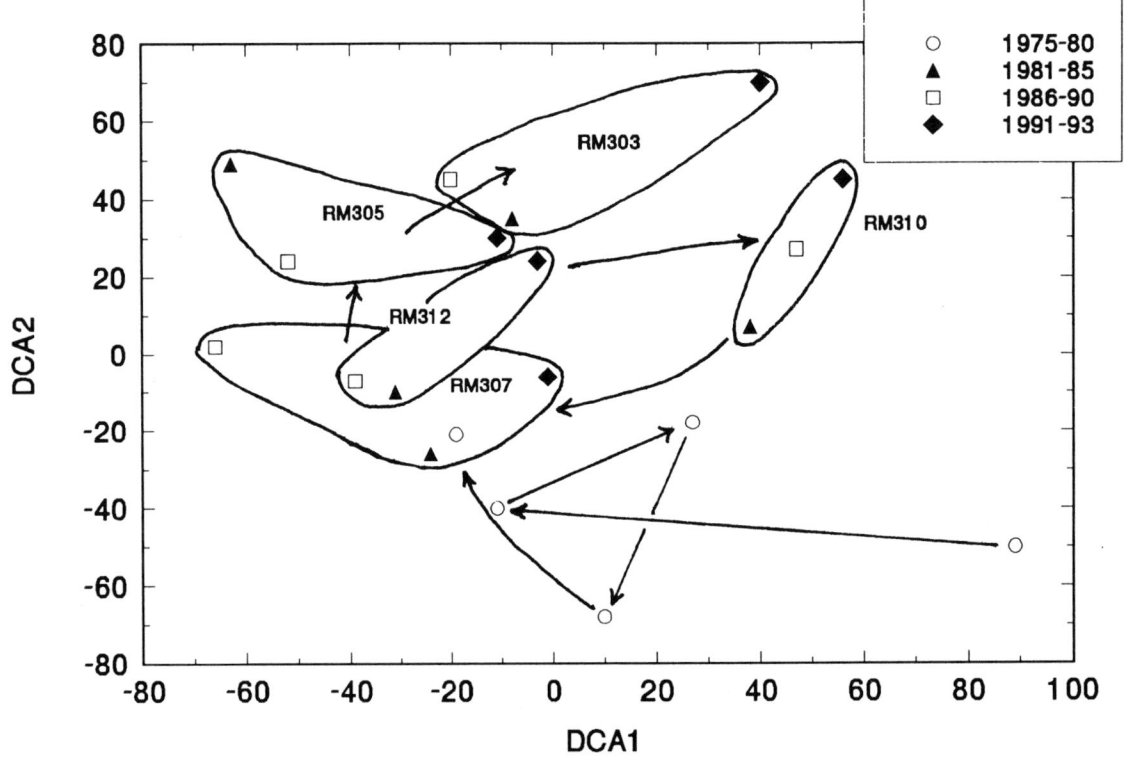

Figure 127: Plots of the first two axes of detrended correspondence analysis (DCA) of species abundance (no/km) for the Wabash River at Lafayette and West Lafayette, Indiana.

Table 19: Mean community values and species abundance during the period 1975-93 at (a) Mascouten, (b) Ouiotenon, and (c) "The Hills" and Collier Island.

Species\Community Index	Mascouten (RM 312)	Ouiotenon (RM 305)	"Hills" & Collier Is. (RM 301 & 299)
Longnose gar	0.816	0.634	0.821
Shortnose gar	0.327	0.697	0.442
Gizzard shad	15.429	14.388	10.442
Goldeye\Mooneye	3.102	0.845	1.179
Channel catfish	2.490	1.437	2.189
Flathead catfish	0.449	0.592	1.137
White bass	1.510	1.225	1.326
Freshwater drum	1.265	1.310	0.905
Carp	6.735	4.627	6.211
Silver redhorse	1.673	0.697	1.326
Golden redhorse	1.673	1.859	1.642
Shorthead redhorse	2.776	1.416	3.874
Northern river carpsucker	4.735	3.317	3.242
Blue sucker	0.490	0.021	0.379
Longear sunfish	0.531	0.676	0.653
Spotted bass	0.367	0.169	0.168
Smallmouth bass	0.286	0.845	0.253
Sauger	0.857	0.761	0.589
Number per kilometer	49.755	38.381	40.400
Kilograms per kilometer	28.374	18.097	28.194
Ave. no. species/collection	8.306	6.663	7.705
S-W Diversity(no.)	1.623	1.458	1.624
Composite Index of Well-being	6.364	5.729	6.256

Total Number of catches =   49   95   95
Total km electrofished =   24.5   47.3   47.5

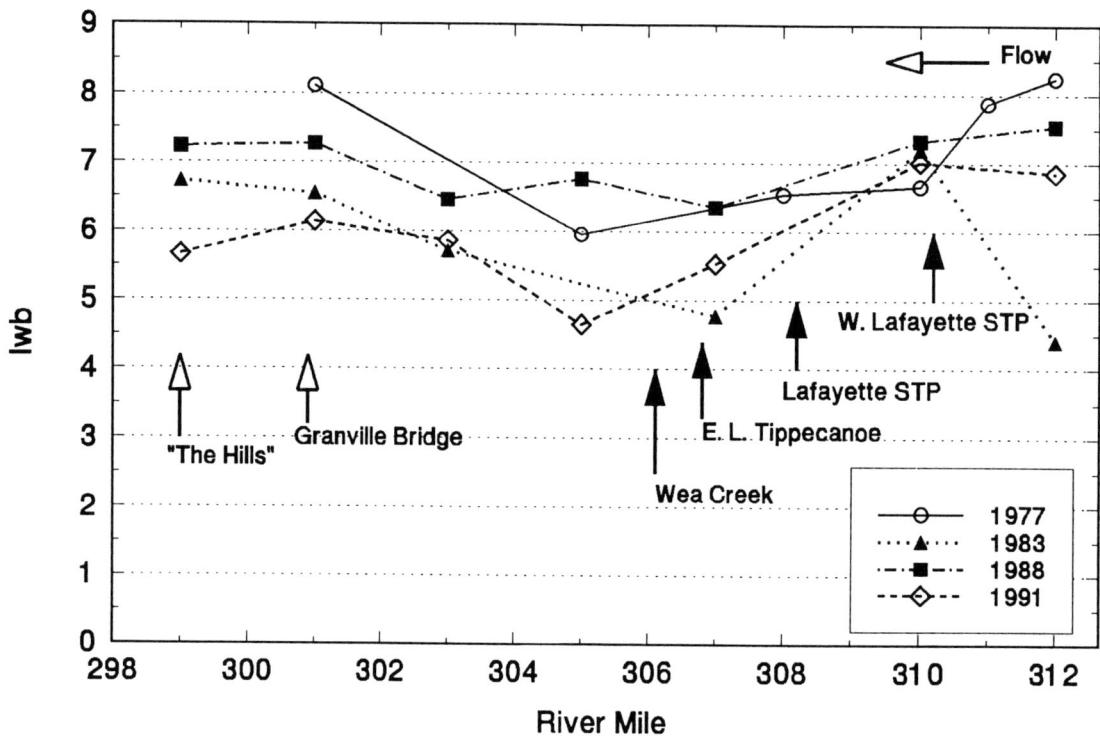

Figure 128: Composite Index of Well-being (Iwb) profiles for the Wabash River at Lafayette\West Lafayette during low-flow summers of 1977, 1983, 1988, and 1991.

THE EFFECT OF THE TERRE HAUTE
METROPOLITAN AREA

The population density of the Terre Haute area is approximately the same as the Lafayette/West Lafayette area 160 km (100 miles) upriver. The Wabash River, which forms the western boundary of Terre Haute, is heated by Cinergy Corp.'s Wabash River EGS 8 kilometers (5 miles) upriver. Many of the city's manufacturing plants are concentrated on about 3 km (2 miles) of "high ground" on the Wabash River's east shore. Treated domestic wastes from Terre Haute enter the river at about Rkm 330 (RM 205).

Three fish collecting stations were located in the metropolitan area (Figure 129). The most upriver site (RM 211) was situated on the west bank just upstream from the Conrail RR (old N.Y. Central) bridge. This station was intended to be an "ambient" site for the Terre Haute area since it is not subject to nonpoint-source runoff from Terre Haute streets and is located as far downriver as possible from the thermal influence of the Wabash EGS. However, catches of fish at that site were unsatisfactory over the years.

The next closest site downriver was located immediately adjacent to a vertical sea wall (RM 210). It was not considered to be good habitat and was deeper than desirable at high river levels. Nevertheless, catches of fish at this station were surprisingly diverse.

The third site was located several kilometers downstream in the southern part of Terre Haute at RM 207. Originally established downriver from the I-70 bridge, it was relocated to its present position above the

bridge because of increasing water depths and decreasing catches of fish at the old site.

The fourth site in the Terre Haute area was located at about RM 204, a short distance downstream from the Terre Haute STP effluent. It consisted of a good gravel bar with relatively shallow, fast water and was regarded as excellent aquatic habitat.

Other downriver stations included in the analysis which are not shown were RM 203 near the Federal Penitentiary, RM 199 immediately upriver from Honey Creek, RM 197, RM 192, and RM 191. Both of the latter stations are located above Big Creek. Some of these stations were situated quite far apart, but in the best habitat available. For analytic purposes the data were grouped into five-year blocks of time beginning in 1974 when fish were first collected in this area.

The Iwb profiles through this section of river are shown in Figure 130. Bridges, geographical features, and the locations of three larger tributaries are also indicated.

Little Sugar Creek, entering from the west and downriver from the I-70 bridge, entered the Wabash River as an iron-red stream until this decade. It originates in the now rehabilitated Green Valley mine, close to the Indiana-Illinois border. As indicated in the chemical\physical section, this deep coal mine was closed in 1963, but the 100+ acre gob-pile continued to pass acid, heavy metals, and sediment into Little Sugar Creek (Thomas 1978) and the Wabash River until fall of 1994. Honey Creek, which enters the Wabash River at Rkm 320 (RM 198) from the northeast, has a history of pollution from coal strip mines in its headwaters (Whitaker and Wallace 1973).

Iwb profiles indicate that the fish community improved as the river flows past the Terre Haute metropolitan area. The best fish community was consistently found between the I-70 bridge and the Terre Haute STP. The community rapidly deteriorated below the Terre Haute STP, however, and finally recovered 21 km (13 miles) downriver near the mouth of Big Creek.

The DCA indicates the relative overall differences in the fish communities at the fish collection sites over time (Figure 131). The spatial trajectory is indicated for the periods 1984-88 and 1989-93. The best fish community was found during 1989-93 at RM 207 and the worst at RM 203. During the period 1979-83 when the Iwb at all sites was depressed there is a clustering of sites at the upper left area.

More specific comparisons were made of the fish communities at three sites for the period 1984-1993: (a) RM 207 upriver from the Terre Haute STP, (b) RM 203 three km (2 miles) downriver from the Terre Haute STP, and (c) RM 191-192 just above Big Creek. As in the Lafayette/West Lafayette area, all of the community values were lower at the most impacted site at RM 203 compared to upriver and downriver sites (Table 20). The biomass of fish was particular low, 10.927 kg/km, compared to 32.202 kg/km above the STP and 20.904 kg/km near Big Creek primarily because the average size of many species was much lower at RM 203 than in flanking collecting sites.

Species populations whose catch rates (no/km) were lower at RM 203 than at flanking areas included gizzard shad, channel catfish, flathead catfish, white bass, freshwater drum, carp, northern river carpsuckers, redhorse, and sauger and walleye.

Iwb profiles of drought years (1977, 1983, 1988, and 1991) were lower relative to the 5-year Iwb profiles (Figure 132) with an even deeper depression downriver from the Terre Haute STP.

The overall impact of the Terre Haute metropolitan area appears to be measureably less after 1983 than before. Furthermore, there appears to be a small, but consistent improvement in the recovery of the fish community downriver from the Terre Haute STP during the period 1989-93 compared to the previous five-year period 1984-88. Nevertheless, the zone of recovery is long, with evidence for negative impacts over a distance of approximately 17 km (11 miles) downriver from the Terre Haute STP.

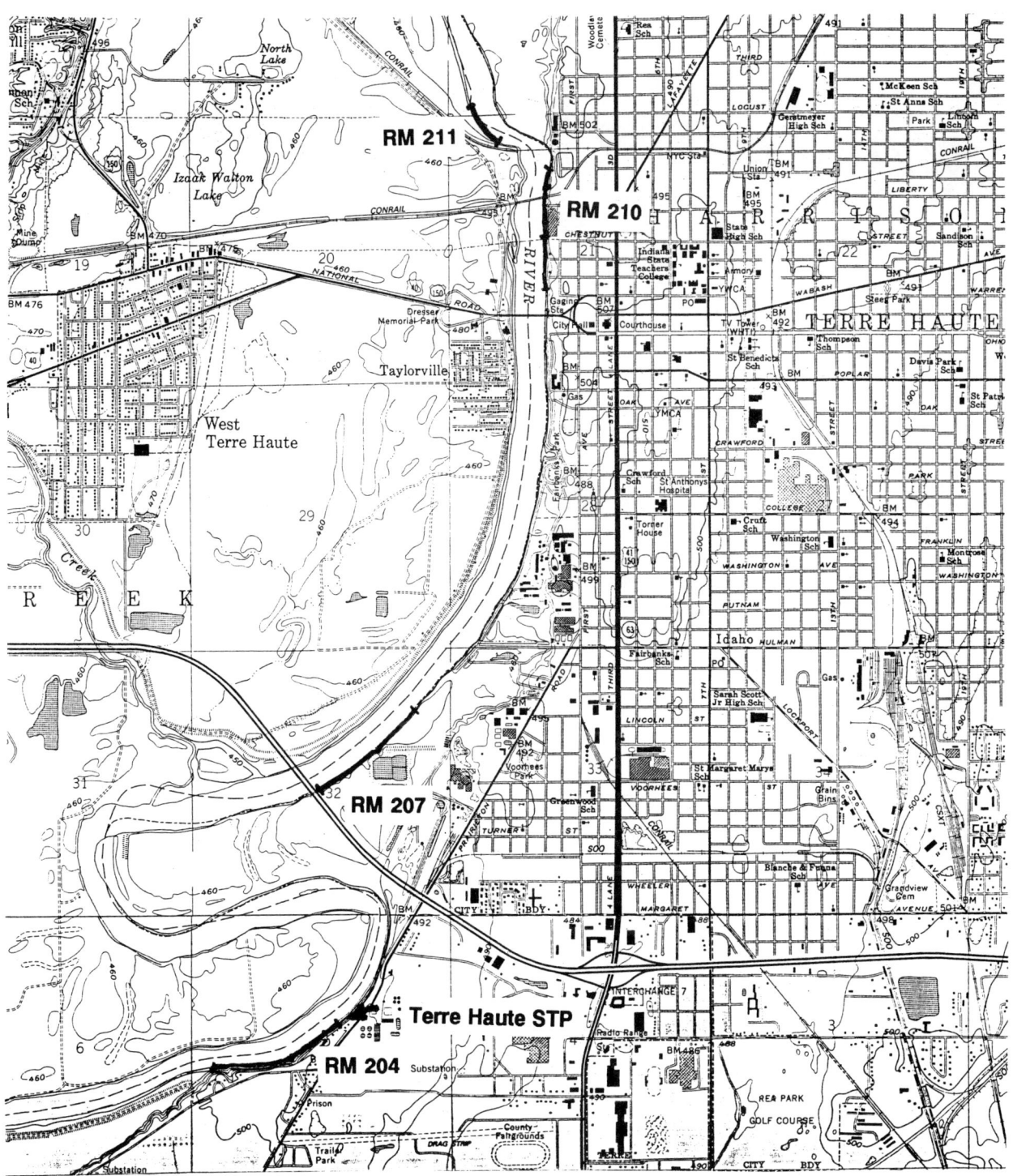

Figure 129: Location of collecting sites at Terre Haute, Indiana.

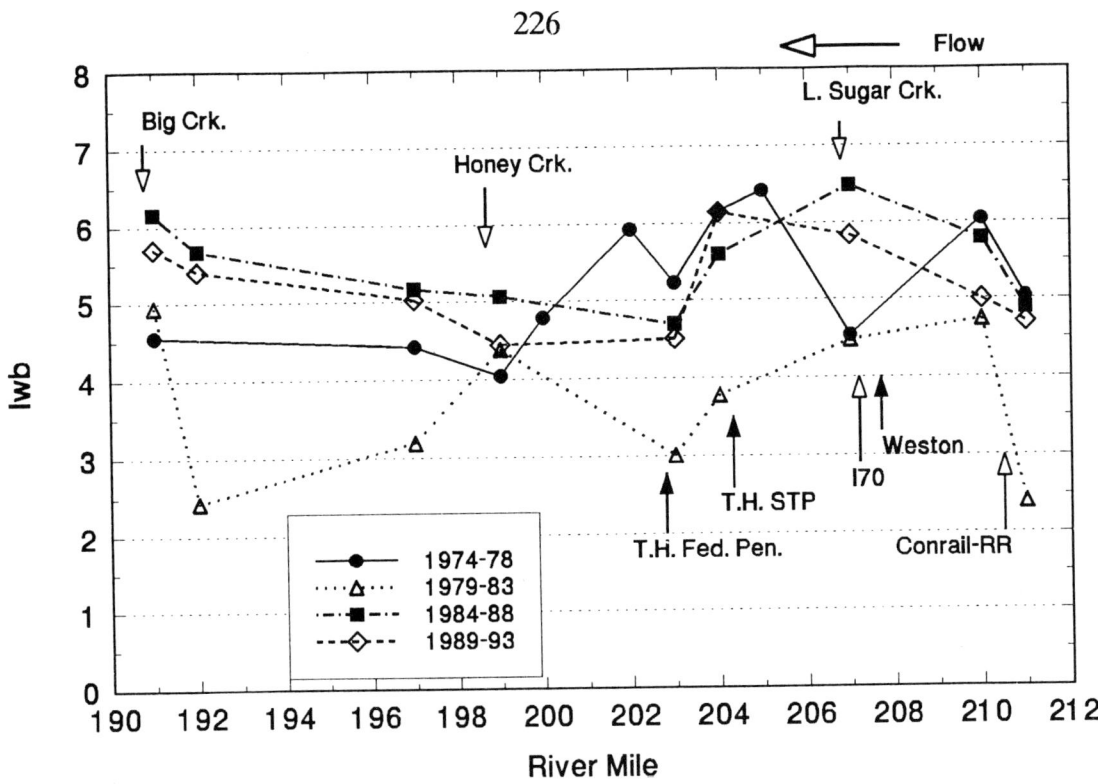

Figure 130: Composite Index of Well-being (Iwb) profiles for the Wabash River at Terre Haute, Indiana grouping the data into five year blocks, each of which includes one low-flow summer.

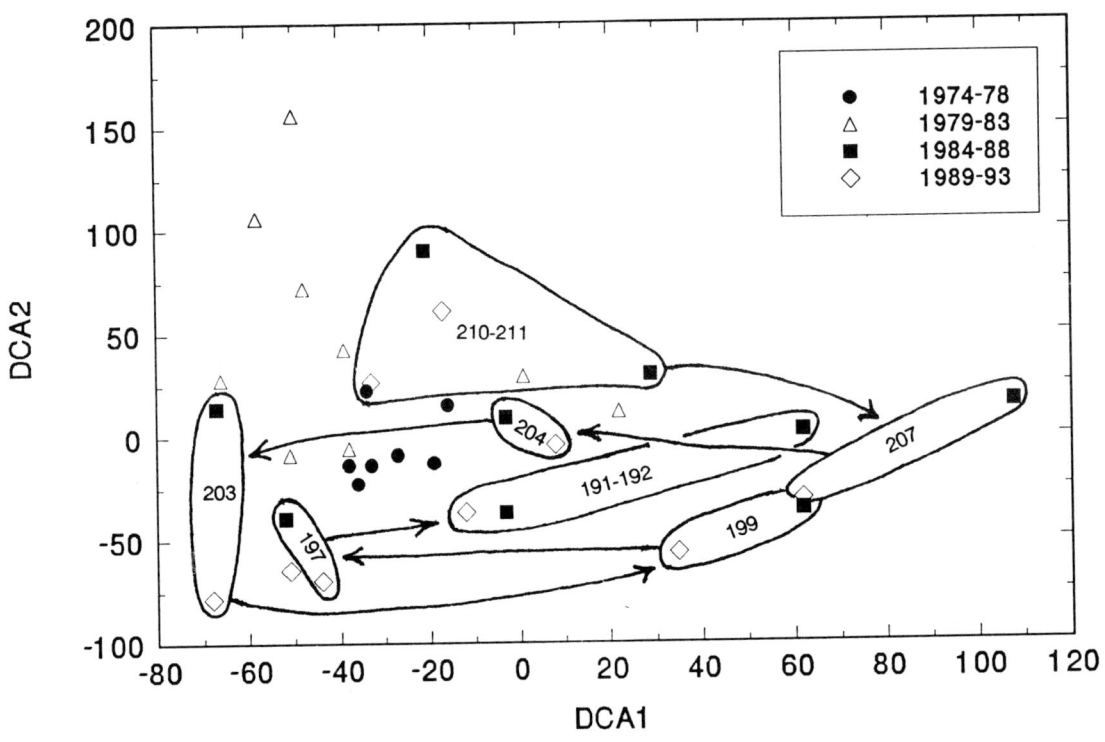

Figure 131: Plots of the first two axes of detrended correspondence analysis (DCA) using species abundance (No/km) at collecting stations in the Terre Haute, Indiana area.

Table 20: Mean community values and species abundance during the period 1984-1993 at (a) RM 207, (b) RM 203, and (c) RM 191-192 combined.

Species\Community Index	Above T.H. STP (RM 207)	2 miles below STP (RM 203)	Above Big Creek (RM 191-192)
Shovelnose sturgeon	-	-	0.185
Longnose gar	0.222	1.111	1.926
Shortnose gar	1.259	5.037	4.407
Gizzard shad	26.519	12.593	14.185
Goldeye\Mooneye	0.518	1.211	0.481
Channel catfish	12.444	0.444	5.185
Flathead catfish	4.148	2.370	3.519
White bass	1.556	0.296	0.556
Freshwater drum	3.333	0.074	1.037
Carp	8.074	0.815	5.037
Redhorse species	0.150	-	0.259
No. river carpsucker	4.444	0.222	1.000
Blue sucker	0.074	0.148	0.704
Buffalofish	0.220	0.444	0.444
Centrarchid bass	-	0.074	0.222
Sauger\Walleye	0.520	-	0.111
Number per kilometer	64.593	25.259	40.185
Kilograms per kilometer	32.202	10.927	20.904
Ave. no. species\coll.	6.593	4.074	5.944
S-W Diversity(no.)	1.337	1.090	1.370
Index of Well-being	6.192	4.608	5.753

Total Number of catches =	27	27.0	54
Total km electrofished =	13.5	13.5	27

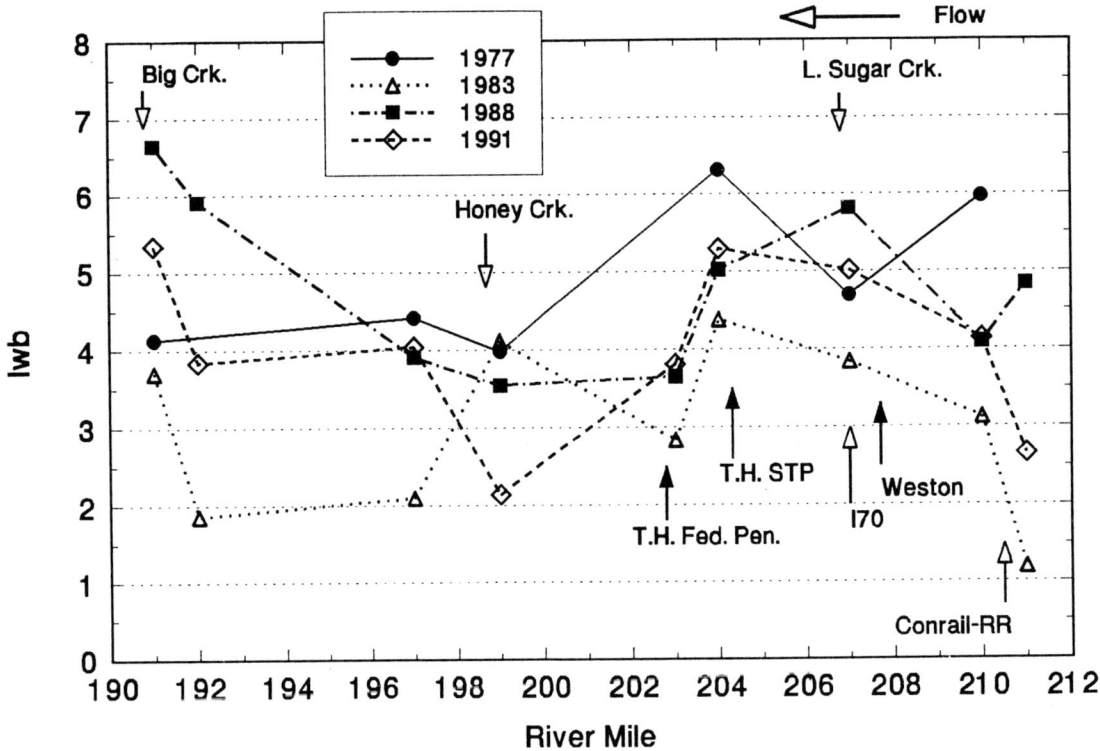

Figure 132: Iwb profiles for the Wabash River at Terre Haute, Indiana during low-flow summers of 1977, 1983, 1988, and 1991.

VALIDATION OF THE IWB AND DCA PATTERNS

Collecting sites throughout the middle Wabash River were placed in habitats which included relatively fast current, moderate depths, and a gravel bottom. This kind of habitat was most often found on the outside of gradual bends in the river. Electrofishing here yielded about 50% more individuals and 25% more species than slow-moving, deeper, or midstream locations during 1968-70 at the Cayuga EGS construction site (Gammon 1973). An additional positive factor was that fast-water habitats also tend to be avoided by the ubiquitous and numerous gizzard shad which favor the still-water habitats of backwaters, eddys, and inside river bends.

This criterion was developed as the outcome of sampling a great range of habitat types during the 1968-70 preoperational electrofishing studies at the Cayuga EGS. As indicated previously, the eight zones studied at Cayuga EGS included a) the shallow, slow, and sandy inside of the river bend (Zone 1), b) the deep cutbank habitat on the outside of the river bend (Zone 2), c) relatively slow current with moderate depths over a sandy bottom (Zones 4 and 6), d) moderately fast current with shallow water over a sandy bottom (Zone 8), and e) fast current, moderate depths, and a gravel bottom (Zones 3, 5, and 7). Zones 3, 5, and 7 were much more productive when electrofished and served as examples of the type of ideal habitat sought for all of the collecting zones established when the "long distance" surveys began in 1973.

In the absence of a methodology for evaluating habitat in large rivers such as the

Wabash, it might be supposed that the depressions in the fish communities down-river from the Lafayette\West Lafayette and Terre Haute metropolitan areas could have resulted from locating the collecting sites in areas of poor habitat. The lack of suitable habitat was one of the reasons some collecting sites were located so far apart. Nevertheless, it is possible that habitat alone could generate the observed Iwb and DCA patterns at Lafayette\West Lafayette, Indiana.

The influence of habitat type on electrofishing catches and the resulting DCA patterns was tested by analyzing two data sets simultaneously: a) the electrofishing catches in 8 zones at preoperational Cayuga EGS during 1968-70, and b) the electrofishing catches in 8 collecting sites in the Lafayette\West Lafayette area (RM 299 to RM 315) during the years 1986-90.

Habitat diversity was much greater within the Cayuga EGS Reach than it was within the collecting sites in the Lafayette\West Lafayette segment. However, the Cayuga EGS zones were distributed over a relatively short distance (5.2 km = 3.25 miles) of river while the sites at Lafayette\West Lafayette were dispersed over a longer stretch of river. Nevertheless, several of the collecting sites in the immediate vicinity of Lafayette\West Lafayette (RM 307, 305, and 303) were located very close together.

The results of the DCA are illustrated in Figure 133. The direction of river flow between adjacent collecting sites is indicated by arrows. The composition of the fish community at the RM 310 site was substantially different from that of flanking collecting stations RM 312 and RM 307. However, even if this is ignored, there is still much less similarity among the fish communities at RM 307, 305, 303, and 301 than among the fish communities in Zones 1, 3, 5, and 7 at Cayuga. Therefore, it seems reasonable to conclude that the dissimilarities are caused by differences in water quality and not variability of habitat.

This DCA pattern also illustrates the similarities of fish communities between upriver sites RM 315 and RM 312 and downriver sites RM 301 and RM 299 from Lafayette\West Lafayette, indicating that the fish communities have recovered from metropolitan influences by the time the river reaches the "Hills" (RM 301) and Collier Island (RM 299).

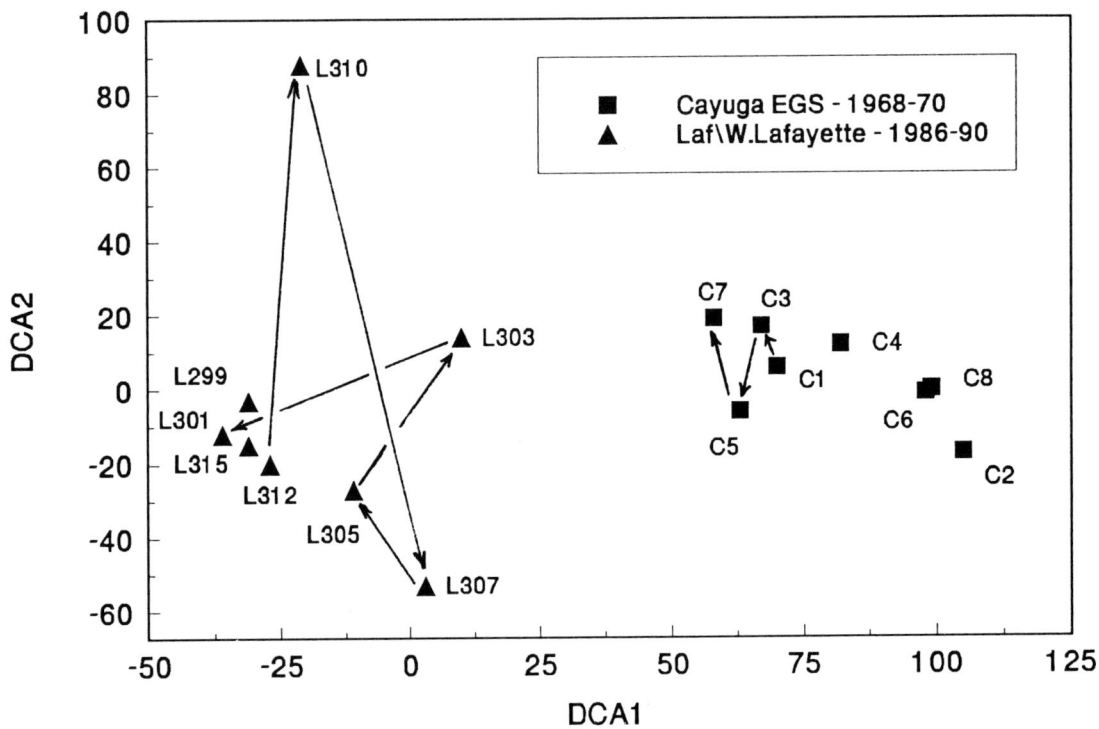

Figure 133: Plots of the first two axes of detrended correspondence analysis (DCA) of species abundance (no/km) at preoperational Cayuga EGS and near Lafayette\West Lafayette, Indiana.

DISCUSSION

Most rivers and streams in Indiana are negatively affected by human activities. Nevertheless, some riverine fish communities are surprisingly diverse and healthy, even in waters which appear to be less than optimum. Most of the fish species still live today in the Wabash River just as they did 175 years ago when Rafinesque (1820) and Lesueur studied them. Some species have disappeared, such as the lake sturgeon (*Acipenser fulvescens*), muskellunge (*Esox masquinongy ohiensis*), and stargazing darter (*Percina uranidea*). Other species are rare; the northern pike (*Esox lucius*), alligator gar (*Lepisosteus spatula*), and crystal darter (*Ammocrypta asprella*), for example. Still others, such as the walleye pike (*Stizostedion vitreum*) persist only because tributary reservoirs are regularly stocked. However, increased collecting activities in recent years have disclosed remnant populations of some species previously thought to have been extirpated.

The initial decline in quality of the river environment was associated with a rapidly increasing human population which cleared native vegetation from the land and converted it to agriculture. Excess agricultural production soon required navigable waters and the state of Indiana obliged by constructing canals over nearly four decades beginning at Fort Wayne in 1828. The canals were finally abandoned in the mid-1860's because of high upkeep expenses, gross mismanagement, and growing competition from railroads, but not before inflicting serious and widespread ecological damage to Indiana's aquatic systems.

Despite the extensive changes in landscape and new industrial impacts in the 1880's, Jordan (1890) found the "fish fauna of the Lower Wabash . . to be unexpectedly rich,. ." especially in the abundance and number of species of darters at the five sites he visited. However, an increase in the human population and industrial activities soon led to gross organic pollution and adverse human health effects (Craven 1912).

Four decades of investigative neglect followed Jordan's departure to Stanford University before Gerking (1945) systematically collected fish throughout Indiana, including 16 sites on the Wabash River. He found fewer species of darters at the sites sampled by Jordan which he attributed to increased siltation from soil erosion as well as the influence of "city sewage, cannery waste, coal mine drainage, paper mill waste, and dairy-products factory waste."

The Wabash River was again ignored for two more decades following Gerking's benchmark study during which time many of the state's reservoirs were constructed. The reservoirs not only reduced the incidence of downstream flooding, but they also provided recreational opportunities such as boating, swimming, and fishing as well as real estate development.

The Wabash River is too small and its flow too unreliable to serve as an important transportation corridor. Nevertheless, the Army Corps of Engineers has been prevailed upon to assess its potential in this regard nearly every decade. Neither is there sufficient recreational boating and fishing activity to support much marina development along its length at the present time.

The murky, aesthetically displeasing appearance of the Wabash River almost certainly inhibited its recreational useage in the past and continues to do so today. This disturbing quality and the problem of difficult access for boats limits its use by anglers as well as the scientific community.

It may surprise some that the Wabash River constitutes more than 40 square miles of surface area, thereby making it the largest body of water contained entirely within the state. Only the Lake Michigan portion of Indiana is larger. As such, the Wabash River represents an underused resource of great potential value, especially since over 10% of Indiana's population lives within easy reach of the river. That aspect alone, however, has failed to spark discernable interest on the part of the public or governmental units.

Perhaps the view of the Wabash River as an overlooked, but potentially valuable, asset is overly optimistic. Indeed, until the last decade there was little reason for optimism. As a newcomer to Indiana over 30 years ago, I regularly drove over this "dirty" river on regular visits to Wisconsin. The contrasts of the Wabash River to the clear waters of the Chippewa, Flambeau, and Namekagon Rivers of northern Wisconsin were hardly flattering. However, I found that superficial appearances were deceptive and that a rich and abundant aquatic life actually thrived in the Wabash River. In 1965 Bob Poppe and I spent an eventful October afternoon collecting fish near Montezuma, Indiana with a hastily constructed electrofishing apparatus. We found that an astounding abundance and diversity of fishes

lived and apparently flourished beneath the clouded surface. From that time to the present the Wabash River has grudgingly and sporadically yielded its mysteries.

We have monitored the fish community of the Wabash River mainstem annually since 1967, but not until 1984 did the fish community conspicuously improve, perhaps for the very first time since Indiana was settled. Significant population increases for most species followed during the next few years. Prior to that time only a few sport fish were found in the electrofishing catches and the catch rate averaged slightly more than two game-fish per kilometer. The average catch rate quadrupled after 1984 as most sport fish populations enlarged. Even small-mouth bass became abundant in parts of the upper Wabash River. Commercial and sport fishing pressure remained relatively light during this period and did not appear to change much over the period of study.

During the sudden and substantial population growth, some fishes extended their range into areas of the river in which they did not or could not live previously. The density of carp, however, remained unchanged and the gizzard shad population actually declined sharply because of increased predator pressure on young-of-the-year. The average size of many species increased, possibly because of greater longevity or perhaps due to faster growth or maybe because of a combination of both factors.

What elements in the river environment were altered to have kindled this positive reorganization? An examination of changes in key chemical constituents of the river provides few clues and a simple answer to that question is unlikely. The improvement probably was the result of a combination of events including a gradual, but substantial (50%) reduction in BOD (organic) loading to the river because of cumulative improvements in point-source waste treatment in conjunction with other events.

The multivariate statistical analysis of catch data and spatial and temporal patterns of the Composite Index of Well-being (Iwb) both indicate that negative local impacts have decreased during the past 10 years at Terre Haute, Indiana. Conditions have also improved over the past 15 years at Lafayette\West Lafayette, Indiana. Furthermore, it is likely that positive local improvements have occurred at population centers throughout the Wabash River basin because of upgraded treatment of human and industrial wastes. It was also demonstrated that negative thermal impacts were reduced at the Cayuga Electric Generating Station because of a relatively small change in operational procedures.

There is no doubt that the concentration of dissolved oxygen (DO) in water profoundly affects aquatic life, especially the fauna, and it appears that a positive change in dissolved oxygen may have played a role in the faunal improvement. The dissolved oxygen model proposed for the middle Wabash River by HydroQual (1984) provides some insights into the basic problem.

The story begins with the amount of living and dead organic matter in the river (biological oxygen demand or BOD) and its strong influence on the oxygen status of the river. The greater the amount of BOD, the greater the removal of dissolved oxygen from the water. More oxygen is consumed by in-river processes than is generated by

photosynthesis on the average summer day in the Wabash River and a dissolved oxygen deficit or DOD is the result.

Summertime is the critical period where dissolved oxygen is concerned. When the summer flow of the Wabash River is low and stable the HydroQual model projected DOD values of 2.0 to 2.5 mg/l from Delphi to Montezuma and then sharply increased values of about 4.0 mg/l from Terre Haute to Merom. During the summer months, then, the DO concentration is two to four milligrams per liter less than the river could hold at saturation. This spatial model is the exact reverse of the pattern exhibited by the Iwb profile and suggests an inverse relationship between the two.

Furthermore, HydroQual estimated that phytoplankton respiration was the largest single user of DO and was responsible for an estimated 50-60% of the DOD in the upper reaches and about 70% downriver from Terre Haute, Indiana.

BOD entering from multiple point sources along the river was the second largest user of oxygen. By far the largest proportion of BOD entering the middle Wabash River, 61% to 77%, was estimated to originate from nonpoint, agricultural sources.

Decomposing BOD accounted for about 10% of the DOD in the upper reaches and over 15% downriver from Terre Haute. Sediment oxygen demand or SOD, which is the removal of oxygen because of decomposition of organic materials by the bottom sediments, was estimated to be the third largest oxygen user. Other oxygen utilizing processes such as respirational uptake by fish and benthos contributed only to a negligible extent.

Most of the BOD in the Wabash River consists of decomposable organic matter and phytoplankton. Although the average annual BOD concentration has declined significantly, monthly average levels are as high as ever from July through September because of high densities of phytoplankton. In the upper Wabash River during this period phytoplankton density gradually increases as the river flows southward. The highest algal densities occur between Layayette and Terre Haute, Indiana but then remain fairly constant further downriver.

The low-flow summers of 1983, 1988, and 1991 produced good reproduction and survival of young fish through the first year of life. Perhaps coincidently, or maybe significantly, there was a 25% reduction in agricultural loadings to the river in 1983 during the Payment-in-Kind (PIK) program when farmers were paid not to grow corn and soybeans. From 1985 through 1992 river discharge during the summer was less than average. Less rainfall presumably resulted in less than usual nonpoint-source pollution.

The relative importance of these various contributing factors have not been assessed, but weather during the decade of the 1990s has profoundly affected fish populations throughout the Wabash River. The Wabash River did not flood as severely as the Mississippi River in 1993, but prolonged high water during midsummer nevertheless essentially eliminated young year-classes of most resident species. The summers of 1992 and 1996 also elevated the river for shorter periods of time. High discharge often occurs in early summer, eg. 1968, 1973-74, 1980-82, 1984, 1986, and 1989. However, prolonged high discharge during mid-summer only occurred during 1992 and 1993. The impact on most larger species of resident Wabash River fishes was devastating. Not only was there a direct negative effect on most species populations, but also there was virtually no reproductive success. Populations of larger species in the Wabash River were at their lowest ebb in 1994 and not until 1997 was recovery convincingly evident.

What must be done to promote continued improvement of the Wabash River ecosystem? One very troublesome part of the overall problem is that of limiting the delivery of nutrients to the Wabash River, thereby reducing densities of algae. If this can be accomplished to some degree then the resulting effect should be an improvement in the dissolved oxygen and dissolved oxygen deficit balance. Less algae would also result in less turbidity of the water which would be highly beneficial to sight-feeding game-fish. The river's aesthetic qualities would also be enhanced if algal densities can be diminished during the summer recreational season.

Agricultural practices as a whole are of paramount importance when agriculture dominates river basins as they do throughout the midwest. An agricultural revolution of sorts has been underway since about 1990, a revolution which has the potential to benefit our aquatic ecosystems. Conservation tillage practices such as chisel plowing, ridge-till, and no-till have rapidly been adopted throughout the U.S., but particularly in the cornbelt states. All types of conservation tillage reduce soil erosion by minimizing physical disturbance of the soil and maintaining a protective surface coating of crop residue. The adoption of no-till farming, in particular, may prove to be an important element in improving our river ecosystems because it reduces soil erosion.

No-till practices have spread rapidly since 1990 when only 7.5% of corn and soybean fields were grown under no-till practices in Putnam County, Indiana. By 1993 no-till farming was practiced on 50% of the corn and soybean fields and this increased to 60% in 1994. If this trend continues it should lead to less soil erosion in the future leading, in turn, to a reduction in sediment and phosphate pollution. It may possibly even have some influence on reducing nitrate levels since the practice fosters a return of earthworms and other soil organisms to previously plowed fields.

Although agriculture occupies a majority of the Wabash River basin, it is not the only nonpoint-source contributor. Active and derelict mines, both old deep mines and old surface mines, may play a significant negative role in their impact on the river ecosystem. It may be just a coincidence that the best fish communities in the middle Wabash River lie upriver from the northern border of coal mines and that community quality sharply declines in the Covington\Perrysville area (Reach 5).

Likewise, the stretch of river from Clinton to Terre Haute, Indiana (Reach 8) appeared to be unusually depressed in past years. Perhaps the improvement here since 1983 is also linked to less nonpoint-source runoff from mining areas as well as agricultural fields.

There are other elements in the Wabash River basin which contribute to an overall understanding of the ecosystem and its problems. Surveys of the status of the river's riparian corridor were undertaken in 1983 and 1994 because of accelerated cutting of trees and enlargement of fields in many stretches of the river during the 1970s. The extent of eroding banks in 1993 was on the order of 5-7% of the river length overall. Another 5-7% of the river corridor has only one or two trees separating the bank from tilled fields. These figures are lower than those estimated in 1983 perhaps because of the natural establishment of willow groves in the interim.

The riparian status of tributaries such as the lower 12 miles of Sugar Creek together with its extreme upper section and upper Big Raccoon Creek is even less encouraging. Clearly, there is an urgent need to maintain and, in many areas, reestablish an effective riparian buffer along our rivers and streams. Riparian forests are critically important to maintaining a healthy river ecosystem.

While seeking to restore an effective vegetative buffer the entire river corridor should also be carefully examined for opportunities to develop and integrate currently rare backwater areas into the river mainstem.

Evaluations of the separate and collective negative impacts due to man must be conducted within changeable and some-times damaging natural processes such as droughts and floods. The droughts of 1988 and 1991 had far less effect on the fish populations than the 1993 flood, but they did produce a depression in our catches. However, the extent of the depression differed somewhat. In 1991 populations were depressed throughout the middle Wabash River. In 1988 populations were depressed only in the lower eight Reaches. We have no clues as to what might have caused these differences.

LITERATURE CITED

Aldridge, A.W., B.S. Payne, and A.C. Miller. 1987. The effects of intermittent exposure to suspended solids and turbulence on three species of freshwater mussels. Environmental Pollution 45:17-28.

Anderson, R. M. 1994. Personal communication. Indiana DNR.

Anonymous. 1907. The Wabash and Erie Canal. No. III in **Indiana Magazine of History**, Vol. III, No. 3:100-107.

Arvin, D. V. 1989. Statistical summary of streamflow data for Indiana. U.S. Geological Survey, Open-File Report 89-62:964 pp.

Ball, R. L. and B. M. Schoenung. 1996. Status of mussel populations in the primary harvest areas. 1995 final report, Indiana Department of Natural Resources, Division of Fish and Wildlife, Indianapolis, Indiana. 72 pp. mimeo.

Bell, J. 1983. Sediment Oxygen Demand in the Middle Wabash River. Unpublished ms. Purdue University, Department of Environmental Engineering, West Lafayette, Indiana.

Bingham, R. L. 1968. Reproductive seasons of eight freshwater mussels from the Wabash, White, and East Fork of the White Rivers of Indiana. Report to the Indiana Department of Natural Resources, Division of Fish and Game. 102 pp.

Bird, G.A. and N.K. Kaushik. 1981. Coarse particulate organic matter in streams. pp. 41-68 in **Perspectives in Running Water Ecology**, edited by M.A. Lock and D.D. Williams. Plenum Press, New York and London.

Blackwell, R. L. 1991. Summary of harvest estimates and license sales for Indiana's inland and Ohio River commercial fisheries, 1990. Fish. Sec., Indiana Dept. Nat. Res., Div. Fish & Wildlife:20 pp mimeo.

Blatchley, W. S. and L. E. Daniels. 1902. On some mussels known to occur in Indiana. Annual Report, Indiana Geological Survey 26:557-628.

Bonnemains, J. 1984. Charles-Alexandre Lesueur en Amerique du Nord (1816-1837). Annales du Musreum du Havrey. 3 volumes.

Braun, E. R. 1990. A survey of upper Wabash River fishes. Indiana Department of Natural Resources, Division of Fish and Wildlife.

Bray, W.M., E.H. Swanson, and I.S. Farrington. 1972. The Making of the Past-The New World. Elsevier. Phaidon.

Bridges, C.L. 1997. Personal communication. Indiana DEM.

Bridges, C.L., H.L. BonHomme, G.R. Bright, D.E. Clark, J.K. Ray, J.L. Rud, and J.L. Winters. 1986. Summer primary productivity and associated ata for four Indiana rivers. Proceedings of the Indiana Academy of Science. Vol. 96, 1986:309-314.

Call, R. E. 1900. A descriptive illustrated catalogue of the Mollusca of Indiana. Indiana Department of Geology and Natural Resources, 24th Annual Report:335-535.

Cammack, E.A. 1954. Notes on Wabash River Steamboating: Early Lafayette. Indiana Magazine of History, Vol. L, March 1954, No. 1:35-50.

Clark, C.D. (ed.). 1980. The Indiana Water Resource. Indiana Department of Natural Resources. 508 pp.

Condit, D. and D. Roseboom. 1989. Stream bank stabilization and the Illinois River Soil Conservation Task Force. in Proceedings Second Conference on the Management of the Illinois River System: The 1990s and Beyond. Special Report No. 18, Water Res. Center, Univ. of Illinois, 205 N. Mathews Ave., Urbana, IL 61801.

Cope, E.D. 1971. A partial synopsis of the fishes of the fresh waters of North Carolina. Proc. Amer. Phil. Soc. 1869-70, 11:448-495.

Corbett, D.M. 1969. Acid mine-drainage problem of the Patoka River watershed, southwestern Indiana. Indiana Water Resources Research Center Report of Investigations No. 4:173 pp.

Corbett, D.M. and A.F. Agnew. 1968. Coal mining effect on Busseron Creek watershed, Sullivan County, Indiana. Indiana Water Resources Research Center Report of Investigations No. 2:187 pp.

Cox, S.C. 1860. Recollections of the Early Settlement of the Wabash Valley. Courier Steam Book & Job Print. House, Lafayette, Ind. 160 pp.

Craven, J. 1912. The protection of our rivers from pollution. Proc. Indiana Acad. Sci. Vol. 22:47-50.

Culbertson, G. 1908. Deforestation and its effects among the hills of southern Indiana. Proc. Indiana Acad. Sci. Vol. 18:27-37.

Cummins, K. W., J.R. Sedell, F.J. Swanson, G.W. Minshall, S.G. Fisher, C.E. Cushing, R.C. Petersen, and R.L. Vannote. 1983. Organic matter budgets for stream ecosystems: problems in their evaluation. pp 299-253 in **Stream Ecology-Application and Testing of General Ecological Theory** edited by J. R. Barnes and G. W. Minshall. Plenum Press, New York and London.

Cummings, K. S., C. A. Mayer, L. M. Page and J. M. K. Berlocher. 1987. Survey of the freshwater mussels (Mollusca: Unionidae) of the Wabash River Drainage. Phase I: Lower Wabash River & Tippecanoe River. Indiana Department of Natural Resources, Section of Faunistic Surveys & Insect Identification, Technical Report 1987 (5): 61 pp.

Cummings, K. S., C. A. Mayer and L. M. Page. 1988. Survey of the freshwater mussels (Mollusca: Unionidae) of the Wabash River Drainage. Phase II: Upper and middle Wabash River. Indiana Department of Natural Resources, Section of Faunistic Surveys & Insect Identification, Technical Report 1988(8): 48 pp.

Cummings, K. S., C. A. Mayer, and L. M. Page. 1991. Survey of the freshwater mussels (Mollusca:Unionidae) of the Wabash River Drainage Phase III: White River and Selected Tributaries. Center for Biodiversity, Tech. Rep. 991(3):39 pp.

Cummings, K. S. and C. A. Mayer. 1992. Field Guide to Freshwater Mussels of the Midwest. Illinois Natural History Survey, Manual 5. 194 pp.

Cummings, K.S., J.L. Harris, and R.J. Neves. 1993. Conservation status of freshwater mussels of the United States and Canada. Fisheries 18(9):6-22.

Crawford, C.G. and L.J. Mansue. 1988. Suspended-sediment characteristics of Indiana streams 1952-84. U.S. Geological Survey Open-File Report 87-527:79 pp.

Daniels, L. E. 1903. A check list of Indiana mollusca, with localities. Indiana Dept. of Geol. and Nat. Res., 26th Ann. Rept.:629-652.

Daniels, L. E. 1914. A supplemental check list of Indiana mollusca with localities and notes. Indiana Dept. of Geol. and Nat. Res., 39th Ann. Rept.:318-326.

Dunn, J.P. 1910. Greater Indianapolis - the History, the Industries, the Institutions, and the People of a City of Homes. Lewis Publ. Co., Chicago, Ill. Vol. 1:1-641.

EA Science and Technology. 1988. A 316(b) study and impact assessment for the Cayuga Generating Station. A Report for Public Service Company of Indiana, Plainfield, Ind. 62 pp.

EA Science and Technology. 1989. A 316(b) study and impact assessment for the Wabash River Generating Station. A Report for Public Service Company of Indiana, Plainfield, Ind. 71 pp.

EA Science, and Technology, Inc. 1990. Recalibration of Wabash River dissolved oxygen model. A Report for Public Service Indiana Prepared by EA Mid-Atlantic Regional Operations, EA Eng., Sci., and Tech., Inc., 15 Loveton Circle, Sparks, MD 21152.

Eigenmann, C.H. and C.H. Beeson. 1894. The Fishes of Indiana. Proc. Ind. Acad. Sci. 1893, 9:76-108.

Esarey, L. 1912. Internal Improvements in Early Indiana. Indiana Historical Soc. Publ. Vol V, No. 2:158 pp.

Evermann, B.W. and O.P. Jenkins. Notes on Indiana Fishes. Proc. U.S. Nat. Mus.:43-57.

Fatout, P. 1972. Indiana Canals. Purdue Univ. Studies, West Lafayette, Ind. 212 pp.

Finni, G.R. 1981. Final report Wabash River biological studies long-stretch adult fish surveys, riffle fish sampling, and benthic macroinvertebrates. A report submitted to American Electric Power Service Corporation by WAPORA, Inc. pp. 116-118.

Forbes, S.A. and R.E. Richardson. 1920. The Fishes of Illinois, 2nd ed. Nat. Hist. Surv. Ill. 3;i-cxxi, 1-357.

Gammon, J. R. 1973. The effect of thermal inputs on the populations of fish and macroinvertebrates in the Wabash River. Purdue University, Water Res. Research Center. Tech. Report No. 32. 106 pp.

Gammon, J. R. 1976. The Fish Populations of the Middle 340 km of the Middle Wabash River. Technical Report No. 32, Water Resources Research Center, Purdue University, Lafayette, Ind. 73 pp.

Gammon, J. R. 1980. The Use of Community Parameters Derived from Electrofishing Catches of River Fish as Indicators of Environmental Quality. Seminar on Water Quality Trade-offs, EPA-905/9-80-009, U.S. Environmental Protection Agency, Washington, D. C., pp. 335-363.

Gammon, J. R. 1982. Changes in the fish community of the Wabash River following power plant start-up: Projected and Observed. ASTM, 6th Symposium on Aquatic Toxicology:350-366.

Gammon, J. R. and J.M. Reidy. 1981. The role of tributaries during an episode of low dissolved oxygen in the Wabash River. Am. Fish. Soc. Warmwater Streams Symposium:396-407.

Gammon, C.W. and J.R. Gammon. 1990. Fish communities and habitat of the Eel River in relation to agriculture. A report for the Indiana Department of Environmental Management, Office of Water Management. 74 pp.

Gammon, J.R. and C.W. Gammon. 1993. Changes in the fish community of the Eel River resulting from agriculture. Proc. Indiana Academy of Science. Vol. 102:67-82.

Gammon, J.R. and C.W. Gammon. 1998. The middle Wabash River seine survey 1997. A Report for Eli Lilly and Company and Cinergy Corporation. 32 pp. mimeo.

Gerking, S. C. 1945. Distribution of the fishes of Indiana. Investigations of Indiana Lakes and Streams, Volume III:1-137.

Glander, P. A. 1987. Summary of harvest estimates and license sales for Indiana's inland commercial fishery, 1984-1986. Indiana Department of Natural Resources, Division of Fish and Wildlife. Fish Management Report. 27 pp.

Goodrich, C. and H. Van der Schalie. 1944. A revision of the Mollusca of Indiana. Am. Mid. Nat. 32(2):257-326.

Hay, O.P. 1895. The Lampreys and Fishes of Indiana. Indiana Dept. Geology and Natural Resources 19:146-296.

Hem, J. D. 1985. Study and interpretation of the chemical characteristics of natural water, 3rd ed. U. S. Geological Survey Water-Supply Paper 2254. 263 pp.

Henschen, M. T. 1989. Commercial mussel harvesting in Indiana, 1982-1987. Indiana Department of Natural Resources, Indianapolis.

Horwitz, R.J. 1978. Temporal variability patterns and the distributional patterns of stream fishes. Ecological Monographs 48:307-321.

Homoya, M.A., D.B. Abrell,, J.R. Aldrich, and T.W. Post. 1986. The natural regions of Indiana. Proc. Indiana Academy of Science, 94(1985):245-268.

Hyde, G. E. 1962. Indians of the Woodlands from prehistoric times to 1725. Univ. Oklahoma Press, Norman, Ok. 295 pp.

HydroQual, Inc. 1984. Dissolved oxygen analysis of the Wabash River. A Report for Eli Lilly and Company, Indianapolis, Ind.

Indiana Stream Pollution Control Board. 1977. Indiana 305 (b) Report. Indiana State Board of Health, Stream Pollution Control Board.

Indiana Department of Environmental Management. 1986-1990. Indiana water quality. Monitoring station records - rivers and streams. Ind. Dept. Env. Mgt., Office of Water Management (published annually).

Indiana State Board of Health. 1957-1985. Water quality monitoring, rivers and streams: Indiana State Board of Health Monitor Station Records (published annually).

Jenkins, O. P. 1889. List of fishes collected in Vigo County. Hoosier Naturalist, Valparaiso, Ind., Vol. II:93-96.

Jordan, D.W. 1878. Catalogue of the Fishes of Indiana. Appendix A in **First Annual Report of the Commission of Fisheries of Indiana,** 1883:96-103.

Jordan, D.S. 1889. Address to State Fish & Game Convention on Dec. 19, 1889. In **Report of the Commission of Fish.** 1891:62-69.

Jordan, D.S. 1890. Report of explorations made during the summer and autumn of 1888, in the Alleghany region of Virginia, North Carolina and Tennessee, and in Western Indiana, with an account of the fishes found in of the river basins of those regions. Bull. U.S. Fish. Comm. 1888, 8:97-173.

Karr, J.R. 1981. Assessment of biotic integrity using fish communities. Fisheries (Bethesda) 6(6):21-27.

Karr, J.R. 1987. Biological monitoring and environmental assessment: a conceptual framework. Env. Mgt. 11:249-256.

Karr, J.R., K.D. Fausch, P.L. Angermeier, P.R. Yant, and I.J. Schlosser. 1986. Assessing biological integrity in running waters: a method and its rationale. Ill. Nat. Hist. Surv. Spec. Publ. 5, Urbana.

Karr, J.R., P.R. Yant, K.D. Fausch, and I.J. Schlosser. 1987. Spatial and temporal variability of the index of biotic integrity in three midwestern streams. Trans. Am. Fish. Soc. 116:1-11.

Karr, J.R. 1992. Measuring biological integrity: lessons from streams. pp 83-104 in Ecological Integrity and the Management of Ecosystems edited by S. Woodley, J. Kay, and G. Francis. St. Lucie Press, Delray Beach, Florida.

Kovacic, D.E., L. L. Osborne, and B.C. Dickson. 1990. The influence of riparian vegetation on nutrient losses in a midwestern stream watershed. Report Project No. S-117. Water Resources Center, University of Illinois, Urbana, IL. 91 pp.

Krumholz, L. A., R. L. Bingham and E. R. Meyers. 1970. A survey of the commercially valuable mussels of the Wabash and White Rivers of Indiana. Proc. Ind. Acad. Sci. 1969. Vol. 79:

Lesniak, D.G., M.C. Tavenner, and J.R. Siefker. 1973. Quantitative chemical analysis of specific components of the waters of Lost Creek and the Wabash River, Vigo County, Indiana. Proc. Indiana Academy of Science, Vol. 72:176-179.

Lewis, D.L. and D.K. Gattie. 1991. The ecology of quiescent microbes. ASM News. Volume 57, No. 1:27-32.

Lewis, R.B. 1997. Personal communication. Cinergy Corp.

Lewis, R.B., J.E. Pike, and K.E. Richard. 1997. Fish and benthic macroinvertebrate community analysis near the Cayuga generating station 1995. Technical Report EPW-060, PSI Energy subsidiary of Cinergy Corp., Plainfield, Indiana.

Lindley, H. (ed.). 1916. Indiana as seen by early travelers. Indiana Historical Comm., Indianapolis. 596 pp.

Lindsey, A.A. 1962. Analysis of an original forest of the lower Wabash floodplain and upland. Proc. Indiana Acad. Sci. 72:282-287.

Lindsey, A.A., R.O. Petty, D.K. Sterling and W.V. Asdall. 1961. Vegetation and environment along the Wabash and Tippecanoe Rivers. Ecol. Mono. 31:105-156.

Ludwig, J. A. and J. F. Reynolds. 1988. Statistical Ecology. John Wiley & Sons, N. Y. 337 pp.

Martin, J. D. and C. G. Crawford. 1987. Statistical analysis of surface-water-quality data in and near the coal-mining region of southwestern Indiana, 1957-80.

Maximilian, Prince of Wied. 1843. Travels in the interior of North America. Ackerman and Co., Strand. pp. 74-92.

McComish, T.S. 1967. Food habits of bigmouth and smallmouth buffalo in Lewis and Clark Lake and the Missouri Riveriver. Trans. Amer. Fish. Soc. 96(1):70-74.

243

McCord, S.S. 1970. Travel Accounts of Indiana 1679-1961. Indiana Historical Collections, Vol. XLVII, Ind. Hist. Bureau. 331 pp.

Ohio Environmental Protection Agency (OEPA). 1987. Water quality implementation manual. AQ Manual (3rd update). Fish. Ohio Environmental Protection Agency, Columbus.

Omernik, J.M. 1977. Nonpoint-source stream nutrient level relationships: a nationwide survey. U.S. Environmental Protection Agency, DPA-600/3-77-105. Corvallis, Oregon.

Omernik, J. M. and A. L. Gallant. 1988. Ecoregions of the upper midwest states. U. S. Environmental Protection Agency, Environmental Research Laboratory, Corvallis OR 97333. EPA/600/3-88/037:56 pp.

ORSANCO. 1990a. Assessment of nonpoint source pollution of the Ohio River. Toxic Substances Control Program, Ohio River Valley Water Sanitation Commission, Cincinnati, Ohio. 68 pp.

ORSANCO. 1990b. Long-term trends assessment of fifteen water quality parameters in the Ohio River. Toxic Substances Control Program, Ohio River Valley Water Sanitation Commission, Cincinnati, Ohio. 26 pp.

Osborne, L.L. and D.E. Kovacic. 1993. Riparian vegetated buffer strips in water-quality restoration and stream mangement. Freshwater Biology 29:243-258.

Owens, L.B. and D.W. Nelson. 1973. Relationship of various indices of water quality to denitrification in surface waters. Proc. Indiana Academy of Science, Vol. 72:404-413.

Parke, N. J. 1985. An investigation on phytoplankton sedimentation in the middle Wabash River. M. A. Thesis, DePauw University, Greencastle, Indiana.

Parke, N. J. and J. R. Gammon. 1986. An investigation of phytoplanktonsedimentation in the middle Wabash River. Proc. Indiana Acad. Sci. Vol. 94:279-288.

Pearson, J. 1975. Upper Wabash River fisheries survey report. Indiana Department of Natural Resources, Division of Fish and Wildlife.

Peterjohn, W.T. and D.L. Correll. 1984. Nutrient dynamics in an agricultural watershed: Observations on the role of a riparian forest. Ecol. 65:1466-1475.

Petty, R.O. and M.T. Jackson. 1966. Plant Communities. Chapter 16, pp 264-296 in A.A. Lindsey (ed.) Natural Features of Indiana, Indiana Acad. Sci., Indiana State Library, Indianapolis, Indiana. 591 pp.

Pielou, E. C. 1977. Mathematical Ecology. Wiley, New York.

Pielou, E. C. 1984. The Interpretation of Ecological Data. Wiley, New York.

Peters, J.G. 1981. Effects of surface mining on water quality in a small watershed, Sullivan County, Indiana. U.S. Geological Survey Open-File Report 81-543:61 pp.

Pitzer, D. E. 1989. The original boatload of knowledge down the Ohio River: William Maclure's and Robert Owen's transfer of science and education to the midwest, 1825-1826. pp 128-142 in **The Ohio River Its History and Environment**, The Ohio Journal of Science, Vol. 89(5), W.J. Mitsch, editor.

Rafinesque, C.S. 1820. Icthyologia Ohiensis or Natural History of the Fishes Inhabiting the River Ohio and Its Tributary Streams. Reprint Edition 1970 by Arno Press Inc. 90 pp.

Rafinesque, C.S. 1836. A Life of Travels and Researches in North America and South Europe. F. Turner, No. 367, Market St. Reprinted in Chronica Botanica, Vol. 8, No. 2:291-360.

Robertson, R. N. 1975. Middle Wabash River stream survey report. Indiana Department of Natural Resources, Division of Fish and Wildlife.

Robins, C.R., R.M. Bailey, C.E. Bond, J.R. Brooker, E.A. Lachner, R.N. Lea, and W.G. Scott. 1991. Common and scientific names of fishes from the United States and Canada. Fifth Edition. American Fisheries Society Special Publication 20.

Rogellin, E.M. 1979. The role of seining in the analysis of the middle Wabash River. M. A. Thesis, DePauw University, Greencastle, Indiana.

Seegert, G. L. 1990. Personal Communication.

Sellers, T. 1995. Personal Communication

Siefker, J.R. and D.P. McCleary. 1979. Chemical analysis of water samples from three lakes in the Greene-Sullivan State Forest and the Wabash River at Clinton, Indiana. Proc. Indiana Academy of Science, Vol. 80:293-295.

Simon, T. 1992. Biological criteria development for large rivers with an emphasis on an assessment of the White River Drainage, Indiana. USEPA, Region 5, EPA-905/R-92/006.

Schneider, H. and A.D. Hasler. 1960. Laute und aluterzengung beim susswasser-trommler *Aplodinotus grunniens* Rafinesque (Sciaenidae, Pisces). Zeitschr. Vergleich. Physiol., Berlin 43(5):499-517.

Schneller, M.V. 1955. Oxygen depletion in Salt Creek, Indiana. Investigations of Indiana Lakes and Streams, Vol. IV, No. 6:163-175.

Slack, K.V. 1955. A study of the factors affecting stream productivity by the comparative method. Investigations of Indiana Lakes and Streams, Vol. IV, No. 6:3-47.

Spacey, A. 1990. Personal communication. Purdue University.

Spacey, A. and A.M. Chaney. 1993. Metabolic effects of suspended solids on Unionid mussels. Project Report E-1-5 (Study 16), Endangered Species Program, Indiana Department of Natural Resources, Division of Fish and Wildlife. 52 pp.

Stefanavage, T. C. 1990. Summary of harvest estimates and license sales for Indiana's inland and Ohio River commercial fisheries, 1987-1989. 20 pp.

Stefanavage, T. C. 1992. Reported 1991 commercial mussel harvest from Indiana's rivers. Fish Mgt. Rep, Fish. Sect. Ind. Dept. Nat. Res., Div. Fish & Wildlife: 12 pp mimeo.

Ter Braak, C. J. F. CANOCO--a FORTRAN program for Canonical Community Ordination by [Partial] [Detrended] [Canonical] Correspondence Analysis, Principal Components Analysis and Redundancy Analysis (Version 2.1). Agriculture Mathematics Group, Wageningen.

Thomas, T.C. 1978. An analysis of abandoned coal mines in West Central Indiana and suggested reclamation techniques (with an inventory of active strip mines). West Central Indiana Econimic Development District, Ind., Terre Haute, Ind.:231 pp.

Thomas, T.C. 1981. An interim report of field investigations of delelict coal mined lands in Southwestern Indiana. Indiana Department of Natural Resources, Division of Reclamation, April 1981. 53 pp. mimeo.

Todd, D. K. (ed.). 1970. The water encyclopedia. Water Information Center, Water Research Building, Manhasset Isle, Port Washington, N. Y. 559 pp.

Visher, S.S. 1944. Indiana floods. Proc. Indiana Acad. Sci. 54:134-141.

Wangsness, D.J. and others. 1981a. Hydrology of area 33, eastern region, Interior Coal Province, Indiana and Kentucky. U.S. Geological Survey Water-Resources Investigations. Open-File Report 81-423:84 pp.

Wangsness, D.J. and others. 1981b. Hydrology of area 32, eastern region, Interior Coal Province, Indiana. U.S. Geological Survey Water-Resources Investigations. Open-File Report 81-498:76 pp.

Wangsness, D.J. and others. 1983. Hydrology of area 30, eastern region, Interior Coal Province, Indiana and Illinois. U.S. Geological Survey Water-Resources Investigations. Open-File Report 82-1005:82 pp.

Wangsness, D.J. 1982. Reconnaissance of stream biota and physical and chemical water quality in areas of selected land use in the coal-mining region, southwestern Indiana, 1979-1980. U.S. Geological Survey Open-File Report 81-423:84 pp.

Whitaker, J.O. Jr. 1976. Fish community changes at one Vigo County, Indiana locality over a twelve year period. Proceedings Indiana Academy of Science 85:191-207.

Wilber, W.G., C.G. Crawford, D.E. Renn, S.E. Ragone, and D.J. Wangsness. 1980. Preliminary assessment of the factors affecting water quality in the coal-mining region, southwestern Indiana, March to October 1979. In Warner, R.E. and P.E. Clark (eds.) Water Resources and land-use management in Indiana, A Symposium. Proceedings: Indiana Water Resources Association:215-234.

Wiley, E.O. 1976. The phyolgeny and biogeography of fossil and recent gars (Actinopterygii: Lepisosteidae). University of Kansas, Museum of Natural History. Miscellaneous Publication No. 64:111 pp.Williams, J.D., M.L. Warren, Jr.,

Williams, J.E., J.E. Johnson, D.A. Hendrickson, S. Contreras-Balderas, J.D. Williams, M. Navarro-Mendoza, D.E. McAllister, and J.E. Deacon. 1989. Fishes of North America endangered, threatened, or of special concern: 1989. Fisheries (Bethesda) 14(6):2-20.

Yost, J.C. and S.R. Rives. 1990. Recalibration of Wabash River dissolved oxygen model. EA Mid-Atlantic Regional Operations, EA Engineering, Science, and Technology, Inc., 15 Loveton Circle, Sparks, Maryland 21152. A Report for Public Service Indiana, Plainfield, Indiana.

Zogorski, J.S., D.S. Ramey, P.W. Lambert, J.D. Martin, and R.E. Warner. 1981. Hydrologic evaluation of a hypothetical coal-mining site near Chrisney, Spencer County, Indiana. U.S. Geological Survey Open-File Report 80-ll07:122 pp.

List of Publications on the Wabash River and its Tributaries
produced by research funded by Cinergy Corp. and Eli Lilly and Company

Gammon, J.R. 1970. Aquatic life survey of the Wabash River, with special reference to the effects of thermal effluents on populations of macroinvertebrates and fish: 1967-1969. Prog. Report to Public Service Indiana. 65 pp. mimeo.

Gammon, J.R. 1971. The response of fish populations in the Wabash River to heated effluents. Proc. 3rd Nat. Symp. Radioecology:513-523.

Gammon, J.R. 1973. The effect of thermal inputs on the populations of fish and macroinvertebrates in the Wabash River. Purdue University Water Res. Research Center. Tech. Report No. 32. 106 pp.

Bartolucci, L.A., R.M. Hoffer, and J.R. Gammon. 1973. Effects of altitude and atmospheric windows on remote measurements of thermal effluent in the Wabash River. Proc. 1st Panamerican Symp. Remote Sensing:147-159.

Teppen, T.C. and J.R. Gammon. 1975. Distribution and abundance of fish populations in the middle Wabash River. Thermal Ecology II (ERDA Symp. Series) CONF-75045:272-283.

Yoder, C.O. and J.R. Gammon. 1975. Seasonal distribution and abundance of Ohio River fishes at the J.M. Stuart Electric Generating Station. Thermal Ecology II (ERDA Symp. Series) CONF-750245:284-295.

Gammon, J.R. 1976. Measurement of entrainment and predictions of impact in the Wabash and Ohio Rivers. Third Nat. Workshop on Entrainment and Impingement. L.D. Jensen, Ed.:159-176.

Gammon, J.R. 1976. The fish populations of the middle 340 Km of the Wabash River. Purdue U. Water Res. Res. Center. Tech. Report No. 86:73 pp.

Mancini, E.R., J.R. Gammon, and P.H. Carlson. 1976. Recent collections of Anepeorus simplex (Walsh) (Ephemeroptera: Heptageniidee) from the Wabash River, Indiana. Ent. News, 87:788:237-238.

Gammon, J.R., Spacie, A., Hamelink, J.L., and R.L. Kaesler. 1979. The role of electrofishing in assessing environmental quality of the Wabash River. ASTM Symp. Ecol. Assess. of Effluent Impacts on Communities of Indigenous Aquatic Organisms: 307-324.

Gammon, J.R. and J.M. Reidy. 1981. The role of tributaries during an episode of low dissolved oxygen in the Wabash River: Am. Fish. Soc. Warmwater Streams Symposium:396-407.

Gammon, J.R. 1980. The use of community parameters derived from electrofishing catches of river fish as indicators of environmental quality. EPA Seminar on Water Quality Management Tradeoffs-Point Source vs. Diffuse Source Pollution: EPA-905/9-80-009. pp. 335-363.

Gammon, J.R. 1982. Changes in the fish community of the Wabash River following power plant start-up: Projected and Observed. ASTM, 6th Symposium on Aquatic Toxicology:350-366.

Gammon, J.R. and J.R. Riggs. 1983. Vermilion River and Sugar Creek. Proc. Ind. Acad. Sci. Vol. 92:183-190.

Parke, N.J. and J.R. Gammon. 1986. An investigation of phytoplankton sedimentation in The Middle Wabash River. Proc. Ind. Acad. Sci. Vol. 95:279-288.

Miller, D.L., P.M. Leonard, R.M. Hughes, J.R. Karr, P.B. Moyle, L.H. Schrader, B.A. Thompson. R.A. Daniels, K.D. Fausch, G.A. Fitzhugh, J.R. Gammon, D.B. Halliwell, P.L. Angermeier and D.J. Orth. 1988. Regional Applications of an index of biotic integrity for use in Water Resource Management. Fisheries, Vol. 13:12-20.

Gammon, J. R., C. W. Gammon, and M. K. Schmid. 1990. Landuse influence on fish communities in central Indiana streams. Proc. 1990 Midwest Pollution Control Biologists Meeting. pp. 111-120.

Gammon, J. R. 1991. Biological monitoring in the Wabash River and its tributaries. pp. 105-112 in G. Flock (ed.) Water Quality Standards for the 21st Century. Office of Water, U. S. Environmental Protection Agency. 251 pp.

Gammon, J. R., C. W. Gammon, and C. E. Tucker. 1991. The fish communities of Sugar Creek. Ind. Acad. Sci. 99:141-155.

Gammon, J. R. 1991. The environment and fish communities of the middle Wabash River. A Report for Eli Lilly & Company and PSI-Energy. 129 pp.

Gammon, J. R. and C. W. Gammon. 1991. Agricultural impacts on the fishes of the Eel River, Indiana. pp. 85-99. In T.P. Simon and W.S. Davis (editors). 1992. Proc. 1991 Midwest Pollution Control Biologists Meeting: Environmental Indicators: Measurement and Assessment Endpoints. U.S. EPA, Region V, Env. Sci. Div., Chicago, IL. EPA 905/R-92/003.

Gammon, J. R. and C. W. Gammon. 1993. Changes in the fish community of the Eel River resulting from agriculture. Proc. Indiana Acad. Sci. 102:67-82.

Gammon, J. R. 1993. The Wabash River: Progress and Promise. pp. 142-161 in Proceedings of the Symposium on **Restoration Planning for the Rivers of the Mississippi River Ecosystem** edited by L.W. Hesse, C.B. Stalnaker, N.G. Benson, and J.R. Zuboy. Biological Report 19, National Biological Survey, U.S. Department of the Interior.

Gammon, J. R. 1995. Environmental assessment of fish populations of the Wabash River and its tributaries. Natural Areas Journal 15(3):259-266.

Gammon, J. R. 1995. The Wabash River ecosystem II. A report to PSI-Energy and Eli Lilly and Company. 235 pp.

Gammon, J.R. 1995. The status of riparian wetlands in west-central Indiana streams. Proceedings of the Indiana Academy of Sciences, Volume 103(3-4):195-213.

Gammon, J.R. 1997. "Beneath the Water's Surface: The Fishes" pp. 310-317 in Indiana - A Natural perspective. M. Jackson, ed.

250

Master of Arts Theses

Bell, Susan C. 1969. The Effects of Thermal Pollution on the Macroinvertebrate Population of the Wabash River.

Pierce, Steven T. 1973. The Effects of Thermal Enrichment on the Macroinvertebrate Populations of the Wabash River.

King, James R. 1974. A Study of Power Plant Entrainment Effects on the Drifting Macroinvertebrates of the Wabash River.

Mancini, Eugene R. 1974. Macroinvertebrate Drift of the Wabash River and its Relation to Wabash Generating Station (Terre Haute, Indiana).

Teppen, Terry C. 1975. Distribution and Abundance of Fish Populations in the Middle Wabash River.

Reidy, Joseph M. 1979. The Role of Tributaries in the Recovery of a River from Stress.

Rogellin, E.M. 1979. The Role of Seining in the Analysis of the Middle Wabash River.

Rud, Jerome L. 1982. The Diets and Interspecific Relationships of Twelve Species of Game Fish from the Middle Wabash River, West-central Indiana.

Parke, Neil. 1985. An Investigation on Phytoplankton Sedimentation in The Middle Wabash River.